T0282645

# CAMBRIDGE LIBRARY COLLECTION

*Books of enduring scholarly value*

## Physical Sciences

From ancient times, humans have tried to understand the workings of the world around them. The roots of modern physical science go back to the very earliest mechanical devices such as levers and rollers, the mixing of paints and dyes, and the importance of the heavenly bodies in early religious observance and navigation. The physical sciences as we know them today began to emerge as independent academic subjects during the early modern period, in the work of Newton and other 'natural philosophers', and numerous sub-disciplines developed during the centuries that followed. This part of the Cambridge Library Collection is devoted to landmark publications in this area which will be of interest to historians of science concerned with individual scientists, particular discoveries, and advances in scientific method, or with the establishment and development of scientific institutions around the world.

## Heat Considered as a Mode of Motion

Professor of natural philosophy at the Royal Institution between 1853 and 1887, the physicist and mountaineer John Tyndall (1820–93) passionately sought to share scientific understanding with the Victorian public. A lucid and highly regarded communicator, he lectured on such topics as heat, light, magnetism and electricity. In this collection of twelve lectures, first published in 1863, Tyndall discusses the general properties of heat and its associated physical processes, such as convection, conduction and radiation. He presents concepts so that they are intelligible to non-specialists, and helpful illustrations of laboratory equipment accompany his descriptions of experiments and phenomena. Throughout, he explains the research and discoveries of renowned scientists, including Sir Humphry Davy, Julius von Mayer, James Prescott Joule, and Hermann von Helmholtz. Several of Tyndall's other publications, from his lectures on sound to his exploration of alpine glaciers, are also reissued in this series.

Cambridge University Press has long been a pioneer in the reissuing of out-of-print titles from its own backlist, producing digital reprints of books that are still sought after by scholars and students but could not be reprinted economically using traditional technology. The Cambridge Library Collection extends this activity to a wider range of books which are still of importance to researchers and professionals, either for the source material they contain, or as landmarks in the history of their academic discipline.

Drawing from the world-renowned collections in the Cambridge University Library and other partner libraries, and guided by the advice of experts in each subject area, Cambridge University Press is using state-of-the-art scanning machines in its own Printing House to capture the content of each book selected for inclusion. The files are processed to give a consistently clear, crisp image, and the books finished to the high quality standard for which the Press is recognised around the world. The latest print-on-demand technology ensures that the books will remain available indefinitely, and that orders for single or multiple copies can quickly be supplied.

The Cambridge Library Collection brings back to life books of enduring scholarly value (including out-of-copyright works originally issued by other publishers) across a wide range of disciplines in the humanities and social sciences and in science and technology.

# Heat
## Considered as
## a Mode of Motion

*Being a Course of Twelve Lectures*
*Delivered at the Royal Institution of Great Britain*
*in the Season of 1862*

JOHN TYNDALL

CAMBRIDGE
UNIVERSITY PRESS

# CAMBRIDGE
## UNIVERSITY PRESS

University Printing House, Cambridge, CB2 8BS, United Kingdom

Published in the United States of America by Cambridge University Press, New York

Cambridge University Press is part of the University of Cambridge.
It furthers the University's mission by disseminating knowledge in the pursuit of
education, learning and research at the highest international levels of excellence.

www.cambridge.org
Information on this title: www.cambridge.org/9781108068901

© in this compilation Cambridge University Press 2014

This edition first published 1863
This digitally printed version 2014

ISBN 978-1-108-06890-1 Paperback

This book reproduces the text of the original edition. The content and language reflect
the beliefs, practices and terminology of their time, and have not been updated.

Cambridge University Press wishes to make clear that the book, unless originally published
by Cambridge, is not being republished by, in association or collaboration with, or
with the endorsement or approval of, the original publisher or its successors in title.

# HEAT

AS

# A MODE OF MOTION.

LONDON

PRINTED BY SPOTTISWOODE AND CO.

NEW-STREET SQUARE

# HEAT

CONSIDERED AS

# A MODE OF MOTION:

BEING

## A COURSE OF TWELVE LECTURES

DELIVERED AT

## THE ROYAL INSTITUTION OF GREAT BRITAIN

IN THE SEASON OF 1862.

BY

## JOHN TYNDALL, F.R.S. &c.

PROFESSOR OF NATURAL PHILOSOPHY IN THE ROYAL INSTITUTION.

WITH ILLUSTRATIONS.

LONDON:

LONGMAN, GREEN, LONGMAN, ROBERTS, & GREEN.

1863.

TO THE MEMBERS

<small>OF THE</small>

ROYAL INSTITUTION OF GREAT BRITAIN

THIS BOOK IS DEDICATED.

FEBRUARY 1863.

# PREFACE.

———•◦•◦—

IN the following Lectures I have endeavoured to bring
the rudiments of a new philosophy within the reach
of a person of ordinary intelligence and culture.

The first seven Lectures of the course deal with *thermometric heat*; its generation and consumption in mechanical
processes; the determination of the mechanical equivalent
of heat; the conception of heat as molecular motion; the
application of this conception to the solid, liquid, and
gaseous forms of matter; to expansion and combustion; to
specific and latent heat; and to calorific conduction.

The remaining five Lectures treat of *radiant heat*; the
interstellar medium, and the propagation of motion through
this medium; the relations of radiant heat to ordinary
matter in its several states of aggregation; terrestrial,
lunar, and solar radiation; the constitution of the sun;
the possible sources of his energy; the relation of this
energy to terrestrial forces, and to vegetable and animal
life.

My aim has been to rise to the level of these questions
from a basis so elementary, that a person possessing any
imaginative faculty and power of concentration, might
accompany me.

Wherever additional remarks, or extracts, seemed likely
to render the reader's knowledge of the subjects referred

to in any Lecture more accurate or complete, I have introduced such extracts, or remarks, as an Appendix to the
Lecture.

For the use of the Plate at the end of the volume, I
am indebted to the Council of the Royal Society; it was
engraved to illustrate some of my own memoirs in the
' Philosophical Transactions.' For some of the Woodcuts
I am also indebted to the same learned body.

To the scientific public, the names of the builders of
this new philosophy are already familiar. As experimental
contributors, Rumford, Davy, Faraday, and Joule, stand
prominently forward. As theoretic writers (placing them
alphabetically), we have Clausius, Helmholtz, Kirchhoff,
Mayer, Rankine, Thomson; and in the memoirs of these
eminent men the student who desires it, must seek a
deeper acquaintance with the subject. MM. Regnault and
Séguin also stand in honourable relationship to the Dynamical Theory of Heat, and M. Verdet has recently published two lectures on it, marked by the learning for which
he is conspicuous. To the English reader it is super-
fluous to mention the well-known and highly-prized work
of Mr. Grove.

I have called the philosophy of Heat a new philosophy,
without, however, restricting the term to the subject of
Heat. The fact is, it cannot be so restricted; for the
connection of this agent with the general energies of the
universe is such, that if we master it perfectly, we master
all. Even now we can discern, though but darkly, the
greatness of the issues which connect themselves with
the progress we have made—issues which were probably
beyond the contemplation of those, by whose industry
and genius the foundations of our present knowledge
were laid.

In a Lecture on the 'Influence of the History of Science on Intellectual Education,' delivered at the Royal Institution, Dr. Whewell has shown 'that every advance in intellectual education has been the effect of some considerable scientific discovery, or group of discoveries.' If the association here indicated be invariable, then, assuredly, the views of the connection and interaction of natural forces—organic as well as inorganic—vital as well as physical—which have grown, and which are to grow, out of the investigation of the laws and relations of Heat, will profoundly affect the intellectual discipline of the coming age.

In the study of Nature two elements come into play, which belong respectively to the world of sense and to the world of thought. We observe a fact and seek to refer it to its laws,—we apprehend the law, and seek to make it good in fact. The one is Theory, the other is Experiment; which, when applied to the ordinary purposes of life, becomes Practical Science. Nothing could illustrate more forcibly the wholesome interaction of these two elements, than the history of our present subject. If the steam-engine had not been invented, we should assuredly stand below the theoretic level which we now occupy. The achievements of Heat through the steam-engine have forced, with augmented emphasis, the question upon thinking minds — 'What is this agent, by means of which we can supersede the force of winds and rivers — of horses and of men? Heat can produce mechanical force, and mechanical force can produce Heat; some common quality must therefore unite this agent and the ordinary forms of mechanical power.' This relationship established, the generalising intellect could pass at once to the other energies of the universe, and it now perceives

the principle which unites them all. Thus the triumphs
of practical skill have promoted the developement of philosophy. Thus, by the interaction of thought and fact,
of truth conceived and truth executed, we have made our
Science what it is,—the noblest growth of modern times,
though as yet but partially appealed to as a source of
individual and national might.

As a means of intellectual education its claims are still
disputed, though, once properly organised, greater and
more beneficent revolutions await its employment here,
than those which have already marked its applications in
the material world.   Surely the men whose noble vocation
it is to systemize the culture of England, can never allow
this giant power to grow up in their midst without endeavouring to turn it to practical account.   Science does
not need their protection, but it desires their friendship
on honourable terms : it wishes to work with them towards
the great end of all education, — the bettering of man's
estate.   By continuing to decline the offered hand, they
invoke a contest which can have but one result.   Science
must grow.   Its developement is as necessary and as irresistible as the motion of the tides, or the flowing of the
Gulf Stream.   It is a phase of the energy of Nature, and
as such is sure, in due time, to compel the recognition, if
not to win the alliance, of those who now decry its influence and discourage its advance.

ROYAL INSTITUTION :
    *February* 1863.

# CONTENTS.

—◆◇◆—

Fusion of an Alloy by this Heat — Measurement of the Amount of Heat generated by a given Expenditure of Force — Dr. Mayer and Mr. Joule — The Mechanical Equivalent of Heat — Definition of the Term 'Foot-pound' — Heat developed increases as the height of the fall, and is proportional to the Square of the Velocity — Calculation of Heat gene-rated by the impact of Projectiles — Heat equivalent to the Stoppage of the Earth in its Orbit — Heat equivalent to the Falling of the Earth into the Sun — Preliminary Statement of Meteoric Theory of the Sun's Heat — Analysis of Combustion — Ignition of Diamond — Its Combustion in Oxygen due to the Showering of the Atoms of the Gas against the Surface of the Diamond — Structure of Flame — Candle and Gas Flames — Combustion on Mont Blanc — The Light of Flames is materially dimin-ished by the Rarefaction of the Air, though the Quantity of Combustible Matter consumed remains the same — Frankland's Experiments — All Cases of Combustion are due to the Collision of Atoms which have been urged together by their mutual Attractions    .    .    .    .    PAGE 23

## LECTURE III.

Expansion of Bodies by Heat — Liberation of Particles from the thrall of Cohesion — The Liquid and Gaseous States of Matter defined — Illustra-tions of the Expansion of Air by Heat — Ascent of Fire Balloon — Gases expand by a constant Increment for every Degree above 32° Fahr.— Coefficient of Expansion — Heating of Gas at a constant Pressure — Heating of Gas at a constant Volume — In the former case Work is done by the Gas — In the latter case no Work is done — In the former case an Excess of Heat equivalent to the Work done must be imparted — Calculation of the Mechanical Equivalent of Heat — Mayer and Joule's Determinations — Absolute Zero of Temperature — Expansion without Refrigeration — Expansion of Liquids — Exceptional Deportment of Water and Bismuth — Energy of Atomic Forces — Pyrometers — Strains and Pressures superinduced by sudden Cooling — Chilling of Metallic Wires by Stretching — Heating of India-rubber by Stretching— Contraction of stretched India-rubber by Heat    .    .    .    .    59

## LECTURE IV.

### APPENDIX TO LECTURE IV.

## LECTURE V.

## LECTURE VI.

## LECTURE VII.

## LECTURE VIII.

# HEAT

CONSIDERED AS

# A MODE OF MOTION.

———•◇•———

## LECTURE I.

[January 23, 1862.]

INSTRUMENTS — GENERATION OF HEAT BY MECHANICAL ACTION —
CONSUMPTION OF HEAT IN WORK.

APPENDIX : — NOTES ON THE THERMO-ELECTRIC PILE AND GALVANOMETER.

THE chief characteristic of Natural Knowledge is its
growth; each fact is vital, and every new discovery
forms a starting point for fresh investigation.    Thus it
seems destined to advance, until the phenomena and laws
of the material universe are entirely subdued by the intellect of man.    But though each department of natural
knowledge has been adding to its store, at a rate unknown
in former times, no branch of it has expanded so rapidly
of late, as that which, in these lectures, is to occupy our
attention.    In scientific manuals but scanty reference has
as yet been made to the modern ideas of Heat, and thus
the public knowledge regarding it is left below the attainable level.    But the reserve is natural, for the subject is
still an entangled one, and, in entering upon it, we must
be prepared to encounter difficulties.    In the whole range

of Natural Science, however, there are none more worthy
of being overcome,—none whose subjugation secures a
greater reward to the worker. For by mastering the laws
and relations of Heat, we make clear to our minds the
interdependence of natural forces generally. Let us then
commence our labours with heart and hope; let us familiarise ourselves with the latest facts and conceptions
regarding this all-pervading agent, and seek diligently
the links of law which underlie the facts and give unity
to their most diverse appearances. If we succeed here
we shall satisfy, to an extent unknown before, that love of
order and of beauty which, I am persuaded, is implanted
in the mind of every person here present. From the
heights at which we aim, we shall have nobler glimpses of
the system of Nature than could possibly be obtained, if I,
while acting as your guide in the region which we are
now about to enter, were to confine myself to its lower
levels and already trodden roads.

It is my first duty to make you acquainted with some
of the instruments which I intend to employ in the examination of this question. I must devise some means of
making the indications of heat and cold visible to you all,
and for this purpose an ordinary thermometer would be
useless. You could not see its action ; and I am anxious
that you should see, with your own eyes, the facts on which
our subsequent philosophy is to be based. I wish to give
you the material on which an independent judgement may
be founded; to enable you to reason as I reason if you deem
me right, to correct me if I go astray, and to censure me
if you find me dealing unfairly with my subject. To
secure these ends I have been obliged to abandon the use
of a common thermometer, and to resort to the little instrument A B (fig. 1), which you see before me on the table,
and which is called a *thermo-electric pile.**

* A brief description of the thermo-electric pile is given in the Appendix
to this Lecture.

By means of this instrument I cause the heat which it receives to generate an electric current. You know, or ought to know, that such a current has the power of deflecting a freely suspended magnetic needle, to which it flows parallel. Before you I have placed such a needle *m n* (fig. 1), surrounded by a covered copper wire, the

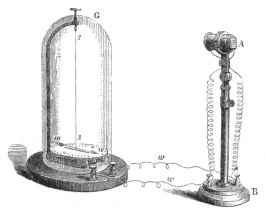

free ends of which, *w w*, are connected with the thermo-electric pile. The needle is suspended by a fibre, *s s*, of unspun silk, and protected by a glass shade, G, from any disturbance by currents of air. To one end of the needle I have fixed a piece of red paper, and to the other end a piece of blue. All of you see these pieces of paper, and when the needle moves, its motion will be clearly visible to the most distant person in this room.*

At the present moment the needle is quite at rest, and points to the zero mark on the graduated disc underneath it. This shows that there is no current passing. I now

---

* In the actual arrangement the galvanometer here described stood on a stool in front of the lecture table, the wires *w w*, being sufficiently long to reach from the table to the stool; for a further description of the galvanometer see the Appendix to this Lecture.

breathe for an instant against the naked face A of the pile
— a single puff of breath is sufficient for my purpose —
observe the effect. The needle starts off and passes through
an arc of 90°. It would go further did I not limit its swing
by fixing, edgeways, a thin plate of mica at 90°. Take
notice of the direction of the deflection; the red end of
the needle moved from me towards you, as if it disliked
me, and had been inspired by a sudden affection for
you. This action of the needle is produced by the small
amount of warmth communicated by my breath to the
face of the pile, and no ordinary thermometer could give
so large and prompt an indication. We will let the heat
thus communicated waste itself; it will do so in a very
short time, and you notice, as the pile cools, that the needle
returns to its first position. Observe now the effect of
*cold* on the face of the pile. I have here some ice, but
I do not wish to wet my instrument by touching it with
ice. Instead of doing so, I will cool this plate of metal
by placing it on the ice; then wipe the chilled metal, and
touch, with it, the face of the pile. You see the effect; a
moment's contact suffices to produce a prompt and energetic deflection of the needle. But mark the direction
of the deflection. When the pile was warmed, the red
end of the needle moved from me towards you; now its
likings are reversed, and the red end moves from you
towards me. Thus you see that cold and heat cause the
needle to move in opposite directions. The important
point here established is, that from the direction in which
the needle moves we can, with certainty, infer whether cold
or heat has been communicated to the pile; and the
energy with which the needle moves — the promptness
with which it is driven aside from its position of rest —
gives us some idea of the comparative quantity of heat
or cold imparted to it in different cases. In a future
lecture I shall explain how we may express the relative

quantities of heat with numerical accuracy; but for the
present a general knowledge of the action of our instruments will be sufficient.

My desire now is to connect heat with the more familiar
forms of force, and I will, therefore, in the first place, try
to furnish you with a store of facts illustrative of the
generation of heat by mechanical processes. I have placed
some pieces of wood in the next room, which my assistant
will now hand to me. Why have I placed them there?
Simply that I may perform my experiments with that sin-
cerity of mind and act which science demands from her cul-
tivators. I know that the temperature of that room is slightly
lower than the temperature of this one, and that hence
the wood which is now before me, must be slightly colder
than the face of the pile with which I intend to test the
temperature of the wood. Let us prove this. I place the
face of the pile against this piece of wood; the red end of
the needle moves from you towards me, thus showing that
the contact has chilled the pile. I now carefully rub the
face of the pile along the surface of the wood; I say 'carefully,' because the pile is a brittle instrument, and rough
usage would destroy it;—mark what occurs. The prompt
and energetic motion of the needle towards you declares
that the face of the pile has been heated by this small
amount of friction. The needle, you observe, goes quite
up to 90° on the side opposite to that towards which it
moved before the friction was applied.

Now these experiments, which illustrate the developement of heat by mechanical means, must be to us what a
boy's school exercises are to him. In order to fix them on
our minds, and obtain due mastery over them, we must
repeat and vary them in many ways. In this task I ask
you to accompany me. Here is a flat piece of brass with
a stem attached to it; I take the stem in my fingers, preserving the brass from all contact with my warm hand, by

enveloping the stem in cold flannel. I place the brass in contact with the face of my pile; the needle moves, showing that the brass is cold. I now rub the brass against the surface of this cold piece of wood, and lay it once more against my pile. I withdraw it instantly, for it is so hot that if I allowed it to remain in contact with the instrument, the current generated would dash my needle violently against its stops, and probably derange its magnetism. You see the strong deflection which even an instant's contact can produce. Indeed, when a boy at school, I have often blistered my hand by the contact of a brass button which I had rubbed energetically against a form. Here, also, is a razor, cooled by contact with ice; and here is a hone, without oil, along which I rub my cool razor as if to sharpen it. I now place the razor against the face of the pile, and you see that the steel, which a minute ago was cold, is now hot. Similarly, I take this knife and knife-board, which are both cold, and rub the knife along the board. I place the knife against the pile, and you observe the result; a powerful deflection which declares the knife to be hot. I pass this cold saw through this cold piece of wood, and place, in the first instance, the surface of the wood against which the saw has rubbed, in contact with the pile. The needle instantly moves in a direction which shows the wood to be heated. I allow the needle to return to zero, and now apply the saw to the pile. It also is hot. These are the simplest and most common-place examples of the generation of heat by friction, and I choose them for this reason. Mean as they appear, they will lead us by degrees into the secret recesses of Nature, and lay open to our view the polity of the material universe.

Let me now make an experiment to illustrate the developement of heat by compression. I have here a piece of deal, cooled below the temperature of the room, and giving, when placed in contact with our pile, the deflection which indicates cold. I place this wood between the plates of a small hydraulic press, and squeeze it forcibly. The plates of the press are also, you will observe, cooler than the air of the room. After compression, I bring the wood into contact with the pile; see the effect. The galvanometer declares that heat has been developed by the act of compression. Precisely the same occurs when I place this lead bullet between the plates of the press and squeeze it thus to flatness.

And now for the effect of percussion. I have here a cold lead bullet, which I place upon this cold anvil, and strike it with a cold sledge-hammer. The sledge descends with a certain mechanical force, and its motion is suddenly destroyed by the bullet and anvil; apparently the force of the sledge is lost. But let us examine the lead; you see it is heated, and could we gather up all the heat generated by the shock of the sledge, and apply it without loss mechanically, we should be able, by means of it, to lift this hammer to the height from which it fell.

I have here arranged another experiment, which is almost too delicate to be performed by the coarse apparatus necessary in a lecture, but which I have made several times before entering this room to-day. Into this small basin I pour a quantity of mercury which has been cooled in the next room. I have coated one of the faces of my thermo-electric pile with varnish, so as to defend it from the mercury which would otherwise destroy the pile; and thus protected I can, as you observe, plunge the pile into the liquid metal. The deflection of the needle shows you that the mercury is cold. Here are two glasses, A and B (fig. 2), swathed thickly round by listing, which will effectually prevent the warmth of my hands from reaching the mercury. Well, I pour the cold mercury from the one glass into the other, and back. It

falls with a certain mechanical force, its motion is destroyed, but heat is developed.    The amount of heat generated by a single pour-
ing out is extremely
small ; I could tell you
the exact amount, but
shall    defer    quanti-
tative    considerations
till  our  next  lecture ;
so I pour the mercury
from glass to glass ten
or fifteen times.   Now
mark the  result when
the pile is plunged into
the   mercury.    The

Fig. 2.

needle moves, and its motion declares that the mercury,
which at the beginning of the  experiment was cooler than
the pile, is now warmer than the pile.  We here introduce
into the lecture-room an effect which occurs in nature at
the base of every waterfall.   There are friends before me
who have stood amid the foam of Niagara.   Had they,
when  there,  dipped sufficiently sensitive thermometers
into the water at the top and bottom of the cataract, they
would have  found  the  latter a little  warmer than  the
former. The sailor's tradition, also, is theoretically correct;
the sea is rendered warmer through the agitation produced
by a storm, the mechanical dash of its billows being ultimately converted into heat.

Whenever friction is overcome, heat is produced, and
the heat produced is the measure of the force expended in
overcoming the friction.   The heat is simply the primitive
force in another form, and if we wish to avoid this conversion we must abolish the friction.  We usually put oil upon
the surface of a hone, we grease a saw, and are careful to
lubricate the axles of our  railway carriages.   What are we

really doing in these cases?   Let us get general notions
first; we shall come to particulars afterwards.   It is
the object of a railway engineer to urge his train bodily
from one place to another; say from London to Edinburgh, or from London to Oxford, as the case may be; he
wishes to apply the force of his steam, or of his furnace,
which gives tension to the steam, to this particular purpose.   It is not his interest to allow any portion of that
force to be converted into another form of force which
would not further the attainment of his object.   He
does not want his axles heated, and hence he avoids as
much as possible expending his power in heating them.
In fact he has obtained his force from heat, and it is
not his object to reconvert the force thus obtained into
its primitive form.   For, for every degree of temperature
generated by the friction of his axles, a definite amount
would be withdrawn from the urging force of his engine.
There is no force lost absolutely.   Could we gather up all
the heat generated by the friction, and could we apply it
all mechanically, we should, by it, be able to impart to the
train the precise amount of speed which it had lost by the
friction.   Thus every one of those railway porters whom
you see moving about with his can of yellow grease, and
opening the little boxes which surround the carriage axles,
is, without knowing it, illustrating a principle which forms
the very solder of Nature.   In so doing, he is unconsciously
affirming both the convertibility and the indestructibility
of force.   He is practically asserting that mechanical
energy may be converted into heat, and that when so converted it cannot still exist as mechanical energy, but that
for every degree of heat developed, a strict and proportional equivalent of the *locomotive force* of the engine
disappears.   A station is approached, say at the rate of
thirty or forty miles an hour; the brake is applied, and
smoke and sparks issue from the wheel on which it

presses. The train is brought to rest — How? Simply by converting the entire moving force which it possessed, at the moment the brake was applied, into heat.

So also with regard to the greasing of a saw by a carpenter. He applies the muscular force of his arm with the express object of getting through the wood. He wishes to tear the wood asunder, to overcome its mechanical cohesion by the teeth of his saw. When the saw moves stiffly, on account of the friction against its flat surface, the same amount of force may produce a much smaller effect than when the implement moves without friction. But in what sense smaller? Not absolutely so, but smaller as regards the act of sawing. The force not expended in the sawing is not lost, it is converted into heat, and I gave you an example of this a few minutes ago. Here again, if we could collect the heat engendered by the friction, and apply it to urge the saw, we should make good the precise amount of work which the carpenter, by neglecting the lubrication of his implement, had simply converted into another form of power.

We warm our hands by rubbing, and in the case of frostbite we thus restore the necessary heat to the injured parts. Savages have the art of producing fire by the skilful friction of well-chosen pieces of wood. It is easy to char wood in a lathe by friction. From the feet of the labourers on the roads of Hampshire sparks issue copiously on a dark night, the collision of their iron-shod shoes against the flints producing the effect. In the common flint and steel the particles of the metal struck off are so much heated by the collision that they take fire and burn in the air. But the heat precedes the combustion. Davy found that when a gunlock, with a flint, was discharged in vacuo, no sparks were produced, but the particles of steel struck off, when examined under the microscope, showed signs of fusion.* Here is a large rock-crystal, I have only to draw

* Works of Sir H. Davy, vol. ii. p. 8.

this small one briskly along it, to produce a stream of light; here are two quartz-pebbles, I have only to rub them together to make them luminous.

A bullet, in passing through the air, is warmed by the friction, and the most probable theory of shooting stars is that they are small planetary bodies, revolving round the sun, which are caused to swerve from their orbits by the attraction of the earth, and are raised to incandescence by friction against our atmosphere. Mr. Joule has shown that the atmospheric friction is competent to produce the effect; and he may be correct in believing that the greater portion of the aërolites are dissipated by heat, and the earth thus spared a terrible bombardment.* These bodies move with planetary velocity; the orbital velocities of the four interior planets are as follows: —

|         |   |   |   |   |   | Miles per Second |
|---------|---|---|---|---|---|-----------------|
| Mercury | . | . | . | . | . | 30·40 |
| Venus   | . | . | . | . | . | 22·24 |
| Earth   | . | . | . | . | . | 18·91 |
| Mars .  | . | . | . | . | . | 15·32 |

while the velocity of the aërolites varies from 18 to 36 miles a second.† The friction engendered by this enormous speed is certainly competent to produce the effects ascribed to it.

More than sixty-four years ago Count Rumford, who was one of the founders of the Royal Institution, executed a series of experiments on the generation of heat by friction, which, viewed by the light of to-day, are of the highest interest and importance. Indeed the services which this Institution has rendered, in connection with this question of the brotherhood of natural forces, can never be forgotten. Thomas Young, a former professor of this Institution, laid the foundations of the undulatory theory of light, which, in its fullest application, embraces our present theory of heat.

---

* Philosophical Magazine, 4th Series, vol. xxxii. p. 349.
† Galbraith and Houghton's Manual of Astronomy, p. 18.

Davy entertained substantially the same views regarding
heat as those which I am now endeavouring to approach
and elucidate.   Faraday established the laws of equivalence
between chemistry and electricity, and his magneto-
electric discoveries were the very first seized upon by Mr.
Joule, in illustration of the mutual convertibility of heat
and mechanical action.*   Rumford, in a paper of great
power both as regards reasoning and experiment, advocated
in 1798† the doctrine regarding the nature of heat which
the recent experiments of eminent men have placed upon
a secure basis.   While engaged in the boring of cannon at
Munich, he was so forcibly struck by the large amount of
heat developed in the process of boring that he was induced
to devise an apparatus for the special examination of the
generation of heat by friction.   He had constructed a hollow
cylinder of iron, into which fitted a solid plunger, which
was caused to press against the bottom of the cylinder.   A
box which surrounded the cylinder contained 18¾lbs. of
water, in which a thermometer was placed.   The original
temperature of the water was 60°.   The cylinder was turned
by horse-labour, and an hour after the friction had commenced the temperature of the water was 107°, having been
raised 47°.   Half-an-hour afterwards he found the temperature to be 142°.   The action was continued, and at the end
of two hours the temperature was 178°.   At the end of two
hours and twenty minutes it was 200°, and at two hours and
thirty minutes from the commencement *the water actually
boiled* !   Rumford's description of the effect of this experiment on those who witnessed it, is quite delightful.   ' It
would be difficult,' he says, ' to describe the surprise and astonishment expressed in the countenances of the bystanders
on seeing so large a quantity of water heated, and actually
made to boil, without any fire.   Though there was nothing

* Philosophical Magazine, 4th Series, vol. xxiii. pp. 265, 347, 435.

† An abstract of this paper is given in the Appendix to Lecture II.

that could be considered very surprising in this matter, yet
I acknowledge fairly that it afforded me a degree of
childish pleasure which, were I ambitious of the reputation
of a grave philosopher, I ought most certainly rather to
hide than to discover.'*  I am sure that both you and I
can dispense with the application of any philosophy which
would stifle such emotion as Rumford here avowed.  In
connection with this striking experiment, Mr. Joulet† has
estimated the amount of mechanical force expended in
producing the heat, and obtained a result which 'is not
very widely different' from that which greater knowledge
and more refined experiments enabled Mr. Joule himself
to obtain, as regards the numerical equivalence of heat
and work.

It would be absurd on my part to attempt here a repetition of the experiment of Count Rumford with all its
conditions.  I cannot devote two hours and a half to a
single experiment, but I hope to be able to show you substantially the same effect in *two minutes and a half*.  I
have here a brass tube, four inches long, and three quarters

Fig. 3.

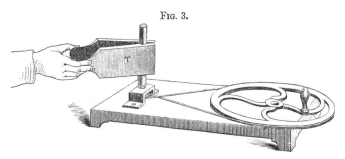

of an inch in interior diameter.  It is stopped at the
bottom, and I thus screw it on to a whirling table, by
means of which I can cause the upright tube to rotate very

* Rumford's Essays, vol. ii. p. 484.
† Philosophical Transactions, vol. cxl. p. 62.

rapidly. I have here two pieces of oak wood, united by a
hinge, and in which are two semicircular grooves, which
are intended to embrace the brass tube. Thus the pieces of
wood form a kind of tongs, T (fig. 3), by gently squeezing
which I can produce friction between the wood and the
brass tube, when the latter rotates. I almost fill the tube
with cold water, and stop it with a cork to prevent the
splashing out of the liquid, and now I put the machine in
motion. As the action continues the temperature of the
water rises, and though the two minutes and a half have
not yet elapsed, those near the apparatus will see steam
escaping from the cork. Three or four times to-day I
have projected the cork by the force of the steam to a
height of twenty feet in the air. There it goes again, and
the steam follows it, producing by its precipitation this
small cloud in the atmosphere.

In all the cases hitherto introduced to your notice, heat
has been *generated* by the expenditure of mechanical
force. Our experiments have gone to show that where
mechanical force is expended, heat is produced, and I wish
now to bring before you the converse experiment, that is,
the *consumption* of heat in mechanical work. And should
you at present find it difficult to form distinct conceptions
as to the bearing of these experiments I exhort you to be
patient. We are engaged on a difficult and entangled
subject, which, I hope, we shall disentangle as we go along.
I have here a strong vessel filled at the present moment
with compressed air. It has been now compressed for
some hours, so that the temperature of the air within the
vessel is the same as that of the air of the room without
it. At the present moment, then, this inner air is pressing
against the sides of the vessel, and if I open this cock a
portion of the air will rush violently out of the vessel.
The word ' rush,' however, but vaguely expresses the true
state of things; the air which rushes out is driven out by the

air behind it; this latter accomplishes the work of urging
forward the stream of air. And what will be the condition
of the *working air* during this process? It will be chilled.
It performs mechanical work, and the only agent which it
can call upon to perform it is the heat which it possesses,
and to which the elastic force with which it presses against
the sides of the vessel, is entirely due. A portion of this
heat will be consumed and the air will be chilled. Observe the experiment which I am about to make. I will
turn the cock *c*, and allow the current of air from the vessel
v (fig. 4), to strike against the face of the pile P. See how
the magnetic needle responds to the act; its red end is

Fig. 4,

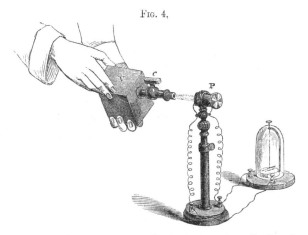

driven towards me, thus declaring that the pile has been
*chilled* by the current of air.

The effect is different when a current of air is urged from
the nozzle of a common bellows against the thermo-electric
pile. In the last experiment the mechanical work of urging
the air forward was performed by the air itself, and a portion of its heat was consumed in the effort. In the case of
the bellows, it is my muscles which perform the work. I
raise the upper board of the bellows and the air rushes in, I

press the boards with a certain force and the air rushes out.
The expelled air strikes the face of the pile, has its motion

FIG. 5.

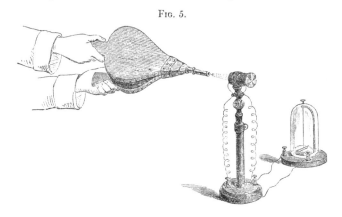

stopped, and an amount of heat equivalent to the destruction of this motion is instantly generated. Thus you observe
that when I urge with the bellows B (fig. 5), a current of
air against the pile, the red end of the needle moves towards
you, thereby showing that the face of the pile has been, in
this instance, *warmed* by the air.   I have here a bottle
of soda water: at present the bottle is slightly warmer
than the pile, as you see by the deflection it produces;
I cut the string which holds the cork and it is driven
out by the elastic force of the carbonic acid gas; the
gas performs work, in so doing consumes heat, and now
the deflection it produces is that of cold.   The truest
romance is to be found in the details of daily life, and here,
in operations with which every child is familiar, we shall
gradually discern the illustration of principles from which
all material phenomena flow.

# APPENDIX TO LECTURE I.

— ◦◇◦ —

NOTE ON THE CONSTRUCTION OF THE THERMO-ELECTRIC
PILE.

LET A B (fig. 6) be a bar of antimony, and B C a bar of bismuth, and
let both bars be soldered together at B. Let the free ends A and C be
united by a piece of wire, A D C. On warming
the place of junction, B, an electric current is
generated, the direction of which is from bismuth
to antimony (B to A, or against the alphabet),
across the junction, and from antimony to bismuth
(A to B, or with the alphabet), through the connecting wire, A D C. The arrow indicates the
direction of the current.

FIG. 6.

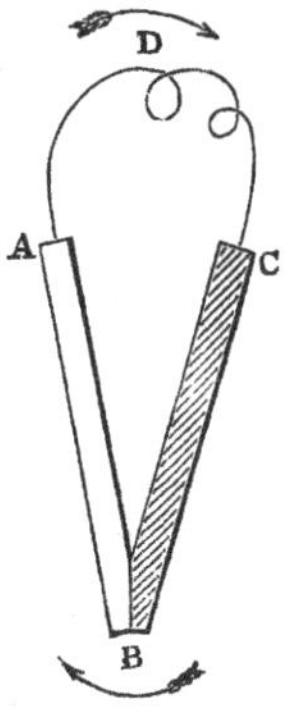

If the junction B be *chilled*, a current is generated opposed in direction to the former. The
figure represents what is called a thermo-electric
pair or couple.

By the union of several thermo-electric pairs
a more powerful current can be generated than
would be obtained from a single pair. Fig. 7, for
example, represents such an arrangement, in which the shaded
bars are supposed to be all of bismuth, and the unshaded
ones of antimony; on warming all the junctions, B, B, &c., a
current is generated in each, and the sum of these currents, all
of which flow in the same direction, will produce a stronger
resultant current than that obtained from a single pair.

The V formed by each pair need not be so wide as it is shown
in fig. 7; it may be contracted without prejudice to the couple.
And if it is desired to pack several pairs into a small compass,

each separate couple may be arranged as in fig. 8, where the black lines represent small bismuth bars, and the white ones small bars of antimony. They are soldered together at the ends, and throughout the length are usually separated by strips of paper

FIG. 7.

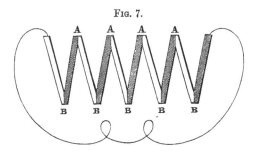

merely. A collection of pairs thus compactly set together constitutes a thermo-electric pile, a drawing of which is given in fig. 9.

The current produced by heat being always from bismuth to antimony across the heated junction, a moment's inspection of fig. 7 will show that when any one of the junctions A, A is heated, a current is generated opposed in direction to that generated when the heat is applied to the junctions B, B. Hence, in the case of the thermo-electric pile, the effect of heat falling upon its two

FIG. 8.                    FIG. 9.

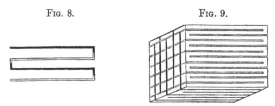

opposite faces is to produce currents in opposite directions. If the temperature of the two faces be alike, they neutralise each other, no matter how high they may be heated absolutely, but if one of them be warmer than the other, a current is produced. The current is thus due to a *difference* of temperature between the two faces of the pile, and within certain limits the strength of the current is exactly proportional to this difference.

From the junction of almost any other two metals, thermo-electric currents may be obtained, but they are most copiously generated by the union of bismuth and antimony.*

### NOTE ON THE CONSTRUCTION OF THE GALVANOMETER.

The existence and direction of an electric current are shown by its action upon a freely suspended magnetic needle.

But such a needle is held in the magnetic meridian by the magnetic force of the earth. Hence, to move a single needle the current must overcome the magnetic force of the earth.

Very feeble currents are incompetent to do this in a sufficiently sensible degree. The following two expedients are, therefore, combined to render sensible the action of such feeble currents: —

The wire through which the current flows is coiled so as to surround the needle several times ; the needle must swing freely within the coil. The action of the single current is thus multiplied.

The second device is to neutralise the directive force of the earth, without prejudice to the magnetism of the needle. This is accomplished by using two needles instead of one, attaching them to a common vertical stem, and bringing their opposite poles over each other, the north end of the one needle, and the south end of the other, being thus turned in the same direction. The double needle is represented in fig. 10.

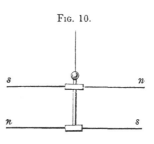

FIG. 10.

It must be so arranged that one of the needles shall be within the coil through which the current flows, while the other needle

---

* The discovery of thermo-electricity is due to Thomas Seebeck, Professor in the University of Berlin. Nobili constructed the first thermo-electric pile but in Melloni's hands it became an instrument so important as to supersede all others in researches on radiant heat. To this purpose it will be applied in future lectures.

swings freely above the coil, the vertical connecting piece passing
through an appropriate slit in the coil. Were both the needles
within, the same current would urge them in opposite directions,
and thus one needle would neutralise the other. But when
one is within and the other without, the current urges both
needles in the same direction.

The way to prepare such a pair of needles is this. Magnetise
both of them to saturation; then suspend them in a vessel, or
under a shade, so as to protect them from air-currents. The
system will probably set in the magnetic meridian, one needle
being in almost all cases stronger than the other; weaken the
stronger needle carefully by the touch of a second smaller
magnet. When the needles are precisely equal in strength, they
will set at *right angles to the magnetic meridian.*

It might be supposed that when the needles are equal in
strength, the directive force of the earth would be completely
annulled, that the double needle would be perfectly *astatic,*
and perfectly neutral as regards direction; obeying simply the
torsion of its suspending fibre. This would be the case if the
magnetic axes of both needles could be caused to lie with mathe-

FIG. 11.

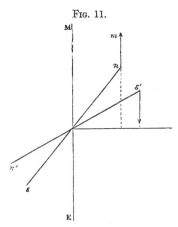

matical accuracy in the same vertical plane. In practice this is
next to impossible; the axes always cross each other. Let *n s,*
*n′ s′* (fig. 11) represent the axes of two needles thus crossing, the
magnetic meridian being parallel to M E; let the pole *n* be drawn

by the earth's attractive force in the direction $n\, m$; the pole $s'$ being urged by the repulsion of the earth in a precisely opposite direction. When the poles $n$ and $s'$ are of exactly equal strength it is manifest that the force acting on the pole $s'$, in the case here supposed, would have the advantage as regards leverage, and would therefore overcome the force acting on $n$. The crossed needles would therefore turn away still further from the magnetic meridian, and a little reflection will show that they cannot come to rest until the line which bisects the angle enclosed by the needles is at right angles to the magnetic meridian.

This is the test of perfect equality as regards the magnetism of the needles; but in bringing the needles to this state of perfection we have often to pass through various stages of obliquity to the magnetic meridian. In these cases the superior strength of one needle is compensated by an advantage, as regards leverage, possessed by the other. By a happy accident a touch is sometimes sufficient to make the needles perfectly equal; but many hours are often expended in securing this result. It is only of course in very delicate experiments that this perfect equality is needed; but in such experiments it is essential.

Another grave difficulty has beset experimenters, even after the perfect magnetisation of their needles has been accomplished. Such needles are sensitive to the slightest magnetic action, and the covered copper wire, of which the galvanometer coils are formed, usually contains a trace of iron sufficient to deflect the prepared needle from its true position. I have had coils in which this deflection amounted to 30 degrees; and in the splendid instruments used by Professor Du Bois Raymond, in his researches on animal electricity, the deflection by the coil is sometimes even greater than this. Melloni encountered this difficulty, and proposed that the wires should be drawn through agate holes, thus avoiding all contact with iron or steel. The disturbance has always been ascribed to a trace of iron contained in the copper wire. Pure silver has also been proposed instead of copper.

To pursue his beautiful thermo-electric researches in a satisfactory manner, Professor Magnus, of Berlin, obtained pure copper, by a most laborious electrolytic process, and after the metal had been obtained, it required to be melted eight times in succession before it could be drawn into wire. In fact the impurity of the coil entirely vitiated the accuracy of the instruments, and almost any amount of labour would be well expended in removing this great defect.

My own experience of this subject is instructive. I had a beautiful instrument constructed a few years ago by Sauerwald, of Berlin, the coil of which, when no current flowed through it, deflected my double needle full 30 degrees from the zero line. It was impossible to attain quantitative accuracy with this instrument.

I had the wire removed by Mr. Becker, and English wire used in its stead; the deflection fell to 3 degrees.

This was a great improvement, but not sufficient for my purpose. I commenced to make enquiries about the possibility of obtaining pure copper, but the result was very discouraging, when, almost despairing, the following thought occurred to me :— The action of the coil must be due to the admixture of iron with the copper, for pure copper is diamagnetic, it is feebly *repelled* by a strong magnet. The magnet therefore occurred to me as a means of instant analysis; I could tell by it, in a moment, whether any wire was free from the magnetic metal or not.

The wire of M. Sauerwald's coil was strongly attracted by the magnet. The wire of Mr. Becker's coil was also attracted, though in a much feebler degree.

Both wires had been covered by green silk; I removed this, but the Berlin wire was still attracted; the English wire, on the contrary, when presented *naked* to the magnet was feebly *repelled ;* it was truly diamagnetic, and contained no sensible trace of iron. Thus the whole annoyance was fixed upon the green silk; some iron compound had been used in the dyeing of it, and to this the deviation of the needle from zero was manifestly due.

I had the green coating removed and the wire overspun with white silk, clean hands being used in the process. A perfect galvanometer is the result; the needle, when released from the action of the current, returns accurately to zero, and is perfectly free from all magnetic action on the part of the coil. In fact, while we have been devising agate plates and other learned methods to get rid of the nuisance of a magnetic coil, the means of doing so are at hand. Let the copper wire be selected by the magnet, and no difficulty will be experienced in obtaining specimens magnetically pure

# LECTURE II.

[January 30, 1862.]

APPENDIX :—EXTRACTS FROM BACON AND RUMFORD.

IN our last lecture the developement of heat by mechanical action was illustrated by a series of experiments, which showed that heat was easily produced by friction, by compression, and by percussion. But facts alone cannot satisfy the human mind; we desire to know the inner and invisible cause of the fact; we search after the principle by the operation of which the phenomena are produced. Why should heat be generated by mechanical action, and what is the real nature of the agent thus generated? Two rival theories have been offered in answer to these questions. Till very lately, however, one of these—*the material theory* —had the greater number of adherents, being opposed by only a few eminent men. Within certain limits this theory involved conceptions of a very simple kind, and this simplicity secured its general acceptance. The material theory supposes heat to be a kind of matter—a subtle fluid stored up in the inter-atomic spaces of bodies. The laborious Gmelin, for example. in his Handbook of Chemistry, defines heat to be 'that substance whose entrance into our bodies causes the sensation of warmth, and its egress the sensation of cold.'* He also speaks of heat combining with bodies as one ponderable substance does with another; and many other eminent chemists treat the subject from the same point of view.

The developement of heat by mechanical means, inasmuch as its generation seemed unlimited, was a great difficulty with the materialists; but they were acquainted with the fact (which I shall amply elucidate in a future lecture), that different bodies possessed different powers of holding heat, if I may use such a term. Take, for example, the two liquids, water and mercury, and warm up a pound of each of them, say from fifty degrees to sixty. The absolute quantity of heat required by the water to raise its temperature 10° is fully thirty times the quantity required by the mercury. Technically speaking, the water is said to have a greater *capacity* for heat than the mercury has, and this term 'capacity' is sufficient to suggest the views of those who invented it. The water was supposed to possess the power of storing up the caloric or matter of heat; of hiding it, in fact, to such an extent that it required thirty measures of this caloric to produce the same sensible effect on it, that one measure would produce upon mercury.

All substances possess, in a greater or less degree, this apparent power of storing up heat. Lead, for example, possesses it; and the experiment with the lead bullet, in which you saw heat generated by compression, was explained by those who held the material theory in the following way. The uncompressed lead, they said, has a higher capacity for heat than the compressed substance; the size of its atomic storehouse is diminished by compression, and hence, when the lead is squeezed, a portion

* English Translation, vol. i. p. 22.

of that heat which, previous to compression, was hidden, must make its appearance, for the compressed substance can no longer hold it all.   In some similar way the experiments on friction and percussion were accounted for. The idea of calling *new heat* into existence was rejected by the believers in the material theory.   According to their views, the quantity of heat in the universe is as constant as the quantity of ordinary matter, and the utmost we can do by mechanical and chemical means, is to store up this heat or to drive it from its lurking places into open light of day.

The *dynamical theory*, or, as it is sometimes called, the *mechanical theory* of heat, discards the idea of materiality as applied to heat.   The supporters of this theory do not believe heat to be matter, but an accident or condition of matter ; namely, *a motion of its ultimate particles*.   From the direct contemplation of some of the phenomena of heat a profound mind is led almost instinctively to conclude that heat is a kind of motion.   Bacon held a view of this kind,* and Locke stated a similar view with singular felicity. ' Heat,' he says, ' is a very brisk agitation of the insensible parts of the object, which produce in us that sensation from whence we denominate the object hot : so what in our sensation is *heat*, in the object is nothing but *motion*.' In our last lecture I referred to the experiments of Count Rumford † on the boring of cannon ; he showed that the hot chips cut from his cannon did not change their capacity for heat ; he collected the scales and powder produced by the abrasion of his metal, and holding them up before his opponents, demanded whether they believed that the vast amount of heat which he had generated had been all

* See Appendix to this Lecture.

† I have particular pleasure in directing the reader's attention to an abstract of Count Rumford's memoir on the Generation of Heat by Friction, contained in the Appendix to this lecture.   Rumford, in this memoir, annihilates the material theory of heat.   Nothing more powerful on the subject has since been written.

squeezed out of that modicum of crushed metal ? 'You have not,' he might have added, 'given yourselves the trouble to enquire whether any change whatever has occurred in the capacity of the metal for heat by the act of friction. You are quick in inventing reasons to save your theory from destruction, but slow to enquire whether these reasons are not merely the finespun fancies of your own brains.' Theories are indispensable, but they sometimes act like drugs upon the mind. Men grow fond of them as they do of dram-drinking, and often feel discontented and irascible when the stimulant to the imagination is taken away.

At this point an experiment of Davy comes forth in its true significance.* Ice is solid water, and the solid has only one-half the capacity for heat that liquid water possesses. A quantity of heat which would raise a pound of ice ten degrees in temperature, would raise a pound of water only five degrees. Further, to simply liquefy a mass of ice, an enormous amount of heat is necessary, this heat being so utterly absorbed or rendered 'latent' as to make no impression upon the thermometer. The question of 'latent heat' shall be fully discussed in a future lecture; what I am desirous of impressing on you at present is, that *liquid water*, at its freezing temperature, possesses a vastly greater amount of heat than *ice* at the same temperature.

Davy reasoned thus: 'If I, by friction, liquefy ice, I produce a substance which contains a far greater absolute amount of heat than the ice; and in this case, it cannot with any show of reason be affirmed that I merely render sensible the heat hidden in the ice, for that quantity is only a small fraction of the heat contained in the water.' He made the experiment, and liquefied the ice by pure friction; and the result has been regarded as the first which proved the immateriality of heat.

* Works of Sir H. Davy, vol. ii. p. 11.

When a hammer strikes a bell the motion of the hammer is arrested, but its force is not destroyed; it has thrown the bell into vibrations, which affect the auditory nerve as sound. So, also, when our sledge hammer descended upon our lead bullet, the descending motion of the sledge was arrested : but it was not destroyed. *Its motion was transferred to the atoms of the lead,* and announced itself to the, proper nerves as heat. The theory, then, which Rumford so powerfully advocated, and Davy so ably supported,* was, that heat is a kind of molecular motion; and that by friction, percussion and compression, this motion may be generated, as well as by combustion. This is the theory which must gradually develope itself during these lectures, until your minds attain to perfect clearness regarding it. And remember, we are entering a jungle and must not expect to find our way clear. We are striking into the brambles in a random fashion at first; but we shall thus become acquainted with the general character of our work, and with due persistence shall, I trust, cut through all entanglement at last.

In our first lecture I showed you the effect of projecting a current of compressed air against the face of the thermo-electric pile. You saw that the instrument was chilled by the current of air. Now *heat* is known to be developed when air is compressed; and, since last Thursday, I have been asked how this heat was disposed of in the case of the condensed air. Pray listen to my reply. Supposing the vessel which contained the compressed air to be formed of a substance perfectly impervious to heat, and

---

* In Davy's first scientific memoir, he calls heat a repulsive motion, which he says may be augmented in various ways. 'First, by the transmutation of mechanical into repulsive motion; that is, by friction or percussion. In this case the mechanical motion lost by the masses of matter in friction is the repulsive motion gained by their corpuscles:' an extremely remarkable passage. I have given further extracts from this paper in the Appendix to Lecture III.

supposing all the heat developed by my arm, in compressing the air, to be retained within the vessel, *that* quantity
of heat would be exactly competent to undo what I had
done, and to restore the compressed air to its original volume and temperature.    But this vessel v (fig.
12), is not impervious to heat, and it was not my
object to draw upon the heat developed by my arm; I

FIG. 12.

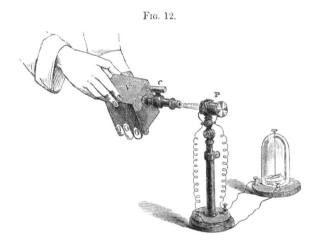

therefore, after condensing the air, allowed the vessel to
rest, till all the heat generated by the condensation had
been dissipated, and the temperature of the air within
and without the vessel was the same.    When, therefore,
the air rushed out, it had not the heat to draw upon,
which had been developed during compression.    The heat
from which it derived its elastic force was only sufficient
to keep it at the temperature of the surrounding air.    In
doing its work a portion of this heat, equivalent to the
work done, was consumed, and the issuing air was consequently chilled.    Do not be disheartened if this reasoning
should not appear quite clear to you.    We are now in

comparative darkness, but as we proceed light will gradually appear, and irradiate retrospectively our present gloom.

I wish now to make evident to you that heat is developed by the compression of air. Here is a strong cylinder of glass T U (fig. 13), accurately bored, and quite smooth within. Into it this piston fits air-tight, so that, by driving the piston down, I can forcibly compress the air underneath it; and when the air is thus com-  FIG. 13. pressed, heat is suddenly generated. Let me prove this. I take a morsel of cotton wool, and wet it with this volatile liquid, the bisulphide of carbon. I throw this bit of wetted cotton into the glass syringe, and instantly eject it. It has left behind it a small residue of vapour. I compress the air suddenly, and you see a flash of light within the syringe. The heat developed by the compression has been sufficient to ignite the vapour. It is not necessary to eject the wetted cotton; I replace it in the tube, and urge the piston downwards; you see the flash as before. If, with this narrow glass tube, I blow out the fumes generated by the combustion of the vapour, I can, without once removing the cotton from the syringe, repeat the experiment twenty times.*

I have here arranged an experiment intended to give you another illustration of the thermal effect produced in air by its own mechanical action. Here is a tin tube, stopped at both ends, and connected with this air-pump. The tin tube is at present full of air, and I bring the face of my pile up against the curved surface of the tube. The instrument declares that the face of the pile in contact with

* The accident which led to this form of the experiment is referred to in the Appendix to this Lecture.

the tin tube has been warmed by the latter. I was quite
prepared for this result, having reason to know that the
air within the tube is slightly warmer than that without.
Now what you are to observe is this:— My assistant shall
work the pump; the cylinders of the machine will be emp-
tied of air, and the air within this tin tube will be driven
into the exhausted cylinders by its own elastic force. I
have already demonstrated the chilling effect of a current
of compressed air on the thermo-electric pile. In the present
experiment I will not examine the thermal condition of the
current at all, but of the vessel in which the work has been
performed. As this tin tube is exhausted I expect to see
the needle, which is now deflected so considerably in the
direction of heat, descend to zero, and pass quite up to
90° in the direction of cold. The pump is now in action,
and observe the result. The needle falls as predicted, and
its advance in the direction of cold is only arrested by its
concussion against the stops.

Three strokes of the pump suffice to chill the tube so as
to send the needle up to 90°,* let it now come to rest. It
would require more time than we can afford to allow the
tube to assume the temperature of the air around it; but
the needle is now sensibly at rest at a good distance on the
cold side of zero. I will now allow a quantity of air to
enter the tube, equal to that which was removed from it a
moment ago by the air-pump. I can turn on this cock,
the air will enter, and each of its atoms will hit the inner
surface of the tube like a projectile. The mechanical
motion of the atom will be thereby annihilated, but an
amount of heat equivalent to this motion will be generated.
Thus as the air enters it will develope an amount of heat

* The galvanometer used in this experiment was that which I employ in my
original researches: it is an exceedingly delicate one. When introduced in
the lectures its dial was illuminated by the electric light; and an image of it,
two feet in diameter, was projected on the screen.

sufficient to re-warm the tube, to undo the present deflection, and to send the needle up on the heat side of zero.
The air is now entering, and you see the effect: the
needle moves, and goes quite up to 90° on that side which
indicates the heating the pile.*

I have now to direct your attention to an interesting
effect connected with this chilling of the air by rarefaction.
I place over the plate of the air-pump a large glass
receiver, which is now filled with the air of this room.
This air, and, indeed, all air, unless it be dried artificially,
contains a quantity of aqueous vapour which, as vapour, is perfectly invisible. A certain temperature is
requisite to maintain the vapour in this invisible state, and
if the air be chilled so as to bring it below this temperature, the vapour will instantly condense, and form a visible
cloud. Such a cloud, which you will remember is not *vapour*, but *liquid water* in a state of fine division, will
form within this glass vessel R (fig. 14), when the air is
pumped out of it; and to make this effect visible to everybody present, to those right and left of me, as well as to
those in front, these six little gas jets are arranged in a
semicircle which half surrounds the receiver. Each person
present sees one or more of these jets on looking through
the receiver, and when the cloud forms, the dimness which
it produces will at once declare its presence. The pump

* In this experiment a mere line along the surface of the tube was in contact with the face of the pile, and the heat had to propagate itself through
the tin envelope to reach the instrument. Previous to adopting this arrangement I had the tube pierced, and a separate pile, with its naked face turned
inwards, cemented air-tight into the orifice. The pile came thus into direct
contact with the air, and its entire face was exposed to the action. The
effects thus obtained were very large; sufficient, indeed, to swing the needle
quite round. My desire to complicate the subject as little as possible induced me to abandon the cemented pile, and to make use of the instrument
with which my audience had already become familiar. With the arrangement actually adopted the effects were, moreover, so large, that I drew
only on a portion of my power to produce them.

is now quickly worked; a very few strokes suffice to precipitate the vapour; there it spreads throughout the entire
receiver, and many of you see a colouring of the cloud, as

Fig. 14.

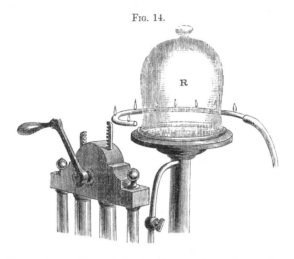

the light shines through it, similar to that observed sometimes, on a large scale, around the moon. When I allow
the air to re-enter the vessel, it is heated, exactly as in the
experiment with our tin tube; the cloud melts away, and
the perfect transparency of the air within the receiver is
restored. Again I exhaust and again the cloud forms;
once more the air enters and the cloud disappears; the
heat developed being more than sufficient to preserve it in
the state of pure vapour.

Sir Humphry Davy refers, in his 'Chemical Philosophy,'
to a machine at Schemnitz, in Hungary, in which air was
compressed by a column of water 260 feet in height.
When a stopcock was opened so as to allow the air to
escape, a degree of cold was produced which not only precipitated the aqueous vapour diffused in the air, but
caused it to congeal in a shower of snow, while the pipe

from which the air issued became bearded with icicles.
'Dr. Darwin,' writes Davy, 'has ingeniously explained the
production of snow on the tops of the highest mountains,
by the precipitation of vapour from the rarefied air which
ascends from plains and valleys. The Andes, placed almost
under the line, rise in the midst of burning sands; about
the middle height is a pleasant and mild climate; the
summits are covered with unchanging snows.'

I would now request your attention to another experiment, in which heat will be developed by what must appear
to many of you a very mysterious agency, and indeed
the most instructed amongst us know, in reality, very little
about the subject. I wish to develope heat by what might
be regarded as friction *against pure space*. And indeed
it may be, and probably is, due to a kind of friction against
that inter-stellar medium, to which we shall have occasion
to refer more fully by and by.

I have here a mass of iron—part of a link of a huge
chain cable—which is surrounded by these multiple coils
of copper wire C C (fig. 15), and which I can instantly
convert into a powerful magnet by sending an electric
current through the wire. You see, when thus excited,
how powerful it is. This poker clings to it, and these
chisels, screws, and nails cling to the poker. Turned
upside down, this magnet will hold a half hundred
weight attached to each of its poles, and probably a score
of the heaviest people in this room if suspended from the
weights. At the proper signal my assistant will interrupt
the electric current:—'Break!' The iron falls and all
the magic disappears: the magnet now is mere common
iron. At the ends of the magnet I place two pieces of
iron P P — movable poles, as they are called — which,
when the magnet is unexcited, I can bring within any
required distance of each other. When the current
passes, these pieces of iron virtually form parts of the

D

Fig. 15.

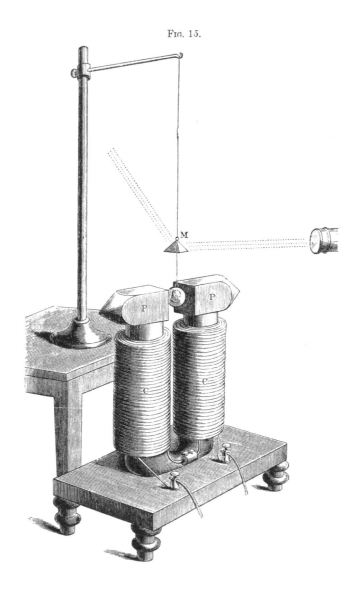

magnet. Between them I will place a substance which
the magnet, even when exerting its utmost power, is
incompetent to attract. This substance is simply a piece
of silver—in fact, a silver medal. I bring it close to the
excited magnet; no attraction ensues. Indeed what little
force—and it is so little as to be utterly insensible in these
experiments—the magnet really exerts upon the silver, is
repulsive instead of attractive.

Well, I suspend this medal between the poles P P
of the magnet, and excite the latter. The medal hangs
there; it is neither attracted nor repelled, but if I seek to
move it I encounter resistance. To turn the medal round
I must overcome this resistance; the silver moves as if it
were surrounded by a viscous fluid. This curious effect
may also be rendered manifest, thus: I have here a rectangular plate of copper, and if I cause it to pass quickly
to and fro like a saw between the poles, when their points
are turned towards it, I seem, though I can see nothing,
to be sawing through a mass of cheese or butter.* Nothing
of this kind is noticed when the magnet is not active: the
copper saw then encounters nothing but the infinitesimal
resistance of the air. Thus far you have been compelled
to take my statements for granted, but I have arranged
an experiment which will make this strange action of the
magnet on the silver medal, strikingly manifest to everybody present.

Above the suspended medal, and attached to it by a bit
of wire, I have a little reflecting pyramid M, formed of four
triangular pieces of looking-glass; both the medal and the
reflector are suspended by a thread which was twisted in
its manufacture, and which will untwist itself when the
weight which it sustains is set free. I place our electric
lamp so as to cast a strong beam of light on this little

* An experiment of Faraday's.

pyramid: you see these long spokes of light passing through the dusty air of the room as the mirror turns.

Let us start it from a state of rest. You now see the beam passing through the room and striking against the white wall. As the mirror commences to rotate, the patch of light moves, at first slowly, over the wall and ceiling. But the motion quickens, and now you can no longer see the distinct patches of light, but instead of them you have this splendid luminous band fully twenty feet in diameter drawn upon the wall by the quick rotation of the reflected beams. At the word of command the magnet will be excited, and the motion of the medal will be instantly stopped. 'Make!' See the effect: the medal seems struck dead by the excitement of the magnet, the band suddenly disappears, and there you have the single patch of light upon the wall. This strange mechanical effect is produced without any visible change in the space between the two poles. Observe the slight motion of the image on the wall: the tension of the string is struggling with an unseen antagonist and producing that slight motion. It is such as would be produced if the medal, instead of being surrounded by air, were immersed in a pot of thick treacle. I destroy the magnetic power, and the viscous character of the space between the poles instantly disappears; the medal begins to twirl as before; there are the revolving beams, and there is now the luminous band. I again excite the magnet: the beams are struck motionless and the band disappears.

By the force of my hand I can overcome this resistance and turn the medal round; but to turn it I must expend force. Where does that force go? It is converted into heat. The medal, if forcibly compelled to turn, will become heated. Many of you are acquainted with the grand discovery of Faraday, that electric currents are developed where a conductor of electricity is set in motion

between the poles of a magnet.   We have these currents
doubtless here, and they are competent to heat the medal.
But what *are* these currents? how are they related to the
space between the magnetic poles—how to the force of my
arm which is expended in their generation?   We do not
yet know, but we shall know by and by.   It does not in
the least lessen the interest of the experiment if the force
of my arm, previous to appearing as heat, appears in
another form—in the form of electricity.   The ultimate
result is the same: the heat developed ultimately is the
exact equivalent of the quantity of strength required to
move the medal in the excited magnetic field.

I wish now to show you the developement of heat by
this action.   I have here a solid metal cylinder, the core
of which is, however, composed of a metal more easily
melted than its outer case.   The outer case is copper, and
this is filled by a hard but fusible alloy.   I set this cylinder
upright between the conical poles P P (fig. 16) of the

Fig. 16.

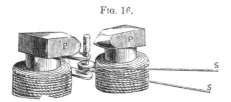

magnet.   A string s s passes from the cylinder to a whirling
table, and by turning the latter the cylinder is caused to
spin round.   It might turn till doomsday, as long as
the magnet remains unexcited, without producing the
effect sought; but when the magnet is in action, I
hope to be able to develope an amount of heat sufficient
to melt the core of that cylinder, and, if successful, I will
pour the liquid metal out before you.   Two minutes will
suffice for this experiment.   The cylinder is now rotating,
and its upper end is open.   I shall leave it thus open

until the liquid metal is seen spattering over the poles of
the magnet. I already see the metallic spray, though a
minute has scarcely elapsed since the commencement
of the experiment. I now stop the motion for a moment, and cork up the end of the cylinder, so as to prevent the scattering about of the metal. Let the action
continue for half a minute longer; the entire mass of the
core is, I am persuaded, now melted. I withdraw the
cylinder, remove the cork, and here is the liquefied mass,
which I thus pour out before you.*

It is now time to consider more closely than we have
hitherto done the relation of the heat developed by mechanical action to the force which produces it. Doubtless
this relation floated in many minds before it received
either distinct enunciation or experimental proof. Those
who reflect on vital processes—on the changes which occur
in the animal body—and the relation of the forces involved
in food, to muscular force, are led naturally to entertain
the idea of interdependence between these forces. It is
therefore not a matter of surprise that the man who first
raised the idea of the equivalence between heat and mechanical energy to philosophic clearness in his own mind,
was a physician. Dr. Mayer of Heilbronn in Germany,
enunciated the exact relation which subsists between heat
and work, giving the number which is now known as the
' mechanical equivalent of heat,' and following up the
statement of the principle by its fearless application.† It
is, however, to Mr. Joule of Manchester that we are
almost wholly indebted for the *experimental* treatment of

* The developement of heat by causing a conductor to revolve between
the poles of a magnet was first effected by Mr. Joule (Phil. Mag. vol. xxiii.
3rd Series, year 1843, pp. 355 and 439), and his experiment was afterwards
revived in a striking form by M. Foucault. The artifice above described,
of fusing the core out of the cylinder, renders the experiment very effective
in the lecture-room.

† See Lectures III. and XII.

this important subject.   Entirely independent of Mayer,
with his mind firmly fixed upon a principle, and undismayed by the coolness with which his first labours appear
to have been received, he persisted for years in his attempts
to prove the invariability of the relation which subsists
between heat and ordinary mechanical force.   He placed
water in a suitable vessel, and agitated that water by
paddles, driven by measurable forces, and determined both
the amount of heat, developed by the stirring of the liquid,
and the amount of labour expended in the process.   He did
the same with mercury and with sperm oil.   He also caused
disks of cast iron to rub against each other, and measured
the heat produced by their friction, and the force expended
in overcoming it.   He also urged water through capillary
tubes, and determined the amount of heat generated by
the friction of the liquid against the sides of the tubes.
And the results of his experiments leave no shadow of
doubt upon the mind that, under all circumstances, the
quantity of heat generated by the same amount of force is
fixed and invariable.   A given amount of force, in causing
the iron disks to rotate against each other, produced precisely the same amount of heat, as when it was applied to
agitate water, mercury, or sperm oil.   Of course at the end
of an experiment, the *temperatures* in the respective cases
would be very different; that of the water, for example,
would be $\frac{1}{30}$th of the temperature of the mercury, because,
as we already know, the capacity of water for heat is
30 times that of mercury.   Mr. Joule took this into
account in reducing his experiments, and found, as I
have stated, that, however the temperatures might differ,
in consequence of the different capacity for heat of the
substances employed, *the absolute amount of heat* generated by the same expenditure of power, was in all cases
the same.

In this way it was found that the quantity of heat which

would raise one pound of water one degree Faht. in temperature, is exactly equal to what would be generated if a pound weight, after having fallen through a height of 772 feet, had its moving force destroyed by collision with the earth. Conversely, the amount of heat necessary to raise a pound of water one degree in temperature, would, if all applied mechanically, be competent to raise a pound weight 772 feet high, or it would raise 772 lbs. one foot high. The term ' foot-pound' has been introduced to express, in a convenient way, the lifting of one pound to the height of a foot. Thus the quantity of heat necessary to raise the temperature of a pound of water one degree being taken as a standard, 772 foot-pounds constitute what is called *the mechanical equivalent* of heat.

In order to imprint upon your minds the thermal effect produced by a body falling from a height, I will go through the experiment of allowing a lead ball to fall from our ceiling upon this floor. The lead ball is at the present moment slightly colder than the air of this room. I prove this by bringing it into contact with the thermo-electric pile, and showing you that the deflection of the needle indicates cold. Here on the floor I have placed a slab of iron, on which I intend the lead to fall, and which you observe is also cooler than the air of the room. At the top of the house I have an assistant, who will heave up the ball after I have attached it to this string. He will not touch the ball, nor will he allow it to touch anything else. He will now let it go; it falls, and is received upon the plate of iron. The height is too small to get much heat by a single fall; I will therefore have the ball drawn up, and dropped three or four times in succession. Observe, there is a length of covered wire attached to the ball, by which I lift it, so that my hand never comes near the ball. There is the fourth collision, and I think I may now examine the temperature of the lead.

I place the ball, which at the commencement was cold, again upon the pile, and the immediate deflection of the needle, in the opposite direction, declares that now the ball is heated ; this heat is due entirely to the destruction of the moving energy which the ball possessed when it struck the plate of iron. According to our theory, the common mechanical motion of the ball as a mass, has been transferred to the atoms of the mass, producing among them the agitation which we call heat.

What was the total amount of heat thus generated ? The space fallen through by the ball in each experiment is twenty-six feet. The heat generated is proportional to the height through which the body falls. Now a ball of lead in falling through 772 feet would generate heat sufficient to raise its own temperature 30°, its ' capacity' being $\frac{1}{30}$th of that of water : hence, in falling through 26 feet, which is in round numbers $\frac{1}{30}$th of 772, the heat generated would, if all concentrated in the lead, raise its temperature one degree. This is the amount of heat generated by a single descent of the ball, and four times this amount would, of course, be generated by four descents. The heat generated is not, however, all concentrated in the ball; it is divided between the ball and the iron on which it falls.

It is needless to say, that if motion be imparted to a body by other means than gravity, the destruction of this motion also produces heat. A rifle bullet when it strikes a target is intensely heated. The mechanical equivalent of heat enables us to calculate with the utmost accuracy the amount of heat generated by the bullet, when its velocity is known. This is a point worthy of our attention, and in dealing with it I will address myself to those of my audience who are unacquainted even with the elements of mechanics. Everybody knows that the greater the height is from which a body falls, the greater is the force with

which it strikes the earth, and that this is entirely due to
the greater velocity imparted to the ball, in falling from
the greater height. The velocity imparted to the body is
not, however, proportional to the height from which it
falls. If the height be augmented four-fold, the velocity
is augmented only two-fold; if the height be augmented
nine-fold, the velocity is augmented only three-fold; if
the height be augmented sixteen-fold, the velocity is
augmented only four-fold; or, expressed generally, the
height augments in the same proportion as the square of
the velocity.

But the heat generated by the collision of the falling
body increases simply as the height; consequently, the
heat generated *increases as the square of the velocity.*

If, therefore, we double the velocity of a projectile, we
augment the heat generated, when its moving force is destroyed, four-fold; if we treble its velocity, we augment the
heat nine-fold; if we quadruple the velocity, we augment
the heat sixteen-fold; and so on.

The velocity imparted to a body by gravity in falling
through 772 feet is, in round numbers, 223 feet a second,
that is to say, immediately before the body strikes the
earth, this is its velocity. Six times this quantity, or 1,338
feet a second, would not be an inordinate velocity for a rifle
bullet.

But a rifle bullet, if formed of lead, moving at a velocity
of 223 feet a second, would generate on striking a target
an amount of heat which, if concentrated in the bullet,
would raise its temperature 30°; with 6 times this velocity
it will generate 36 times this amount of heat; hence 36
times 30, or 1,080°, would represent the augmentation of
temperature of a rifle ball on striking a target with a
velocity of 1,338 feet a second, if all the heat generated
were confined to the bullet itself. This amount of heat
would be far more than sufficient to fuse the lead; but in

reality a portion only of the heat generated is lodged in the ball, the total amount being divided between it and the target.    Were the ball iron instead of lead, the heat generated, under the conditions supposed, would be competent to raise the temperature of the ball only by about $\frac{1}{3}$rd of 1,080°, because the capacity of iron for heat is about three times that of lead.

From these considerations I think it is manifest that if we know the velocity and weight of any projectile, we can calculate, with ease, the amount of heat developed by the destruction of its moving force.    For example, knowing, as we do, the weight of the earth, and the velocity with which it moves through space, a simple calculation would enable us to determine the exact amount of heat which would be developed, supposing the earth to be stopped in her orbit.    We could tell, for example, the number of degrees which this amount of heat would impart to a globe of water equal to the earth in size.    Mayer and Helmholtz have made this calculation, and found that the quantity of heat generated by this colossal shock, would be quite sufficient, not only to fuse the entire earth, but to reduce it, in great part, to vapour.    Thus by the simple stoppage of the earth in its orbit ' the elements ' might be caused ' to melt with fervent heat.'    The amount of heat thus developed would be equal to that derived from the combustion of fourteen globes of coal, each equal to the earth in magnitude.    And if, after the stoppage of its motion, the earth should fall into the sun, as it assuredly would, the amount of heat generated by the blow, would be equal to that developed by the combustion of 5,600 worlds of solid carbon.

Knowledge, such as that which you now possess, has caused philosophers, in speculating on the mode in which the sun is nourished, and his supply of light and heat kept up, to suppose the heat and light to be caused by the

showering down of meteoric matter upon the sun's surface.*
Some philosophers suppose the Zodiacal Light to be a cloud
of meteorites, and from it, it is imagined, the showering
meteoric matter may be derived. Now whatever be the
value of this speculation, it is to be borne in mind that
the pouring down of meteoric matter, in the way indicated,
would be competent to produce the light and heat of the
sun. With regard to the probable truth or fallacy of the
theory, it is not necessary that I should offer an opinion; I
would only say that the theory deals with a cause which, if
in sufficient operation, would be competent to produce the
effects ascribed to it.

Let me now pass from the sun to something less,—in
fact, to the opposite pole of nature. And here that divine
power of the human intellect which annihilates mere
magnitude in its dealings with *law*, comes conspicuously
into play. Our reasoning applies not only to suns and
planets, but equally so to the very ultimate atoms of
which matter is composed. Most of you know the scientific
history of the diamond, that Newton, antedating intellectually the discoveries of modern chemistry, pronounced it
to be an unctuous or combustible substance. Everybody
now knows that this brilliant gem is composed of the
same substance as common charcoal, graphite, or plumbago.
A diamond is pure carbon, and carbon burns in oxygen.
I have here a diamond, held fast in a loop of platinum
wire; I will heat the gem to redness in this flame, and
then plunge it into this jar, which contains oxygen gas.
See how it brightens on entering the jar of oxygen,
and now it glows, like a little terrestrial star, with a
pure white light. How are we to figure the action here

---

* Mayer propounded this hypothesis in 1848, and worked it fully out.
It was afterwards enunciated independently by Mr. Waterston, and
developed by Professor William Thomson (Transactions of the Royal Soc.
of Edinb., 1853). See Lecture XII.

going on? Exactly as you would present to your minds
the conception of meteorites showering down upon the
sun. The conceptions are, in quality, the same, and to the
intellect the one is not more difficult than the other. You
are to figure the atoms of oxygen showering against this
diamond on all sides. They are urged towards it by what
is called chemical affinity, but this force, made clear,
presents itself to the mind as pure attraction, of the same
mechanical quality, if I may use the term, as gravity.
Every oxygen atom as it strikes the surface, and has its
motion of translation destroyed by its collision with
the carbon, assumes the motion which we call heat: and
this heat is so intense, the attractions exerted at these
molecular distances are so mighty, that the crystal is kept
white-hot, and the compound, formed by the union of its
atoms with those of the oxygen, flies away as carbonic acid
gas.

Let us now pass on from the diamond to ordinary flame.
I have here a burner from which I can obtain an ignited
jet of gas. Here is the flame: what is its constitution?
Within the flame we have a core of pure unburnt gas, and
outside the flame we have the oxygen of the air. The external surface of the core of. gas is in contact with the air,
and here it is that the atoms clash together and produce
light and heat by their collision. But the exact constitution
of the flame is worthy of our special attention, and for our
knowledge of this we are indebted to one of Davy's most
beautiful investigations. Coal-gas is what we call a
hydro-carbon; it consists of carbon and hydrogen in a
state of chemical union. From this transparent gas
escapes the soot and lampblack which we notice when the
combustion of the gas is incomplete. Soot and lampblack
are there now, but they are compounded with other substances to a transparent form. Here, then, we have a
surface of this compound gas, in presence of the oxygen

of our air: we apply heat, and the attractions are instantly so intensified that the gas bursts into flame.   The
oxygen has a choice of two partners, or, if you like, it is
in the presence of two foes; it closes with that which it
likes best, or hates most heartily, as the case may be.   It
first closes with the hydrogen, and sets the carbon free.
Solid particles of carbon thus scattered in numbers innumerable in the midst of burning matter, are raised to a
state of intense incandescence; they become white-hot,
and mainly to them the *light* of our lamps is due.   The
carbon, however, in due time, closes with the oxygen, and
becomes, or ought to become, carbonic acid; but in passing
from the hydrogen with which it was first combined, to the
oxygen, with which it enters into final union, it exists, for
a time, in the single state, and,
as a bachelor, it gives us all the
splendour of its light.

Fig. 17.

The combustion of a candle
is in principle the same as that
of a jet of gas.   Here you have
a rod of wax or tallow (fig. 17),
through which is passed the
cotton wick.   You ignite the
wick; it burns, melts the tallow
at its base, the liquid ascends
through the wick by capillary
attraction, it is converted by the
heat into vapour, and this vapour
is a hydro-carbon, which burns exactly like the gas.   Here
also you have unburnt vapour within, common air without,
while between both is a shell which forms the battle-
ground of the clashing atoms, where they develope their
light and heat.   There is hardly anything in nature more
beautiful than a burning candle; the hollow basin partially
filled with melted matter at the base of the wick, the

creeping up of the liquid; its vaporisation; the structure of
the flame; its shape tapering to a point, while converging
air-currents rush in to supply its needs. Its beauty, its
brightness, its mobility, have made it a favourite type of
spiritual essences, and its dissection by Davy, far from
diminishing the pleasure with which we look upon a
flame, has rendered it more than ever a miracle of beauty
to the enlightened mind,

You ought now to be able to picture clearly before your
minds the structure of a candle-flame. You ought to see
the unburnt core within and the burning shell which en-
velopes this core. From the core, through this shell, the constituents of the candle are incessantly passing and escaping
to the surrounding air, In the case of a candle you
have a hollow cone of burning matter. Imagine this cone
cut across horizontally; you would then expose a burning
ring. I will practically cut the flame of a candle thus across.
I have here a piece of white paper which I will bring down
upon the candle; pressing it down upon the flame until it
almost touches the wick. Observe the upper surface of
that paper; it becomes charred, but how? Exactly in
correspondence with the burning ring of the candle, we
have a charred ring upon the paper (fig. 18). I might

FIG. 18,

operate in the same manner with a jet of gas. I will do
so. Here is the ring which it produces. Within the ring
you see there is no charring of the paper, for at this place
the unburnt vapour of the candle, or the unburnt gas of
the jet, impinges against the surface, and no charring can
be produced.

To the existence, then, of solid carbon particles the

light of our lamps is mainly due.   But the existence
of these particles, in the single state, implies the absence
of oxygen to seize hold of them.  If, at the moment
of their liberation from the hydrogen with which they
are first combined, oxygen were present to seize upon
them, their state of bachelorhood would be extinguished,
and we should no longer have their light.   Thus when we
mix a sufficient quantity of air with the gas issuing from a
jet, when we mix it so that the oxygen penetrates to the
very heart of the jet, we find the light destroyed.   Here
is a burner, invented by Prof. Bunsen, for the express
purpose of destroying the light by causing the quick combustion of the carbon particles.   The burner from which
the gas escapes is introduced into a tube; this tube is perforated nearly on a level with the gas orifice, and through
these perforations the air enters, mingles with the gas, and
the mixture issues from the top of the tube.   Fig. 19 represents a form of this burner; the gas is

FIG. 19.

discharged into the perforated chamber $a$,
where air mingles with it, and both ascend
the tube $a\,b$ together: $d$ is a rose-burner
which may be used to vary the shape of the
flame.  I ignite the mixture, but the flame
produces hardly any light.   Heat is the
thing here aimed at, and this lightless
flame is much hotter than the ordinary
flame, because the combustion is much
quicker, and therefore more intense.  If I stop the
orifices in $a$ I cut off the supply of air, and the flame at once
becomes luminous : we have now the ordinary case of a
core of unburnt gas surrounded by a burning shell.   The
illuminating power of a gas may, in fact, be estimated by
the quantity of air necessary to prevent the precipitation
of the solid carbon particles ; the richer the gas, the more
air will be required to produce this effect.

An interesting observation may be made on almost any windy Saturday evening in the streets of London, on the sudden, and almost total extinction of the light of the huge gas jets, exposed chiefly in butchers' shops. When the wind blows, the oxygen is carried mechanically to the very heart of the flame, and the white light instantly vanishes to a pale and ghastly blue. During festive illuminations the same effect may be observed; the absence of the light being due, as in the case of Bunsen's burner, to the presence of a sufficient amount of oxygen to consume, instantly, the carbon of the flame.

To determine the influence of height upon the rate of combustion, was one of the problems which I had set before me, in my journey to the Alps in 1859. Fortunately for science, I invited Dr. Frankland to accompany me on the occasion, and to undertake the experiments on combustion, while I proposed devoting myself to observations on solar radiation. The plan pursued was this: six candles were purchased at Chamouni and carefully weighed; they were then allowed to burn for an hour in the Hotel de l'Union, and the loss of weight was determined. The same candles were taken to the summit of Mont Blanc, and, on the morning of Aug. 21, were allowed to burn for an hour in a tent, which perfectly sheltered them from the action of the wind. The aspect of the six flames at the summit surprised us both. They seemed the mere ghosts of the flames which the same candles were competent to produce in the valley of Chamouni. Pale, small, feeble, and suggesting to us a greatly diminished energy of combustion. The candles being carefully weighed on our return, the unexpected fact was revealed, that the quantity of stearine consumed above was almost precisely the same as that consumed below. Thus, though the light-giving power of the flame was diminished in an extraordinary degree by the elevation, the energy of

E

the combustion was the same above as it was below.  This curious result is to be ascribed mainly to the mobility of the air at this great height.  The particles of oxygen could penetrate the flame with comparative freedom, thus destroying its light, and making atonement for the smallness of their number by the promptness of their action.

Dr. Frankland has made these experiments the basis of a most interesting memoir.*  He shows that the quantity of a candle consumed in a given time is, within wide limits, independent of the density of the air ; and the reason is, that although by compressing the air we augment the number of active particles in contact with the flame, we almost, in the same degree, diminish their mobility, and retard their combustion.  When an excess of air, moreover, surrounds the flame, its chilling effect will tend to prolong the existence of the carbon particles in a solid form, and even to prevent their final combustion.  One of the beautiful experimental results of Dr. Frankland's investigation is, that by condensing the air around it, the pale and smoke-less flame of a spirit-lamp may be rendered as bright as that of coal gas, and, by pushing the condensation sufficiently far, the flame may actually be rendered smoky, the sluggish oxygen present being incompetent to effect the complete combustion of the carbon.

But to return to our theory of combustion : it is to the clashing together of the oxygen of the air and the constituents of our gas and candles, that the light and heat of our flames are due.  I scatter steel filings in this flame, and you see the star-like scintillations produced by the combustion of the steel.  Here the steel is first heated, till the attraction between it and the oxygen becomes sufficiently strong to cause them to combine, and these rocket-like flashes are the result of their collision.  It is the impact of the atoms of oxygen against the atoms of sulphur which

* Philosophical Transactions for 1861.

produces the flame observed when sulphur is burned in oxygen or in air; to the collision of the same atoms against phosphorus are due the intense heat and dazzling light which result from the combustion of phosphorus in oxygen gas. It is the collision of chlorine and antimony which produces the light and heat observed where these bodies are mixed together; and it is the clashing of sulphur and copper which causes the incandescence of the mass when these substances are heated together in a Florence flask. In short, all cases of combustion are to be ascribed to the collision of atoms which have been urged together by their mutual attractions.

# APPENDIX TO LECTURE II.

WHEN I say of motion that it is the genus of which heat is a
species, I would be understood to mean, not that heat generates
motion, or that motion generates heat (though both are true in
certain cases), but that heat itself, its essence and quiddity, is
motion and nothing else; limited, however, by the specific differences which I will presently subjoin, as soon as I have added
a few cautions for the sake of avoiding ambiguity. . . . .

Nor again, must the communication of heat, or its transitive
nature, by means of which a body becomes hot when a hot body
is applied to it, be confounded with the form of heat. For heat
is one thing, and heating is another. Heat is produced by the
motion of attrition without any preceding heat. . . . .

Heat is an expansive motion, whereby a body strives to dilate
and stretch itself to a larger sphere or dimension than it had previously occupied. This difference is most observable in flame,
where the smoke or thick vapour manifestly dilates and expands
into flame.

It is shown also in all boiling liquid, which manifestly swells,
rises, and bubbles, and carries on the process of self-expansion,
till it turns into a body far more extended and dilated than the
liquid itself, namely, into vapour, smoke, or air.

\*       \*       \*       \*       \*       \*       \*

The third specific difference is this, that heat is a motion of
expansion, not uniformly of the whole body together, but in the
smaller parts of it; and at the same time checked, repelled and
beaten back, so that the body acquires a motion alternative, per-

petually quivering, striving and struggling, and irritated by reper-
cussion, whence springs the fury of fire and heat.

Again, it is shown in this that when the air is expanded in a
calender glass, without impediment or repulsion, that is to say
uniformly and equably, there is no perceptible heat. Also when
wind escapes from confinement, although it burst forth with the
greatest violence, there is no very great heat perceptible ; because
the motion is of the whole, without a motion alternating in the
particles.

And this specific difference is common also to the nature of
cold; for in cold contractive motion is checked by a resisting
tendency to expand, just as in heat the expansive action is checked
by a resisting tendency to contract. Thus whether the particles
of a body work inward or outward, the mode of action is the
same.

*    *    *    *    *    *    *

Now from this our first vintage it follows, that the form or true
definition of heat (heat that is in relation to the universe, not
simply in relation to man) is in a few words as follows: *Heat is
a motion, expansive, restrained, and acting in its strife upon the
smaller particles of bodies.* But the expansion is thus modified;
*while it expands all ways, it has at the same time an inclination up-
wards.* And the struggle in the particles is modified also; *it is
not sluggish but hurried and with violence.**

---

ABSTRACT OF COUNT RUMFORD'S ESSAY, ENTITLED, ' AN
ENQUIRY CONCERNING THE SOURCE OF THE HEAT WHICH
IS EXCITED BY FRICTION.'

*[Read before the Royal Society, January 25, 1798.]*

Being engaged in superintending the boring of cannon in the
workshops of the military arsenal at Munich, Count Rumford was
struck with the very considerable degree of heat which a brass
gun acquires, in a short time, in being bored, and with the still
more intense heat (much greater than that of boiling water) of
the metallic chips separated from it by the borer, he proposed
to himself the following questions :

* Bacon's Works, vol. iv.: Spedding's Translation.

'Whence comes the heat actually produced in the mechanical operation above mentioned?

'Is it furnished by the metallic chips which are separated from the metal?'

If this were the case, then the *capacity for heat* of the parts of the metal so reduced to chips ought not only to be changed, but the change undergone by them should be sufficiently great to account for *all* the heat produced. No such change however had taken place; for the chips were found to have the same capacity as slices of the same metal cut by a fine saw, where heating was avoided. Hence, it is evident, that the heat produced could not possibly have been furnished at the expense of the latent heat of the metallic chips. Rumford describes those experiments at length, and they are conclusive.

He then designed a cylinder for the express purpose of generating heat by friction, by having a blunt borer forced against its solid bottom, while the cylinder was turned round its axis by the force of horses. To measure the heat developed, a small round hole was bored in the cylinder for the purpose of introducing a small mercurial thermometer. The weight of the cylinder was 113·13 lbs. avoirdupois.

The borer was a flat piece of hardened steel, 0·63 of an inch thick, 4 inches long, and nearly as wide as the cavity of the bore of the cylinder, namely, $3\frac{1}{2}$ inches. The area of the surface by which its end was in contact with the bottom of the bore was nearly $2\frac{1}{2}$ inches. At the beginning of the experiment the temperature of the air in the shade and also that of the cylinder was 60 degrees Fahr. At the end of 30 minutes, and after the cylinder had made 960 revolutions round its axis, the temperature was found to be 130 degrees.

Having taken away the borer, he now removed the metallic dust, or rather scaly matter, which had been detached from the bottom of the cylinder by the blunt steel borer, and found its weight to be 837 grains troy. 'Is it possible,' he exclaims, 'that the very considerable quantity of heat produced in this experiment—a quantity which actually raised the temperature of above 113 pounds of gun metal at least 70 degrees of Fahrenheit's thermometer—could have been furnished by so inconsiderable a quantity of metallic dust, and this merely in consequence of a *change* in its capacity for heat?

'But without insisting on the improbability of this supposition, we have only to recollect that from the results of actual and decisive experiments, made for the express purpose of ascertaining that fact, the capacity for heat of the metal of which great guns are cast is *not sensibly changed* by being reduced to the form of metallic chips, and there does not seem to be any reason to think that it can be much changed, if it be changed at all, in being reduced to much smaller pieces by a borer which is less sharp.'

He next surrounded his cylinder by an oblong deal box, in such a manner that the cylinder could turn water-tight in the centre of the box, while the borer was pressed against the bottom of the cylinder. The box was filled with water until the entire cylinder was covered, and then the apparatus was set in action.. The temperature of the water on commencing was 60 degrees.

'The result of this beautiful experiment,' writes Rumford, 'was very striking, and the pleasure it afforded me amply repaid me for all the trouble I had had in contriving and arranging the complicated machinery used in making it. The cylinder had been in motion but a short time, when I perceived, by putting my hand into the water, and touching the outside of the cylinder, that heat was generated.

'At the end of one hour the fluid, which weighed 18·77 lbs., or $2\frac{1}{2}$ gallons, had its temperature raised 47 degrees, being now 107 degrees.

'In thirty minutes more, or one hour and thirty minutes after the machinery had been set in motion, the heat of the water was 142 degrees.

'At the end of two hours from the beginning, the temperature was 178 degrees.

'At two hours and twenty minutes it was 200 degrees, and at two hours and thirty minutes it ACTUALLY BOILED!'

It is in reference to this experiment that Rumford made the remarks regarding the surprise of the bystanders, which I have quoted in Lecture I.

He then carefully estimates the quantity of heat possessed by each portion of his apparatus at the conclusion of the experiment, and adding all together, finds a total sufficient to raise 26·58 lbs. of ice-cold water to its boiling point, or through 180 degrees Fahrenheit. By careful calculation, he finds this heat equal to that given out by the combustion of 2303·8 grains (= $4\frac{8}{10}$ oz. troy) of wax.

He then determines the '*celerity*' with which the heat was generated, summing up his computations thus: 'From the results of these computations, it appears that the quantity of heat produced equably, or in a continuous stream, if I may use the expression, by the friction of the blunt steel borer against the bottom of the hollow metallic cylinder, was *greater* than that produced in the combustion of nine *wax candles*, each $\frac{3}{4}$ of an inch in diameter, all burning together with clear bright flames.'

'One horse would have been equal to the work performed, though two were actually employed. Heat may thus be produced merely by the strength of a horse, and, in a case of necessity, this heat might be used in cooking victuals. But no circumstances could be imagined in which this method of procuring heat would be advantageous; for more heat might be obtained by using the fodder necessary for the support of a horse as fuel.'

[This is an extremely significant passage, intimating as it does, that Rumford saw clearly that the force of animals was derived from the food; no *creation of force* taking place in the animal body.]

'By meditating on the results of all these experiments we are naturally brought to that great question which has so often been the subject of speculation among philosophers, namely, What is heat—is there any such thing as an *igneous fluid*? Is there any thing that, with propriety, can be called caloric?'

'We have seen that a very considerable quantity of heat may be excited by the friction of two metallic surfaces, and given off in a constant stream or flux *in all directions*, without interruption or intermission, and without any signs of *diminution* or *exhaustion*. In reasoning on this subject we must not forget *that most remarkable circumstance*, that the source of the heat generated by friction in these experiments appeared evidently to be *inexhaustible*. (The italics are Rumford's.) It is hardly necessary to add, that anything which any *insulated* body or system of bodies can continue to furnish *without limitation* cannot possibly be a *material substance*; and it appears to me to be extremely difficult, if not quite impossible, to form any distinct idea of anything capable of being excited and communicated in those experiments, except it be MOTION.'

When the history of the dynamical theory of heat is written, the man who, in opposition to the scientific belief of his time, could experiment and reason upon experiment, as Rumford did in the investigation here referred to, cannot be lightly passed over. Hardly anything more powerful against the materiality of heat has been since adduced, hardly anything more conclusive in the way of establishing that heat is what Rumford considered it to be, *Motion.*

---

## ON THE COMPRESSION OF AIR CONTAINING BISULPHIDE OF CARBON VAPOUR.

' A very singular phenomenon was repeatedly observed during the experiments with bisulphide of carbon. After determining the absorption of the vapour, the tube was exhausted as perfectly as possible, the trace of vapour left behind being exceedingly minute. Dry air was then admitted to cleanse the tube. On again exhausting, after the first few strokes of the pump, a jar was felt and a kind of explosion heard, while dense volumes of blue smoke immediately issued from the pump cylinders. The action was confined to the latter, and never propagated itself backwards into the experimental tube.

' It is only with bisulphide of carbon that this effect has been observed. It may, I think, be explained in the following manner : — To open the valve of the piston, the gas beneath it must have a certain tension, and the compression necessary to produce this appears sufficient to cause the combination of the constituents of the bisulphide of carbon with the oxygen of the air. Such a combination certainly takes place, for the odour of sulphurous acid is unmistakeable amid the fumes.

' To test this idea I tried the effect of compression in the air syringe. A bit of tow or cotton wool moistened with bisulphide of carbon, and placed in the syringe, emitted a bright flash when the air was compressed. By blowing out the fumes with a glass tube, this experiment may be repeated twenty times with the same bit of cotton.

' It is not necessary even to let the moistened cotton remain in

the syringe.   If the bit of tow or cotton be thrown into it, and
out again as quickly as it can be ejected, on compressing the air
the luminous flash is seen.   Pure oxygen produces a brighter
flash than atmospheric air.   These facts are in harmony with the
above explanation.' *

* Phil. Trans., 1861 ; Phil. Mag., Sept. 1861.

# LECTURE III.

## [February 6, 1862.]

EXPANSION: THE SOLID, LIQUID, AND GASEOUS FORMS OF MATTER—HYPO-
THESES REGARDING THE CONSTITUTION OF GASES—COEFFICIENT OF EXPAN-
SION—HEAT IMPARTED TO A GAS UNDER CONSTANT PRESSURE—HEAT
IMPARTED TO A GAS AT CONSTANT VOLUME—MAYER'S CALCULATION
OF THE MECHANICAL EQUIVALENT OF HEAT—DILATATION OF GASES
WITHOUT REFRIGERATION—ABSOLUTE ZERO OF TEMPERATURE—EXPANSION
OF LIQUIDS AND SOLIDS: ANOMALOUS DEPORTMENT OF WATER AND
BISMUTH—ENERGY OF THE FORCE OF CRYSTALLIZATION—THERMAL EFFECT
OF STRETCHING WIRES—ANOMALOUS DEPORTMENT OF INDIA-RUBBER.

APPENDIX:—ADDITIONAL DATA CONCERNING EXPANSION—EXTRACTS FROM
SIR H. DAVY'S FIRST SCIENTIFIC MEMOIR: FUSION OF ICE BY FRICTION, &C.

YOUR reappearance here to-day, after the strain which
has already been put upon your attention, encourages
me to hope that our present experiment will not be entirely unsuccessful.   I need not tell an audience like this
that nothing intellectually great is either accomplished or
appropriated without effort.   Newton ascribed the difference between himself and other men to his patience in
steadily looking at a question, until light dawned upon it,
and if we have firmness to imitate his example, we shall,
no doubt, reap a commensurate reward.

In our first lecture I permitted a sledge-hammer to
descend upon a mass of lead, and we found that the lead
became heated, as soon as the mechanical motion of the
hammer was arrested.   Formerly it was assumed that the
force of the hammer was simply lost by the concussion.
In elastic bodies it was supposed that a portion of the
force was restored by the elasticity of the body, which

caused the descending mass to rebound; but in the collision
of inelastic bodies it was taken for granted that the force
of impact was lost. This, according to our present notions,
was a fundamental mistake; we now admit no loss, but
assume, that when the motion of the descending hammer
ceases, it is simply a case of *transference*, instead of annihilation. The motion of the mass, as a whole, has been
transformed into a motion of the molecules of the mass.
This motion of heat, however, though intense, is executed
within limits too minute, and the moving particles are too
small, to be visible. To discern these processes we must
make use of a finer eye and higher powers, namely, the
eye and powers of the mind. In the case of solid bodies,
then, while the force of cohesion still holds the particles
together, you must conceive a power of vibration, within
certain limits, to be possessed by the particles. You must
suppose them oscillating to and fro across their positions
of rest; and the greater the amount of heat we impart
to the body, or the greater the amount of mechanical action which we invest in it by percussion, compression, or friction, the more intense will be the molecular
vibration, and the wider the amplitude of the atomic
oscillations.

Now, nothing is more natural than that particles thus
vibrating, and ever as it were seeking wider room,
should urge each other apart, and thus cause the body of
which they are the constituents, to expand in volume.
This, in general, is the consequence of imparting heat to
bodies—expansion of volume. We shall closely consider
the few apparent exceptions to this law by and by. By
the force of cohesion, then, the particles are held together; by the force of heat they are pushed asunder:
here are the two antagonist principles on which the molecular aggregation of the body depends. Let us suppose
the communication of heat to continue; every increment

of heat pushes the particles more widely apart; but the
force of cohesion, like all other known forces, acts more
and more feebly, as the distance between the particles
which are the seat of the force is augmented.   As, therefore, the heat strengthens, its opponent grows weak, until,
finally, the particles are so far loosed from the rigid thrall
of cohesion, that they are at liberty, not only to vibrate to
and fro across a fixed position, but also to roll or glide
around each other.   Cohesion is not yet destroyed, but it
is so far modified, that while the particles still offer resistance to being torn directly asunder, their lateral mobility
over each other's surfaces is secured.   *This is the liquid
condition of matter.*

In the interior of a mass of liquid the motion of every
atom is controlled by the atoms which surround it.   But
suppose you develope heat of sufficient power within the
body of a liquid, what occurs?   Why, the particles break
the last fetters of cohesion, and fly asunder to form bubbles of vapour.   If one of the surfaces of the liquid be
quite free, that is to say, uncontrolled either by a liquid
or solid; it is quite easy to conceive that some of the
vibrating superficial particles will be jerked quite away
from the liquid, and will fly with a certain velocity through
space.   *Thus freed from the influence of cohesion, we have
matter in the vaporous or gaseous form.*

My object here is to familiarise your minds with the
general conception of atomic motion.   I have spoken
of the vibration of the particles of a solid as causing
its expansion; the particles have been thought by some
to revolve round each other, and the communication of
heat, by augmenting the centrifugal force of the particles,
was supposed to push them more widely asunder.   I have
here a weight attached to a spiral spring; if I twirl the
weight round in the air it tends to fly away from me, the

spring stretches to a certain extent, and as I augment the
speed of revolution, the spring stretches still more, the distance between my hand and the weight being thus augmented.  It has been thought that the augmentation of
the distance between a body's atoms by heat, may be also
due to a revolution of its particles.  And imagine the motion to continue till the spring snaps ; the ball attached to
it would fly off along a tangent to its former orbit, and
thus represent an atom freed, by heat, from the force of
cohesion, which is rudely represented by our spring.  The
ideas of the most well-informed philosophers are as yet
uncertain regarding the exact nature of the motion of
heat ; but the great point, at present, is to regard it as
motion of some kind, leaving its more precise character to
be dealt with in future investigations.

We might extend the notion of revolving atoms to gases
also, and deduce their phenomena from a motion of this
kind.  But I have just thrown out an idea regarding gaseous
particles, which is at present very ably maintained : *
the idea, namely, that such particles fly in straight lines
through space.  Everybody must have remarked how
quickly the perfume of an odorous body fills a room, and
this fact harmonises with the idea of the direct projection
of the particles.  But it may be proved, that if the theory
of rectilinear motion be true, the particles must move at
the rate of several hundred feet a second.  Hence it might
be objected that, according to the above hypothesis, odours
ought to spread much more quickly than they are observed to do.

The answer to this objection is, that they have to make
their way through a crowd of air particles, with which
they come into incessant collision.  On an average, the

* By Joule, Krönig, Maxwell; and, in a series of extremely able papers,
by Clausius.

distance through which an odorous particle can travel in common air, without striking against a particle of air, is infinitesimal, and hence the propagation of a perfume through air is enormously retarded by the air itself. It is well known that when a free communication is opened between the surface of a liquid and a vacuum, the vacuous space is much more speedily filled to saturation with the vapour of the liquid, than when air is present.

According to this hypothesis, then, we are to figure a gaseous body as one whose particles are flying in straight lines through space, impinging like little projectiles upon each other, and striking against the boundaries of the space which they occupy. Mr. Anderson will place this bladder, half filled with air, under the receiver of the air-pump; he will now work the pump, and remove the air that surrounds the bladder. The bladder swells; the air within it appears quite to fill it, so as to remove all its folds and creases. How is this expansion of the bladder produced? According to our present theory, it is produced by the shooting of atomic projectiles against its interior surface, which drive the envelope outwards, until its tension is able to cope with their force. When air is admitted into the receiver, the bladder shrivels to its former size; and here we must figure the discharge of the air particles against the outer surface of the bladder, which drive the envelope inwards, causing, at the same time, the particles within to concentrate their fire, until finally the force from within equals that from without, and the envelope remains quiescent. All the impressions, then, which we derive from heated air or vapour are, according to this hypothesis, due to the impact of the gaseous atoms. They stir the nerves in their own peculiar way, the nerves transmit the motion to the brain, and the brain declares it to be heat. Thus the impression one receives on entering the hot room of a Turkish bath, is caused by the atomic cannonade which is there maintained against the surface of the body.

If, instead of placing this bladder under the receiver of an air-pump, and withdrawing the external air, I augment, by heat, the projectile force of the particles within it, these particles, though comparatively few in number, will strike with such impetuous energy against the inner surface as to cause the envelope to retreat : the bladder swells and becomes apparently filled with air ; I hold the bladder close to the fire, and here it is, you see, with all its creases removed.   But you will retort, perhaps, by saying that this ought not to be the case, inasmuch as the air outside the bladder is also near the fire, and therefore animated with a like projectile energy, which tends to drive the envelope in.   True, the bladder and the air in contact with it are equally near the fire ; but in a future lecture you will learn that the air outside the bladder allows the rays of heat to pass through it with very little augmentation of temperature, while the bladder intercepts the radiant heat ; the *envelope* becomes first warmed and then communicates its heat, by contact, to the air within.   The air, moreover, in contact with the bladder on the outside, though heated by the bladder, has free space to dilate in, and is therefore incompetent to resist the expansion of the confined air which the bladder contains.

This, then, is a simple illustration of the expansive force of heat, and I have here an apparatus intended to show you the same fact in another manner.   Here is a flask, F (fig. 20), empty, except as regards air, which I intend to heat by this little spirit-lamp underneath.   From the flask a bent tube passes to this dish, containing a coloured liquid. In the dish, a 2-foot glass tube, $tt$, is inverted, closed at the top, but with its open end downwards ; you know that the pressure of the atmosphere is competent to keep the column of liquid in this tube, and here you have it quite

filled to the top with the liquid. The tube passing from
the flask is caused to turn up exactly underneath the open
end of this upright tube, so that if a bubble of air should
issue from the former, it will ascend the latter. I now
heat the flask, and as I do so the air expands, for the

Fig. 20.

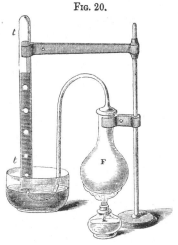

reasons already given; bubbles are driven from the end of
the bent tube, and they ascend in the tube $t\ t$. The air
speedily depresses the liquid column, until now, in the
course of a very few seconds, the whole column of liquid
has been superseded by air.

It is perfectly manifest that the air, thus expanded by
heat, is lighter than the unexpanded air. Our flask, at the
conclusion of this experiment, is lighter than it was at
the commencement, by the weight of the air transferred
from it into the upright tube. Supposing, therefore, a
light bag to be filled with such air, it is plain that the
bag would, with reference to the heavy air outside it,
be like a drop of oil in water; the oil being lighter
than the water will ascend through the latter; so also our
bag, filled with heated air, will ascend in the atmosphere;

F

and this is the principle of the so-called fire-balloon.   Mr.
Anderson will ignite some tow in this vessel, over it he
will place this funnel, and over the funnel I will hold the
mouth of this paper balloon.   The heated air ascending
from the burning tow enters the balloon, causes it to swell;
its tendency to rise is already manifest.   I let it go, and
thus it sails aloft till it strikes the ceiling of the room.

But we must not be content with regarding these phenomena in a general way; without exact quantitative determinations our discoveries would confound and bewilder us.
We must now enquire what is the amount of expansion
which a given quantity of heat is able to produce in a gas?
This is an important point, and demands our special attention.   When we speak of the volume of a gas, we should
have no distinct notion of its real quantity, if its temperature were omitted, the volume varies so largely with the
temperature.   Take, then, a measure of gas at the precise
temperature of water when it begins to freeze, or of ice
when it commences to melt, that is to say, at a temperature of 32° Fahr. or 0° Cent., and raise that volume of gas
one degree in temperature, *the pressure on every square
inch of the envelope which holds the gas being preserved
constant.*   The volume of the gas will become expanded
by a quantity which we may call $a$; raise it another degree
in  temperature, its  volume  will  be  expanded  by  $2a$, a
third  degree  will  cause  an  expansion  of  $3a$, and  so  on.
Thus, we see, that for every degree which we add  to the
temperature of the gas, it is expanded by the same amount.
What is this amount?   No matter  what  the quantity of
gas may be at the freezing temperature, by raising it one
degree *Fahrenheit* we augment its volume by $\frac{1}{490}$th of its
own amount; while by raising it one degree *Centigrade* we
augment the volume by $\frac{1}{273}$rd of its own amount.   A cubic
foot of gas, for example, at 0° C., becomes, on being
heated  to  1°, $1\frac{1}{273}$ cubic foot, or expressed in decimals,

1 vol. at 0° C. becomes      $1 + \cdot 00367$ at 1° C.

at 2° C. it becomes    $1 + \cdot 00367 \times 2$

at 3° C. it becomes    $1 + \cdot 00367 \times 3$, and so on.

The constant number $\cdot 00367$, which expresses the fraction of its own volume, which a gas, at the freezing temperature, expands on being heated one degree, is called the *coefficient of expansion* of the gas. Of course if we use the degrees of Fahrenheit, the coefficient will be smaller in the proportion of 9 to 5.

Fig. 21.

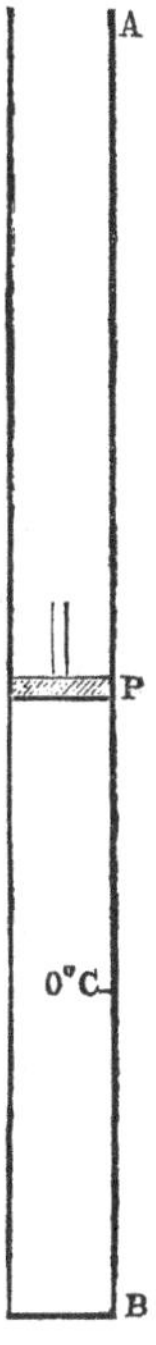

This much made clear, we shall now approach, by slow degrees, an interesting but difficult subject. Suppose I have a quantity of air contained in a very tall cylinder, A B (fig. 21), the transverse section of which is one square inch in area. Let the top A of the cylinder be open to the air, and let P be a piston, which, for reasons to be explained immediately, I will suppose to weigh two pounds one ounce, and which moves air-tight and without friction, up or down in the cylinder. At the commencement of the experiment, let the piston be at the point P of the cylinder, and let the height of the cylinder from its bottom B to the point P be 273 inches, the air underneath the piston being at a temperature of 0° C. Then, on heating the air from 0° to 1° C. the piston will rise one inch; it will now stand at 274 inches above the bottom. If the temperature be raised two degrees, the piston will stand at 275, if raised three degrees, it will stand at 276, if raised ten degrees it will stand at 283, if 100 degrees it will stand at 373 inches above the bottom; finally, if the temperature were raised to 273° C., it is quite manifest 273 inches would be added to the height of the column, or, in other words, by heating the air to 273° C., *its volume would be doubled.*

It is evident that the gas, in this experiment, executes work. In expanding from P upwards it has to overcome the downward pressure of the atmosphere, which amounts to 15 lbs. on every square inch, and also the weight of the piston itself, which is 2 lbs. 1 oz. Hence, the section of the cylinder being one square inch in area, in expanding from P to P′ the work done by the gas is equivalent to the raising a weight of 17 lbs. 1 oz., or 273 ounces, to a height of 273 inches. It is just the same as what it would accomplish, if the air above P were entirely abolished, and a piston weighing 17 lbs. 1 oz. were placed at P.

Let us now alter our mode of experiment, and instead of allowing our gas to expand when heated, let us oppose its expansion by augmenting the pressure upon it. In other words, let us keep *its volume constant* while it is being heated. Suppose, as before, the initial temperature of the gas to be 0° C., the pressure upon it, including the weight of the piston P, being, as formerly, 273 ounces. Let us warm the gas from 0° C. to 1° C.; what weight must we add at P in order to keep its volume constant? Exactly one ounce. But we have supposed the gas, at the commencement, to be under a pressure of 273 ounces, and the pressure it sustains is the measure of its elastic force; hence, by being heated one degree, the elastic force of the gas has augmented by $\frac{1}{273}$rd of what it possessed at 0°. If we warm it 2°, 2 ozs. must be added to keep its volume constant; if 3°, 3 ozs. must be added. And if we raise its temperature 273°, we should have to add 273 ozs.; that is, we should have to *double the original pressure* to keep the volume constant.

It is simply for the sake of clearness, and to avoid fractions in our reflections, that I have supposed the gas to be under the original pressure of 273 ozs. No matter what its pressure may be, the addition of 1° C. to its temperature produces an augmentation of $\frac{1}{273}$rd of the elastic force which the gas possesses at the freezing temperature; and by raising its temperature 273°, while its volume is kept constant, its elastic force is doubled. Let us now compare this experiment with the last one. *There* we heated a certain amount of gas from 0° to 273°, and doubled its volume by so doing, the double volume being attained while the gas lifted a weight of 273 ozs. to a height of 273 inches. *Here* we heat the same amount of gas from 0° to 273°, but we do not permit it to lift any weight. We keep its volume constant. The quantity of matter heated in both cases is the same; the *temperature* to which it is heated is in both cases the same; but are the *absolute quantities* of heat imparted in both cases the same? By no means. Supposing that to raise the temperature of the gas, whose *volume* is kept constant, 273°, 10 grains of combustible matter are necessary; then to raise the temperature of the gas whose *pressure* is kept constant an equal number of degrees, would require the consumption of $14\frac{1}{4}$ grains of the same combustible matter. *The heat produced by the combustion of the additional $4\frac{1}{4}$ grains, in the latter case, is entirely consumed in lifting the weight.* Using the accurate numbers, the quantity of heat applied when the volume is constant, is to the quantity applied when the pressure is constant, in the proportion of

$$1 \text{ to } 1\cdot421.$$

This extremely important fact constitutes the basis from which the mechanical equivalent of heat was first calculated. And here we have reached a point which is worthy of, and which will demand, your entire attention. I will endeavour to make this calculation before you.

Let c (fig. 21*a*) be a cylindrical vessel with a base one

square foot in area.  Let P P mark the upper surface of a
cubic foot of air at a temperature of 32° Fahr.  The height
A P will be then one foot.  Let the air be

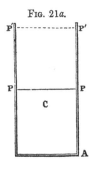

FIG. 21a.

heated till its volume is doubled; to effect
this it must, as before explained, be raised
273° C., or 490° F. in temperature; and,
when expanded, its upper surface will
stand at P′ P′, one foot above its initial
position.  But in rising from P P to P′P′
it has forced back the atmosphere, which
exerts a pressure of 15 lbs. on every square
inch of its upper surface; in other words it
has lifted a weight of $144 \times 15 = 2,160$ lbs.
to a height of one foot.

The ' capacity ' for heat of the air thus expanding is 0·24;
water being unity.  The weight of our cubic foot of air is
1·29 oz. hence the quantity of heat required to raise 1·29 oz.
of air 490° Fahr. would raise a little less than one-fourth
of that weight of water 490°.  The exact quantity of water
equivalent to our 1·29 oz. of air is $1·29 \times 0·24 = 0·31$ oz.

But 0·31 oz. of water, heated to 490°, is equal to 152 ozs.
or $9\frac{1}{2}$ lbs. heated 1°.  Thus the heat imparted to our
cubic foot of air, in order to double its volume, and enable
it to lift a weight of 2,160 lbs. one foot high, would be competent to raise $9\frac{1}{2}$ lbs. of water one degree in temperature.

The air has here been heated *under a constant pressure*,
and we have learned, that the quantity of heat necessary to
raise the temperature of a gas under constant pressure a
certain number of degrees, is to that required to raise the
gas to the same temperature, *when its volume is kept
constant*, in the proportion of 1·42 : 1; hence we have
the statement—

lbs.    lbs.

$1·42 : 1 = 9·5 : 6·7$

which shows that the quantity of heat necessary to·augment the temperature of our cubic foot of air, at constant volume, 490°, would heat 6·7 lbs. of water 1°.

Deducting 6·7 lbs. from 9·5 lbs. we find that the excess of heat imparted to the air, in the case where it is permitted to expand, is competent to raise 2·8 lbs. of water 1° in temperature.

As explained already, this excess is employed to lift the weight of 2,160 lbs. one foot high. Dividing 2,160 by 2·8, we find that a quantity of heat sufficient to raise 1 lb. of water 1° Fahr. in temperature, is competent to raise a weight of 771·4 lbs. a foot high.

This method of calculating the mechanical equivalent of heat was followed by Dr. Mayer, a physician in Heilbronn, Germany, in the spring of 1842.

Mayer's first paper contains merely an indication of the way in which he had found the equivalent; but does not contain the calculation. The paper was evidently a kind of preliminary note, from which date might be taken. In it were enunciated the convertibility and indestructibility of force, and its author referred to the mechanical equivalent of heat, merely in illustration of his principles. Had this first paper stood alone, Mayer's relation to the dynamical theory of heat would be very different from what it now is; but in 1845 he published an Essay on Organic Motion, which, though exception might be taken to it here and there, is, on the whole, a production of extraordinary merit. This was followed in 1848 by an Essay on 'Celestial Dynamics,' in which with remarkable boldness, sagacity, and completeness, he developed the meteoric theory of the sun. Taking him all in all, the right of Mayer to stand, as a man of true genius, in the front rank of the founders of the dynamical theory of heat, cannot be disputed.

On August 21, 1843, Mr. Joule communicated a paper to the British Association, then meeting at Cork, and in the third part of this paper* he describes a series of experiments on magneto-electricity, executed with a view to determine the 'mechanical value of the heat.' The results of this elaborate investigation gave the following weights raised one foot high, as equivalent to the warming of 1lb. of water 1° Fahr.

| | | | |
|---|---|---|---|
| 1. | 896 lbs. | 5. | 1026 lbs. |
| 2. | 1001 „ | 6. | 587 „ |
| 3. | 1040 „ | 7. | 742 „ |
| 4. | 910 „ | 8. | 860 „ |

In 1844 Mr. Joule deduced from experiments on the condensation of air, the following equivalents to 1lb. of water heated 1° Fahr.

$$823 \text{ foot pounds}$$
$$795 \quad \text{,,}$$
$$820 \quad \text{,,}$$
$$814 \quad \text{,,}$$
$$760 \quad \text{,,}$$

As the experience of the experimenter increased, we find that the coincidence of his results becomes closer. In 1845 Mr. Joule deduced from experiments with water, agitated by a paddle-wheel, an equivalent of

$$890 \text{ foot pounds.}$$

Summing up his results in 1845, and taking the mean, he found the equivalent to be

$$817 \text{ foot pounds.}$$

In 1847 he found the mean of two experiments to give as equivalent

$$781 \cdot 8 \text{ foot pounds.}$$

Finally, in 1849, applying all the precautions suggested

---

* Phil. Mag., 1843, vol. xxiii. p. 435.

by seven years' experience, he obtained the following
numbers for the mechanical equivalent of heat :—

| 772·692, | from friction of water, | mean of | 40 | experiments |
| 774·083, | „ „ mercury, | „ | 50 | „ |
| 774·987, | „ „ cast-iron, | „ | 20 | „ |

For reasons assigned in his paper, Mr. Joule fixes the
exact equivalent of heat at

772 foot pounds.

According to the method pursued by Mayer, in 1842,
the mechanical equivalent of heat is

771·4 foot pounds.

Such a coincidence relieves the mind of every shade of
uncertainty, regarding the correctness of our present mechanical equivalent of heat.

Do I refer to these things in order to exalt Mayer, at
the expense of Joule ? It is far from my intention to do
so. The man who through long years, without encourage-
ment, and in the face of difficulties which might well be
deemed insurmountable, could work with such unswerving
steadfastness of purpose to so triumphant an issue, is safe
from depreciation. And it is not the experiments alone, but
the spirit which they incorporate, and the applications
which their author made of them, that entitle Mr. Joule
to a place in the foremost rank of physical philosophers.
Mayer's labours have, in some measure, the stamp of a
profound intuition, which rose, however, to the energy of
undoubting conviction in the author's mind. Joule's labours, on the contrary, are an experimental demonstration.
True to the speculative instinct of his country, Mayer
drew large and weighty conclusions from slender premises,
while the Englishman aimed, above all things, at the firm
establishment of facts. And he did establish them. The
future historian of science will not, I think, place these

men in antagonism.  To each belongs a reputation which
will not quickly fade, for the share he has had, not only
in establishing the dynamical theory of heat, but also in
leading the way towards a right appreciation of the
general energies of the universe.

Let us now check our conclusion regarding the influence which the performance of work has on the
quantity of heat communicated to a gas.  Is it not
possible to allow a gas to expand, without performing
work?  This question is answered by the following important experiment, which was first made by Gay Lussac.
I have here two copper vessels, A, B (fig. 22), of the same
size, one of which, A, is exhausted, and the other, B,

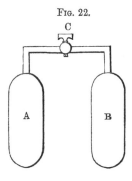

FIG. 22.

filled with air.  I turn the
cock c; the air rushes out of B
into A, until the same pressure
exists in both vessels.  Now the
air in driving its own particles
out of B performs work, and experiments which we have already
made inform us, that the residue
of air which remains in B must
be chilled.  The particles of air
enter A with a certain velocity, to
generate which the heat of the
air in B has been sacrificed; but they immediately strike
against the interior surface of A, their motion of translation
is annihilated, and the exact quantity of heat lost by B
appears in A.   Mix the contents of A and B together,
and you have air of the original temperature.  There is
no work performed, and there is no heat lost.  Mr. Joule
made this experiment by compressing twenty-two atmo-
spheres of air into one of his vessels, while the other was
exhausted.  On surrounding both vessels by water, kept
properly agitated, no augmentation of temperature was

observed in the water, when the gas was allowed to stream
from one vessel into the other.* In like manner, supposing
the top of the cylinder (fig. 20) to be closed, and the half
above the piston a perfect vacuum; and suppose the air
in the lower half to be heated 273°, its volume being kept
constant. If the pressure were removed, the air would
expand and fill the cylinder; the lower portion of the column
would thereby be chilled, but the upper portion would be
heated, and mixing both portions together, we should have
the whole column at a temperature of 273°. In this case
we raise the temperature of the gas from 0° to 273°, and
afterwards allow it to double its volume; the state of the
gas at the commencement, and at the end, is the same as
when the gas expands against a constant pressure, or lifts
a constant weight; but the absolute quantity of heat in
the latter case is 1·421 times that employed in the former,
the difference being due to the fact that the gas, in the one
case, performs mechanical work, and in the other not.

We are taught by this experiment that mere rarefaction is not of itself sufficient to produce a lowering of the
mean temperature of a mass of air. It was, and is still, a
current notion, that the mere expansion of a gas produced
refrigeration, no matter how that expansion was effected.
The coldness of the higher atmospheric regions was accounted for by reference to the expansion of the air. It
was thought that what we have called the 'capacity for
heat' was greater in the case of the rarefied than of the
unrarefied gas. But the refrigeration which accompanies
expansion is, in reality, due to the consumption of heat in
the performance of work by the expanding gas. Where
no work is performed there is no absolute refrigeration.

All this needs reflection to arrive at clearness, but every
effort of this kind which you make will render your subsequent efforts easier, and should you fail, at present, to

* Phil. Mag. 1845, vol. xxvi. p. 378.

gain clearness of comprehension, I repeat my recommenda-
tion of patience.  Do not quit this portion of the subject
without an effort to comprehend it—wrestle with it for a
time, but do not despair if you fail to arrive at clearness.

I have now to direct your attention to one other interesting question.  We have seen the elastic force of our gas
augmented by an increase of temperature.  In an inflexible
envelope we have, for every degree of temperature, a certain
definite increment of elastic force, due to the augmented
energy of the gaseous projectiles.  Reckoning from 0° C.
upwards, we find that every degree added to the temperature
produces an augmentation of elastic force, equal to $\frac{1}{273}$rd of
that which the gas possesses at 0°, and, hence, that by imparting 273° we double the elastic force. Supposing the same law
to hold good when we reckon from 0° *downwards*—that for
every degree of temperature *withdrawn* from the gas we
diminish its elastic force, or the motion which produces it,
by $\frac{1}{273}$rd of what it possesses at 0°, it is manifest that at a
temperature of 273° Centigrade below 0° we should cease
to have any elastic force whatever.  The motion to which
the elastic force is due must here vanish, and we reach
what is called the *absolute zero of temperature.*

No doubt, practically, every gas deviates from the above
law of contraction  before it sinks so low, and it would become solid before reaching — 273° C., or the absolute zero.
This is considerably below any temperature which we have
as yet been able to obtain.

I will not subject your minds to any further strain in
connection with this subject to-day, but will now pass on
to illustrate experimentally the expansion of liquids by heat.

Here is a Florence flask filled with alcohol, and tightly
corked ; through the cork a tube, $t'$ (fig. 23), passes water-
tight, and the liquid rises a foot or so in this tube.  I will
heat this flask, the alcohol will expand, and it will rise in
the tube. But I wish you to see it rising, and to enable you

to do so I will place the tube $t\,t'$ in front of the electric lamp E, and send a strong beam of light across it, at the

FIG. 23.

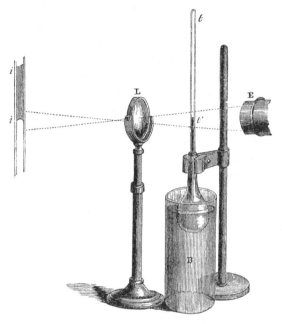

place $t'$, where the liquid column ends; I thus illuminate the tube and column.  In front of the tube I place this lens L, and arrange its distance so that it shall cast an enlarged image $ii$, of the column upon the screen.  You now see clearly where the column ends; you see this quivering of the top of the column, and if it moves you will be able to see its motion.  I now fill this beaker, B, with hot water, and I will raise the beaker so that the hot water shall surround the Florence flask.  It is needless to say that the image upon the screen is inverted, and that when the liquid expands, the top of the column will *descend* along the screen.  Observe the experiment from the commencement; the flask is now in the hot water,

and the head of our column *ascends,* as if the liquid contracted. Now it stops and commences to descend, and it
will continue to do so permanently. But why the first
ascent ? It is not due to the contraction of the liquid,
but *to the momentary expansion of the flask,* to which the
heat is first communicated. The glass expands before the
heat can fairly reach the liquid, and hence the column falls;
but soon the expansion of the liquid exceeds that of the glass,
and the column rises. Two things are here illustrated ; the
expansion of the solid glass by heat, and the fact that the
observed dilatation of the liquid does not give us its true
augmentation of volume, but only the difference of dilatation between the glass and it.

I have here another flask filled with water, exactly similar
in size to the former, and furnished with a similar tube. I
place it in the same position, and repeat with it the experiment made with the alcohol. You see, first of all, the
transitory effect due to the expansion of the glass, and
afterwards, the permanent expansion of the liquid ; but you
can observe that the latter proceeds much more slowly
than in the case of alcohol ; the alcohol expands more
speedily than the water. Now we might go over a hundred liquids in this way, and find them all expanding
by heat, and we might thus be led to conclude that
expansion by heat is a law without exception ; but we
should err in this conclusion. And it is really to illustrate
an exception of this kind that I have introduced this
flask of water. I will cool the flask down by plunging
it into a substance somewhat colder than water, when it
first freezes. This substance I obtain by mixing pounded
ice with salt. You see the column gradually sinking, the
heat is being given up to the freezing mixture, and the
water contracts. This contraction is now very slow, and now
it stops altogether. A slight motion commences in the
opposite direction, and now the liquid *is visibly expanding.*

I stir the freezing mixture, so as to bring colder portions
of it into contact with the flask; the colder the mixture the
quicker is the expansion. Here then we have Nature
stopping in her ordinary course, and reversing her ordinary
habits. The fact is, that the water goes on contracting
till it reaches a temperature of 39° Faht., or 4° Cent., at
which point the contraction ceases. This is the so-called
point of maximum density of the water; from this downwards, to its freezing point, the liquid expands; and when
it is converted into ice, the expansion is large and sudden.
Ice, we know, swims upon water, being lightened by this
expansion. If I now apply heat, the series of changes are
reversed: the column descends, showing the *contraction
of the liquid by heat.* After a time the contraction ceases,
and permanent expansion sets in.

The force with which these molecular changes are
effected is all but irresistible. The changes usually occur
under conditions which allow us no opportunity of observing the energy involved in their accomplishment. But to
give you an example of this energy, I have confined a quantity of water in this iron bottle. The iron is fully half an
inch thick, and the quantity of water is small, though sufficient to fill the bottle. The bottle is closed by a screw firmly
fixed in its neck. I have here a second bottle of the same
kind, and prepared in a similar manner. Both of them I
place in this copper vessel, and surround them with a freezing
mixture. They cool gradually, the water within approaches
its point of maximum density; no doubt, at this moment,
the water does not quite fill the bottle, a small vacuous
space exists within. But soon the contraction ceases, and
expansion sets in; the vacuous space is slowly filled, the
water gradually changes from liquid to solid; in doing so
it requires more room, which the rigid iron refuses to
grant. But its rigidity is powerless in the presence of the
atomic forces. These atoms are giants in disguise; you

hear that sound ; the bottle is shivered by the crystallising molecules — there goes the other; and here are the fragments of the vessels, which show their thickness, and impress you with the might of that energy by which they were thus riven.*

You have now no difficulty in understanding the effect of frosty weather upon the water-pipes of your houses. I have here a number of pieces of such pipes, all rent. You become first sensible of the damage when the thaw sets in, but the mischief is really done at the time of freezing; the pipes are then rent, and through the rents the water escapes, when the solid within is liquefied.

It is hardly necessary for me to say a word on the importance of this property of water in the economy of nature. Suppose a lake exposed to a clear wintry sky; the superficial water is chilled, contracts, becomes thus heavier, and sinks by its superior weight, its place being supplied by the lighter water from below. In time this is chilled, and sinks in turn. Thus a circulation is established, the cold, dense water descending, and the lighter and warmer water rising to the top. Supposing this to continue, even after the first pellicles of ice were formed at the surface; the ice would sink as it was formed,† and the process would not cease until the entire water of the lake would be solidified. Death to every living thing in the water would be the consequence. But just when matters become critical, Nature steps aside from her ordinary proceeding, causes the

* Metal cylinders, an inch in thickness, are unable to resist the decomposing force of a small galvanic battery. M. Gassiot has burst many such cylinders by electrolytic gas.

† Prof. William Thomson has recently raised a point which deserves the grave consideration of theoretic geologists : Supposing the constituents of the earth's crust to *contract* on solidifying, as the experiments thus far made indicate, a breaking in, and sinking of the crust would assuredly follow its formation. Under these circumstances, it is extremely difficult to conceive that a solid shell should be formed, as is generally assumed, round a liquid nucleus.

water to expand by cooling, and the cold water swims like a scum on the surface of the warmer water underneath. Solidification ensues, but the solid is much lighter than the subjacent liquid, and the ice forms a protecting roof over the living things below.

Such facts naturally and rightly excite the emotions; indeed, the relations of life to the conditions of life—the general adaptations of means to ends in Nature, excite, in the profoundest degree, the interest of the philosopher. But in dealing with natural phenomena, the feelings must be carefully watched. They often lead us unconsciously to overstep the bounds of fact. Thus, I have heard this wonderful property of water referred to as an irresistible proof of design, unique of its kind, and suggestive of pure benevolence. 'Why,' it is urged, 'should this case of water stand out isolated, if not for the purpose of protecting Nature against herself?' The fact, however, is, that the case is not an isolated one. You see this iron bottle, rent from neck to bottom; I break it with this hammer, and you see a core of metal within. This is the metal bismuth, which, when it was in a molten condition, I poured into this bottle, and closed the bottle by a screw, exactly as in the case of the water. The metal cooled, solidified, expanded, and the force of its expansion was sufficient to burst the bottle. There are no fish here to be saved, still the molten bismuth acts exactly as the water acts. Once for all, I would say that the natural philosopher, as such, has nothing to do with purposes and designs. His vocation is to enquire *what* Nature is, not *why* she is; though he, like others, and he, more than others, must stand at times rapt in wonder at the mystery in which he dwells, and towards the final solution of which his studies furnish him with no clue.

We must now pass on to the expansion of solid bodies, by heat, and I will illustrate it in this way: I have here two wooden stands, A and B (fig. 24), with plates of brass,

$p\,p'$, riveted against them. I hold in my hand two bars of
equal length, one of brass, the other of iron, and these, as
you observe, are not sufficiently long to stretch from stand

FIG. 24.

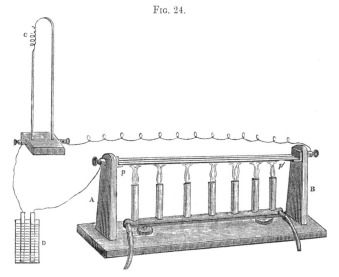

to stand. I will support them on two little projections of wood
attached to the stands at $p$ and $p'$. I connect one of the
plates of brass, $p$, with one pole of a small voltaic battery, D,
and from the other, $p'$, a wire proceeds to the little instrument c, which you see in front of the table; and again from
that instrument a wire returns direct to the other pole of the
battery. The instrument in front consists merely of an
arrangement to support a spiral c of platinum wire, which
will glow with a pure white light when the current from D
passes through it. At the present moment the only break
in the circuit is due to the insufficient length of the bars
of brass and iron to bridge the space from stand to
stand. Underneath the bars is a row of gas jets, which I
will now ignite; the bars are heated, the metals expand,
and I expect that in a few moments they will stretch quite

across from plate to plate; when this occurs, the current
will pass, and the fact of the gap being bridged will
be declared by the sudden glowing of the platinum spiral.
It is still non-luminous, the bridge is not yet complete;
but now it brightens up, showing that one, or both, of these
bars have expanded so as to stretch quite across from
stand to stand. Which of the bars is it? I remove the
iron, but the platinum still glows: I restore the iron,
and remove the brass; the light disappears. It was the
brass that bridged the gap. So that we have here an illustration, not only of the general fact of expansion, but also
of the fact that different bodies expand in different
degrees.

The expansion of both brass and iron is very small: and
various instruments have been devised to measure the expansion. Such instruments go under the general name of
pyrometers. But I have here a means of multiplying
the effect, far more powerful than the ordinary pyrometer.
Here is a solid upright bar of iron two feet long, and on a
mirror connected with the top of the bar I throw a beam
of light from the electric lamp, which beam is reflected to
the upper part of the wall. If the bar shorten, the mirror will
turn in one direction: if it lengthen, the mirror will turn
in the opposite direction. Every movement of the mirror,
however slight, is multiplied by this long index of light;
which, besides its length, has the advantage of moving with
twice the angular velocity of the mirror. Even the breath,
projected against this massive bar of iron, produces a sensible motion of the beam; and if I warm it for a moment
with the flame of a spirit-lamp, the luminous index will
travel downwards, the patch of light upon the wall moving
through a space of full thirty feet. I withdraw the lamp,
and allow the bar to cool; it contracts, and the patch of
light reascends the wall: I hasten the contraction by throwing a little alcohol on the bar of iron, the light moves more

speedily upwards, and now it occupies a place near the ceiling, as at the commencement of the experiment.*

I have stated that different bodies possess different
powers of expansion;† that brass, for example, expands
more, on being heated, than iron.  Here are two rulers,
one of brass and the other of iron, riveted together so as
to form, at this temperature, a straight compound ruler.
But if the temperature be changed, the ruler is no longer
straight.  I heat it, it bends in one direction : I cool it, it
bends in the opposite direction.  When heated, the brass
expands most, and forms the convex side of the curved
ruler.  When cooled, the brass contracts most, and forms the
concave side of the ruler.  Facts like these must, of course,
be taken into account, in structures where it is necessary to
avoid distortion.  The force with which bodies expand
when heated, is quite irresistible by any mechanical ap-
pliances that we can make use of.  All these molecular
forces, though operating in such minute spaces, are almost
infinite in energy.  The contractile force of cooling has
been applied by engineers to draw leaning walls into an
upright position.  If a body be brittle, the heating of one
portion of it, producing expansion, may so press or strain
another portion, as to produce fracture.  Hot water poured
into a glass often cracks it, through the sudden expansion
of the interior.  It may also be cracked by the contraction
produced by intense cold.

I have here some flasks of very thick glass, which, when
blown, were allowed to cool quickly.  The external portions
become first chilled and rigid.  The internal portions
cooled more gradually, but they found themselves, on cool-

* The piece of apparatus with which this experiment was made is intended for a totally different purpose.  I therefore indicate its principle
merely.

† The coefficients of expansion of a few well-known substances are
given in the Appendix to this Lecture.

ing, surrounded, as it were, by a rigid shell, on which they
exerted the powerful strain of their contraction. The consequence is, that the superficial portions of these flasks are
in such a state of tension that the slightest scratch produces
rupture. I throw into this flask a grain of quartz; the
mere dropping of the little bit of hard quartz into the flask
causes the bottom to fly out of it. Here also I have these
so-called Rupert drops, or Dutch tears, produced by glass
being fused to drops, which are suddenly cooled. The
external rigid shell has to bear the strain of the inner
contraction ; but the strain is distributed so equally all over
the surface, that no part gives way. But by simply breaking this filament of glass, which forms the tail of the
drop, the solid mass is instantly reduced to powder. I dip
the drop into a small flask filled with water, and break the
tail of the drop outside the flask ; the drop is shivered
with such force that the shock, transferred through the
water, is sufficient to break the bottle in pieces.

A very curious effect of expansion was observed, and
explained, some years ago by the Reverend Canon Mosely.
The choir of Bristol Cathedral was covered with sheet lead,
the length of the covering being 60 feet, and its depth
19 feet 4 inches. It had been laid on in the year 1851,
and two years afterwards—viz., in 1853—it had moved
bodily down for a distance of eighteen inches. The descent had been continually going on from the time the lead
had been laid down, and an attempt made to stop it by
driving nails into the rafters had failed ; for the force with
which the lead descended was sufficient to draw out the
nails. The roof was not a steep one, and the lead would
have rested on it for ever, without *sliding* down by gravity.
What, then, was the cause of the descent? Simply this.
The lead was exposed to the varying temperatures of day
and night. During the day the heat imparted to it caused
it to expand. Had it lain upon a horizontal surface, it

would have expanded equally all round, but as it lay upon
an inclined surface, it expanded more freely downwards
than upwards. When, on the contrary, the lead contracted at night, its upper edge was drawn more easily
downwards than its lower edge upwards. Its motion was
therefore exactly that of a common earthworm; it pushed
its lower edge forward during the day, and drew its upper
edge after it during the night, and thus by degrees it
crawled through a space of eighteen inches in two years.
Every local change of temperature during the day and
during the night contributed also to the result; indeed
Canon Mosely afterwards found the main effect to be due
to these quicker alternations of temperature.

Not only do different bodies expand differently by heat,
but the same body may expand differently in different directions. In crystals the atoms are laid together according to
law, and along some lines they are more closely packed than
along others. It is also likely that the atoms of many
crystalline bodies oscillate more freely and widely in some
directions than in others. The consequence of this would
be an unequal expansion by heat in different directions.
This crystal I hold in my hand (Iceland spar) has been
proved by Professor Mitscherlich to expand more along its
crystallographic axis than in any other direction. Nay,
while the crystal expands as a whole—that is to say, while
its volume is augmented by heat—it actually contracts in a
direction at right angles to the crystallographic axis.
Many other crystals also expand differently in different
directions; and, I doubt not, most organic structures would,
if examined, exhibit the same fact.

Nature is full of anomalies which no foresight can pre-
dict, and which experiment alone can reveal. From the
deportment of a vast number of bodies, we should be led
to conclude that heat always produces expansion, and that
cold always produces contraction. But water steps in,

and bismuth steps in to qualify this conclusion. If a
metal be compressed, heat is developed: but if a metal
wire be stretched, cold is developed. Mr. Joule and others
have worked at this subject, and found the above fact all
but general.

One striking exception to this rule (I have no doubt
there are many others) has been known for a great
number of years; and I will now illustrate this exception
by an experiment. My assistant will hand me a sheet of
India-rubber, which I have placed in the next room to
keep it quite cold. From this sheet I cut a strip three
inches long, and an inch and a half wide; I turn my
thermo-electric pile upon its back, and upon its exposed
face I lay this piece of India-rubber. From the deflection
of the needle, you see that that piece of rubber is cold.
I now lay hold of the ends of the strip, suddenly stretch it,
and press it, while stretched, on the face of the pile. See
the effect! The needle moves with energy, and showing
that the stretched rubber has heated the pile.

But one deviation from a rule always carries other
deviations in its train. In the physical world, as in the
moral, acts are never isolated. Thus with regard to our
India-rubber; its deviation from the rule referred to is
only part of a series of deviations. In many of his investigations Mr. Joule has been associated with an eminent
natural philosopher—Professor William Thomson—and
when Mr. Thomson was made aware of the deviation of
India-rubber from an almost general rule, he suggested
that the stretched India-rubber might *shorten*, on being
heated. The test was applied by Mr. Joule, and the
shortening was found to take place. This singular experiment, thrown into a suitable form, I will now perform
before you.

I fasten to this arm, *a a* (fig. 25), a length of common
vulcanised India-rubber tubing, and stretch it by a weight,

FIG. 25.

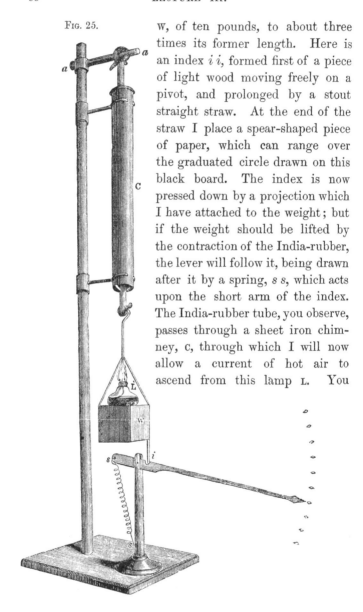

w, of ten pounds, to about three times its former length. Here is an index *i i*, formed first of a piece of light wood moving freely on a pivot, and prolonged by a stout straight straw. At the end of the straw I place a spear-shaped piece of paper, which can range over the graduated circle drawn on this black board. The index is now pressed down by a projection which I have attached to the weight; but if the weight should be lifted by the contraction of the India-rubber, the lever will follow it, being drawn after it by a spring, *s s*, which acts upon the short arm of the index. The India-rubber tube, you observe, passes through a sheet iron chimney, c, through which I will now allow a current of hot air to ascend from this lamp L.　You

see the effect; the index rises, showing that the rubber
contracts, and by continuing to apply the heat for a minute
or so, I cause the end of my index to describe an arc fully
three feet long. I withdraw the lamp, and as the India-
rubber returns to its former temperature, it lengthens; the
index moves downwards, and now it rests even below the
position which it first occupied.

# APPENDIX TO LECTURE III.

## FURTHER REMARKS ON DILATATION.

It is not within the scope of these lectures to dwell in detail on all the phenomena of expansion by heat; but for the sake of my young readers, I will supplement this lecture by a few additional remarks.

The linear, superficial, or cubic coefficient of expansion, is that fraction of a body's length, surface, or volume, which it expands on being heated one degree.

Supposing one of the sides of a square plate of metal, whose length is 1, to expand, on being heated one degree, by the quantity $a$; then the side of the new square is $1 + a$, and its area is

$$1 + 2a + a^2.$$

In the case of expansion by heat, the quantity $a$ is so small, that its square is almost insensible; the square of a small fraction is, of course, greatly less than the fraction itself. Hence without sensible error, we may throw away the $a^2$ in the above expression, and then we should have the area of the new square

$$1 + 2a.$$

$2a$, then, is the superficial coefficient of expansion; hence we infer that by multiplying the linear coefficient by 2, we obtain the superficial coefficient.

Suppose, instead of a square, that we had a cube, having a side $= 1$; and that on heating the cube one degree, the side expanded to $1 + a$; then the volume of the expanded cube would be

$$1 + 3a + 3a^2 + a^3.$$

In this, as in the former case, the square of $a$, and much more the cube of $a$, may be neglected, on account of their exceeding smallness; we then have the volume of the expanded cube

$$= 1 + 3a;$$

that is to say, the cubic coefficient of expansion is found by trebling the linear coefficient.

The following table contains the coefficients of expansion, for a number of well-known substances :—

| Copper | . | . | . 0·000017 | 0·000051 | 0·000051 |
| Lead . | . | . | . 0·000029 | 0·000087 | 0·000089 |
| Tin . | . | . | . 0·000023 | 0·000069 | 0·000069 |
| Iron . | . | . | . 0·0000123 | 0·000037 | 0·000037 |
| Zinc . | . | . | . 0·0000294 | 0·000088 | 0·000089 |
| Glass . | . | . | . 0·000080 | 0·000024 | 0·000024 |

The second column here gives the linear coefficient of expansion for 1° C.; the third column contains this coefficient trebled, which is the cubic expansion of the substance; and the fourth column gives the cubic expansion of the same substance, determined directly by Professor Kopp.* It will be seen that Kopp's coefficients agree almost exactly with those obtained by the trebling of the linear coefficients.

The linear coefficient of glass for 1° C is

$$0·0000080.$$

That of platinum is

$$0·0000088.$$

Hence glass and platinum expand nearly alike. This is of the greatest importance to chemists, who often find it necessary to *fuse* platinum wires into their glass tubes. Were the coefficients different, the fracture of the glass would be inevitable during the contraction.

### *The Thermometer.*

Water owes its liquidity to the motion of heat; when this motion sinks sufficiently, crystallisation, as we have seen, sets in. The temperature of crystallisation is perfectly constant if the water be kept under the same pressure. For example, water crystallises in all climates at the sea-level, at a temperature of

* Phil. Mag., 1852, vol. iii. p. 268.

32° F., or of 0° C. The temperature of condensation from the state of steam is equally constant, as long as the pressure remains the same. The melting of ice and the freezing of water touch each other, if I may use the expression, at 32° F. ; the condensation of steam and the boiling of water touch each other at 212° : 32° then is the freezing point of water, and it is the melting point of ice; 212° is the condensing point of steam and the boiling point of water. Both are invariable as long as the pressure remains the same. Here, then, we have two invaluable standard points of temperature, and they have been used for this throughout the world. The mercurial thermometer consists of a bulb and a stem with capillary bore. The bore ought to be of equable diameter throughout. The bulb and a portion of the stem are filled with mercury. Both are then plunged into melting ice, the mercury shrinking, the column descends, and finally comes to rest. Let the point at which it becomes stationary be marked; it is the *freezing point* of the thermometer. Let the instrument be now removed and thrust into boiling water; the mercury expands, the column rises, and finally attains a stationary height. Let this point be marked; it is the *boiling point* of the thermometer. The space between the freezing point and the boiling point has been divided by Reaumur into 80 equal parts, by Fahrenheit into 180 equal parts, and by Celsius into 100 equal parts, called degrees. The thermometer of Celsius is also called the Centigrade thermometer.

Both Reaumur and Celsius call the freezing point 0°, Fahrenheit calls it 32°, because he started from a zero which he incorrectly imagined was the greatest terrestrial cold. Fahrenheit's boiling point is therefore 212°. Reaumur's boiling point is 80°, while the boiling point of Celsius is 100°.

The length of the degrees being in the proportion of 80 : 100 : 180, or of 4 : 5 : 9 ; nothing can be easier than to convert one into the other. If you want to convert Fahrenheit into Celsius, multiply by 5 and divide by 9 ; if Celsius into Fahrenheit, multiply by 9 and divide by 5. Thus 20° of Celsius are equal to 36° Fahrenheit; but if we would know what temperature by Fahrenheit's thermometer corresponds to 20° of Celsius, we must add 32 to the 36, which would make the temperature 20°, as shown by Celsius, equal the temperature 68°, as shown by Fahrenheit.

EXTRACTS FROM SIR H. DAVY'S FIRST SCIENTIFIC MEMOIR,
BEARING THE TITLE 'ON HEAT, LIGHT, AND THE COMBINA-
TIONS OF LIGHT.'* ·

THE peculiar modes of existence of bodies, solidity, fluidity,
and gazity, depend (according to the calorists) on the quantity of
the fluid of heat entering into their composition. This substance
insinuating itself between their corpuscles, separating them from
each other, and preventing their actual contact, is by them supposed to be the cause of repulsion.

Other philosophers, dissatisfied with the evidences produced in
favour of the existence of this fluid, and perceiving the generation of heat by friction and percussion, have supposed it to be
motion. Considering the discovery of the true cause of the
repulsive power as highly important to philosophy, I have endeavoured to investigate this part of chemical science by experiments; from these experiments (of which I am now about to
give a detail) I conclude that heat or the power of repulsion is
not *matter*.

*The Phenomena of Repulsion are not dependent on a peculiar
elastic fluid for their existence, or Caloric does not exist.*

Without considering the effects of the repulsive power on
bodies, or endeavouring to prove from these effects that it is
motion, I shall attempt to demonstrate by experiments, that it is
not matter; and in doing this, I shall use the method called by
mathematicians, *reductio ad absurdum*.

First, let the increase of temperature produced by friction and
percussion be supposed to arise from a diminution of the capacities of the acting bodies. In this case it is evident some change
must be induced in the bodies by the action, which lessens their
capacities and increases their temperatures.

*Experiment.*— I procured two parallelopipedons of ice,† of the
temperature of 29°, six inches long, two wide, and two-thirds of
an inch thick : they were fastened by wires to two bars of iron.

* Sir Humphry Davy's works, vol. ii.
† The result of this experiment is the same, if wax, tallow, resin, or any
substance fusible at a low temperature, be used; even iron may be fused by
collision.

By a peculiar mechanism, their surfaces were placed in contact, and kept in a continued and most violent friction for some minutes. They were almost entirely converted into water, which water was collected, and its temperature ascertained to be 35°, after remaining in an atmosphere of a lower temperature for some minutes. The fusion took place only at the plane of contact of the two pieces of ice, and no bodies were in friction but ice.

From this experiment it is evident that ice by friction is converted into water, and according to the supposition, its capacity is diminished; but it is a well-known fact, that the capacity of water for heat is much greater than that of ice; and ice must have an absolute quantity of heat added to it, before it can be converted into water. Friction consequently does not diminish the capacities of bodies for heat.

From this experiment it is likewise evident, that the increase of temperature consequent on friction cannot arise from the decomposition of the oxygen gas in contact, for ice has no attraction for oxygen. Since the increase of temperature consequent on friction cannot arise from the diminution of capacity, or oxydation of the acting bodies, the only remaining supposition is, that it arises from an absolute quantity of heat added to them, which heat must be attracted from the bodies in contact. Then friction must induce some change in bodies, enabling them to attract heat from the bodies in contact.

*Experiment.*—I procured a piece of clockwork, so constructed as to be set at work in the exhausted receiver; one of the external wheels of this machine came in contact with a thin metallic plate. A considerable degree of sensible heat was produced by friction between the wheel and plate when the machine worked, uninsulated from bodies capable of communicating heat. I next procured a small piece of ice; * round the superior edge of this a small canal was made, and filled with water. The

* The temperature of the ice and of the surrounding atmosphere at the commencement of the experiment was 32°, that of the machine was likewise 32°. At the end of the experiment the temperature of the coldest part of the machine was near 33°, that of the ice and surrounding atmosphere the same as at the commencement of the experiment; so that the heat produced by the friction of the different parts of the machine was sufficient to raise the temperature of near half a pound of metal at least one degree; and to convert eighteen grains of wax (the quantity employed) into a fluid.

machine was placed on the ice, but not in contact with the water. Thus disposed, the whole was placed under the receiver (which had been previously filled with carbonic acid), a quantity of potash (i. e. caustic vegetable alkali) being at the same time introduced.

The receiver was now exhausted. From the exhaustion and from the attraction of the carbonic acid gas by the potash, a vacuum nearly perfect, was, I believe, made.

The machine was now set to work; the wax rapidly melted, proving an increase of temperature.

Caloric then was collected by friction; which caloric, on the supposition, was communicated by the bodies in contact with the machine. In this experiment, ice was the only body in contact with the machine. Had this ice given out caloric, the water on the top of it must have been frozen. The water on the top of it was not frozen, consequently the ice did not give out caloric. The caloric could not come from the bodies in contact with the ice, for it must have passed through the ice to penetrate the machine, and an addition of caloric to the ice would have converted it into water.

Heat, when produced by friction, cannot be collected from the bodies in contact, and it was proved, by the first experiment, that the increase of temperature consequent on friction cannot arise from diminution of capacity or oxydation. But if it be considered as matter, it must be produced in one of these modes. Since (as is demonstrated by these experiments) it is produced in neither of these modes, it cannot be considered as matter. It has therefore been experimentally demonstrated that caloric, or the matter of heat, does not exist.

Solids, by long and violent friction, become expanded, and if of a higher temperature than our bodies, affect the sensory organs with the peculiar sensation known by the common name of heat.

Since bodies become expanded by friction, it is evident that their corpuscles must move or separate from each other.

Now a motion or vibration of the corpuscles of bodies must be necessarily generated by friction and percussion. Therefore we may reasonably conclude that this motion or vibration is heat, or the repulsive power.

Heat, then, or that power which prevents the actual contact of the corpuscles of bodies, and which is the cause of our peculiar sensations of heat and cold, may be defined a peculiar motion, probably a vibration of the corpuscles of bodies, tending to separate them. It may with propriety be called the repulsive motion.

Since there exists a repulsive motion, the particles of bodies may be considered as acted on by two opposing forces; the approximating power, which may (for greater ease of expression) be called attraction, and the repulsive motion. The first of these is the compound effect of the attraction of cohesion, by which the particles tend to come in contact with each other; the attraction of gravitation, by which they tend to approximate to the great contiguous masses of matter, and the pressure under which they exist, dependent on the gravitation of the superincumbent bodies. The second is the effect of a peculiar motory or vibratory impulse given to them, tending to remove them farther from each other, and which can be generated, or rather increased, by friction or percussion. The effects of the attraction of cohesion, the great approximating cause, on the corpuscles of bodies, is exactly similar to that of the attraction of gravitation on the great masses of matter composing the universe, and the repulsive force is analogous to the planetary projectile force.

In his ' Chemical Philosophy,' pp. 94 and 95, Davy expresses himself thus :—' By a moderate degree of friction, as it would appear from Rumford's experiments, the same piece of metal may be kept hot for any length of time; so that, if the heat be pressed out, the quantity must be inexhaustible. When any body is cooled, it occupies a smaller volume than before; it is evident therefore, that its parts must have approached each other; when the body has expanded by heat, it is equally evident that its parts must have separated from each other. The immediate cause of the phenomenon of heat, then, is motion, and the laws of its communication are precisely the same as the laws of the communication of motion.'

Since all matter may be made to fill a smaller space by cooling, it is evident that the particles of matter must have space between them; and since every body can communicate the power of expansion to a body of a lower temperature—that is, can give an expansive motion to its particles—it is a probable inference that its own particles are possessed of motion; but as there is no change in the position of its parts, as long as its temperature is uniform, the motion, if it exist, must be a vibratory or undulatory motion, or a motion of the particles round their axes, or a motion of the particles round each other.

It seems possible to account for all the phenomena of heat, if it be supposed that in solids the particles are in a constant state of vibratory motion, the particles of the hottest bodies moving with the greatest velocity, and through the greatest space; that in fluids and elastic fluids, besides the vibratory motion, which must be conceived greatest in the last, the particles have a motion round their own axes with different velocity, the particles of elastic fluids moving with the greatest quickness, and that in ethereal substances the particles move round their own axes, and separate from each other, penetrating in right lines through space. Temperature may be conceived to depend upon the velocity of the vibrations; increase of capacity in the motion being performed in greater space; and the diminution of temperature during the conversion of solids into fluids or gases, may be explained on the idea of the loss of vibratory motion, in consequence of the revolution of particles round their axes, at the moment when the body becomes fluid or aëriform, or from the loss of rapidity of vibration in consequence of the motion of the particles through space.

# LECTURE IV.

[February 13, 1862.]

THE TREVELYAN INSTRUMENT — GORE'S REVOLVING BALLS — IN-
FLUENCE OF PRESSURE ON FUSING POINT — LIQUEFACTION AND
LAMINATION OF ICE BY PRESSURE — DISSECTION OF ICE BY A
CALORIFIC BEAM — LIQUID FLOWERS AND THEIR CENTRAL SPOT —
MECHANICAL PROPERTIES OF WATER PURGED OF AIR — THE BOILING
POINT OF LIQUIDS : INFLUENCING CIRCUMSTANCES — THE GEYSERS
OF ICELAND.

APPENDIX : — NOTE ON THE TREVELYAN INSTRUMENT — PHYSICAL
PROPERTIES OF ICE.

BEFORE finally quitting the subject of expansion, I wish to show you an experiment which illustrates in a curious and agreeable way the conversion of heat into mechanical energy. The fact which I wish to reproduce was first observed by a gentleman named Schwartz, in one of the smelting works of Saxony. A quantity of silver which had been fused in a ladle was allowed to solidify, and to hasten its cooling it was turned out upon an anvil. Some time afterwards, a strange buzzing sound was heard in the locality, and was finally traced to the hot silver, which was found quivering upon the anvil. Many years subsequent to this, Mr. Arthur Trevelyan chanced to be using a hot soldering-iron, which he laid by accident against a piece of lead. Soon afterwards, his attention was excited by a most singular sound which, after some searching, was found to proceed from the soldering-iron. Like the silver of Schwartz, the soldering iron was found

in a state of vibration.  Mr. Trevelyan made his discovery
the subject of a very interesting investigation.  He determined the best form to be given to the ' rocker' as the
vibrating mass is now called, and throughout Europe at
present, this instrument is known as 'Trevelyan's Instrument.'    Since that time the subject has engaged the
attention of Prof. J. D. Forbes, Dr. Seebeck, Mr. Faraday,
M. Sondhaus, and myself; but to Trevelyan and Seebeck
we owe most.

Here is such a rocker made of brass.  Its length, A c (fig.
26), is five inches, the width, A B, 1·5 in., and the length of
the handle, which terminates in the knob F, is ten inches.

FIG. 26.

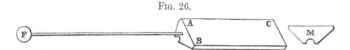

A groove runs at the back of the rocker, along its centre;
the cross section of the rocker and its groove is given
at M.  I heat the rocker to a temperature somewhat higher
than that of boiling water, and lay it on this block of
lead, allowing its
knob to rest upon
the table.  You
hear a quick suc-
cession of forcible
taps.   But you
cannot see the os-
cillations of the
rocker to which
the taps are due.
I therefore place
on it this rod of

FIG. 27.

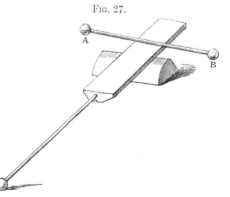

brass, A B (fig. 27), with two balls of brass at its end, the
oscillations are thereby rendered much slower, and you
can easily follow with the eye the pendulous motion of the

H 2

rod and balls. This motion will continue as long as the
rocker is able to communicate sufficient heat to the carrier
on which it rests. Thus we render the vibrations slow,
but I can also render them quick by using a rocker with a
wider groove. The sides of this rocker do not overhang so
much as those of the last; it is virtually a shorter pendulum,
and will vibrate more quickly. Placed upon the lead, as
before, it commences an unsteady and not altogether pleasant music. It is still restless, sometimes seeming to expos-
tulate, sometimes even to objurgate, as if it disliked the
treatment to which it is subjected. Now it becomes
mellow, and fills the room with a clear full note. Its
taps have become periodic and regular, and have linked
themselves together to produce music. Here is a third
rocker, with a still wider groove, and with it I can
obtain a shriller tone. You know of course that the
pitch of note augments with the number of the vibrations;
this wide-grooved rocker oscillates more quickly, and
therefore emits a higher note. By casting a beam of light
upon the rocker I obtain a better index than the rod
and balls. This index is without weight, and therefore
does not retard the motion of the rocker. To the latter
I have fastened, by a single screw at its centre, a small
disk of polished silver; on which the beam of the electric
lamp now falls, and is reflected against the screen. When
the rocker vibrates, the beam vibrates also, but with twice
the angular velocity, and there you see the patch of light
quivering upon the screen.

What is the cause of these singular vibrations and tones?
They are due simply to the sudden expansion by heat of
the body on which the rocker rests. Whenever the hot
rocker comes into contact with its lead carrier, a nipple
suddenly juts from the latter, being produced by the heat
communicated to the lead at the point of contact. The
rocker is tilted up and some other point of it comes into

contact with the lead, a fresh nipple is produced, and the rocker is again tilted. Let A B (fig 28) be the surface of the lead, and R the cross section of the hot rocker; tilted to the right, the nipple is formed as at R; tilted to the left, it is formed as at L. The consequence is that until its temperature

Fig. 28.

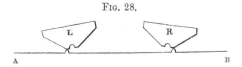

falls sufficiently, the rocker is tossed to and fro, and the quick succession of its taps against the lead produces a musical sound.

I have here fixed two pieces of sheet-lead in a vice; their edges are exposed, and are about half an inch asunder. I balance a long bar of heated brass across the two lead edges. It rests first on one edge, which expands at the point of contact and jerks it upwards; it then falls upon the second edge, which also rejects it; and thus it goes on oscillating, and will continue to do so as long as the bar

Fig. 29.

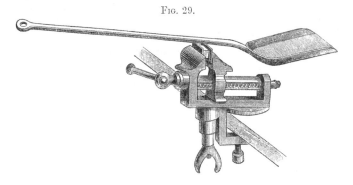

can communicate sufficient heat to the lead. This fire-shovel will answer quite as well as the prepared bar. I balance the heated shovel thus upon the edges of the lead, and it oscillates exactly as the bar did (fig. 29). I may add, that by properly laying either the poker or the fire-shovel upon a block of lead, supporting the handle so as to avoid friction, you may obtain notes as sweet and musical as any which you have heard to-day. A heated hoop placed upon a plate of lead may be caused to vibrate and sing ; and a hot penny-piece or half-crown may be caused to do the same.*

Looked at with an eye to the connection of natural forces, this experiment is interesting. The atoms of bodies must be regarded as all but infinitely small, but then they must be regarded as all but infinitely numerous. The augmentation of the amplitude of any oscillating atom by the communication of heat, is insensible, but the summation of an almost infinite number of such augmentations become sensible. Such a summation, effected almost in an instant, produces our nipple, and tilts the heavy mass of the rocker. Here we have a direct conversion of heat into common mechanical motion. But the tilted rocker falls again by gravity, and in its collision with the block restores almost the precise amount of heat which was consumed in lifting it. Here we have the direct conversion of common gravitating force into heat. Again the rocker is surrounded by a medium capable of being set in motion. The air of this room weighs some tons, and every particle of it is shaken by the rocker, and every tympanic membrane, and every auditory nerve present, is similarly shaken. Thus we have *the conversion of a portion of the heat into sound.* And, finally, every sonorous vibration which speeds through the air of this room, and wastes itself upon the walls, seats, and cushions, is converted into the form with which the cycle of actions commenced — namely, into heat.

Here is another curious effect, for which we are indebted

* For further information see Appendix to this lecture.

to Mr. George Gore, and which admits of a similar explanation. You see this line of rails. Two strips of brass, s s, s's' (fig. 30), are set edgeways, and about an inch asunder. I place this hollow metal ball B upon the rails; if I push it, it rolls along them; but if I do not push it, it stands still. I connect these two rails, by the wires $w\,w'$,

Fig. 30.

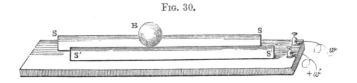

with the two poles of a Voltaic battery. A current now passes down one rail to the metal ball, thence along the ball to the other pail, and finally back to the battery. In passing from the rail to the ball, and from the ball to the other rail, the current encounters resistance, and wherever a current encounters resistance, heat is developed. Heat, therefore, is generated at the two points of contact of the ball with the rails; and this heat produces an elevation of the rail at these points. Observe the effect; the ball which a moment ago was tranquil is now very uneasy. It vibrates a little at first without rolling; now it actually rolls a little way, stops, and rolls back again. It gradually augments its excursion, now it has gone further than I intended: it has quite rolled off the rails, and injured itself by falling on the floor.

Here is another apparatus for which I am indebted to Mr. Gore himself, and in which the rails form a pair of concentric hoops; when the circuit is established, the ball F (fig. 31) rolls round the circle.* Mr. Gore has also obtained the rotation of light balls, by placing them on circular rails of hot copper, the *rolling* force in this case

* Phil. Mag., vol. 15, p. 521.

being the same as the *rocking* force in the Trevelyan
instrument.

In my last lecture I made evident to you the expansion
of water when it passes from the liquid to the solid condition ; with most other substances solidification is accom-

Fig. 31.

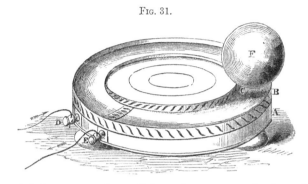

panied by contraction. I have here a round glass dish
into which I pour some hot water. Over the water I pour
from a ladle a quantity of melted bees'-wax. The wax now
forms a liquid layer nearly half an inch thick above
the water. We will suffer both water and wax to
cool, and when they are cool you will find that the wax
which now overspreads the entire surface, and is attached all
round to the glass, will retreat, and we shall finally obtain
a cake of wax of considerably smaller area than the dish.

The wax, then, in passing from the solid to the liquid
state *expands*. To assume the liquid form, its particles
must be pushed more widely apart, a certain play
between the particles being necessary to the condition of
liquidity. Now supposing we resist the expansion of the
wax by an external mechanical force; suppose we have a
very strong vessel completely filled with solid wax, and
which offers a powerful resistance to the expansion of the
mass within it; what would you expect if you sought to
liquefy the wax in this vessel? When the wax is free, the

heat has only to conquer the attraction of its own particles, but in the strong vessel it has not only to conquer the attraction of the particles, but also the resistance offered by the vessel. By a mere process of reasoning, we should thus be led to infer that a greater amount of heat would be required to melt the wax under pressure, than when it is free; or in other words, that the point of fusion of the wax is *elevated* by pressure. This reasoning is completely justified by experiment, not only with wax, but with other substances which contract on solidifying, and expand on liquefying. Messrs. Hopkins and Fairbairn have, by pressure, raised the melting point of some substances which contract considerably on solidifying as much as 20° and 30° Fahr.

These experiments bear on a very remarkable speculation. The earth is known gradually to augment in temperature as we pierce it deeper, and the depth has been calculated at which all known terrestrial bodies would be in a state of fusion. Mr. Hopkins, however, observes that owing to the enormous pressure of the superincumbent layers, the deeper strata would require a far higher temperature to fuse them, than would be necessary to fuse the strata near the earth's surface. Hence he infers that the solid crust must have a considerably greater thickness than that given by a calculation, which assumed the fusing points of the superficial and the deeper strata to be the same.

Now let us turn from wax to ice. Ice on liquefying *contracts*; in the arrangement of its atoms to form a solid, more room is required than they need in the neighbouring liquid state. No doubt this is due to crystalline arrangement; the attracting poles of the molecules are so placed that when the crystallising force comes into play, the molecules unite so as to leave larger inter-atomic spaces in the mass. We may suppose them to attach themselves by their corners; and in turning corner to corner, to cause

a recession of the atomic centres. At all events their
centres retreat from each other when solidification sets in.
By cooling, then, this power of retreat, and of consequent
enlargement of volume, is conferred.  It is evident that
pressure in this case would resist the expansion which is
necessary to solidification, and hence the tendency of
pressure, in the case of water, is to keep it liquid.  Thus
reasoning, we should be led to the conclusion that the
fusing points of substances which expand on solidifying
are *lowered* by pressure.

Professor James Thomson first drew attention to this
fact, and his theoretic reasonings have been verified by the
experiments of his brother Professor William Thomson.

Let us illustrate these principles by a striking experiment.  I have here a square pillar of clear ice an inch
and a half in height and about a square inch in cross
section.  At present the temperature of this ice is 0° C.
But suppose I subject this ice to pressure, I lower its
point of fusion: the ice under pressure will melt at a
temperature under 0° C., and hence the temperature which
it now possesses is in excess of that at which it will melt
under pressure.  I have cut this ice so that its planes of
freezing are perpendicular to the height of the pillar.
The direction of the stratified air-bubbles in the ice from
which this clear piece was taken, enabled me to fix at
once upon its planes of freezing.  Well, I place the column
of ice, L, upright between two slabs of boxwood, B,B′ (fig.
32), and place the whole between the plates of this small
hydraulic press; through the ice I send a beam from the
electric lamp.  In front of the ice I place a lens, and by it
project a magnified image of the ice upon the screen before
you.  The beam which passes through the ice has been
purified beforehand, so that, although it is still hot, its heat
is not of such a quality as can melt the ice; hence the light
passes through the substance without causing fusion.  I

work the arm of the press; the pillar of ice is now gently squeezed between the two slabs of boxwood.  I apply the pressure cautiously, and now you see dark streaks beginning to show themselves across the ice, at right angles to the direction of pressure.  Right in the middle of the mass they

FIG. 32.

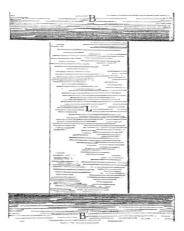

are appearing; and as I continue the pressure, the old streaks expand and new ones appear.   The entire column is now scarred across by these striæ.  What are they?  They are simply *liquid layers* foreshortened, and when you examine this column and look into it obliquely, you see these surfaces.  We have liquefied the ice in planes perpendicular to the pressure, and these liquid planes interspersed throughout the mass give it this strongly pronounced laminated appearance.*

Whether as a solid, a liquid, or a gas, water is one of the most wonderful substances in nature.   Let us consider its wonders a little further.   At all temperatures above

---

* See Appendix to this lecture for further information.

32° Fahr. or 0° C., the motion of heat is sufficient to keep
the molecules of water from rigid union.    But at 0° C.
the motion becomes so reduced that the atoms then seize
upon each other and aggregate to a solid.    This union,
however, is a union according to law.    To many persons
here present this block of ice may seem of no more interest
and beauty than a block of glass; but in the estimation of
science it bears the same relation to glass, that an oratorio
of Handel does to the cries of a market-place.    The ice
is music, the glass is noise; the ice is order, the glass is
confusion.    In the glass, molecular forces constitute an
inextricably entangled skein; in the ice they are woven to
a symmetric web; the miraculous texture of which I will
now try to reveal.

How shall I dissect this ice?    In the solar beam,— or
failing that, in the beam of an electric lamp, we have an
anatomist competent to perform this work.    I remove the
agent by which this beam was purified in the last experiment, and will send the rays direct from the lamp through
this slab of pellucid ice.    It shall pull the crystal edifice to
pieces by accurately reversing the order of its architecture.
Silently and symmetrically the crystallizing force builds
the atoms up, silently and symmetrically the electric beam
will take them down.    I place this slab of ice in front of
the lamp, the light of which now passes through the ice.
Compare the beam before it enters with the beam after its
passage through the substance: to the eye there is no sensible difference; the light is scarcely diminished.    Not so
with the heat.    As a thermic agent, the beam, before entering, is far more powerful than it is after its emergence.    A
portion of the beam has been arrested in the ice, and that
portion is our working anatomist.    Well, what is he doing?
I place a lens in front of the ice, and cast a magnified
image of the slab upon the screen.    Observe that image
(fig. 33), which, in beauty, falls far short of the actual

effect. Here we have a star, and there a star; and as
the action continues, the ice appears to resolve itself into

Fig. 33.

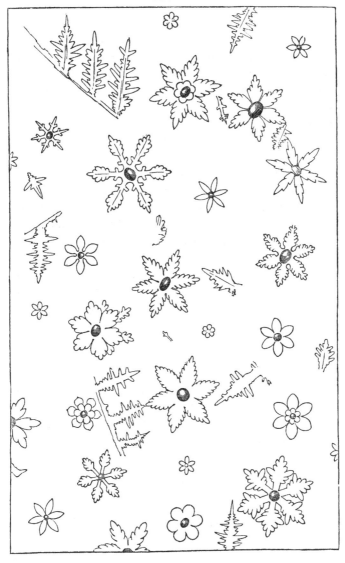

stars, each one possessing six rays, each one resembling a
beautiful flower of six petals.  And as I shift my lens to
and fro, I bring new stars into view, and as the action
continues, the edges of the petals become serrated, and
spread themselves out like fern leaves upon the screen.
Probably few here present were aware of the beauty latent
in a block of common ice.  And only think of lavish Nature
operating thus throughout the world.  Every atom of the
solid ice which sheets the frozen lakes of the North has
been fixed according to this law.  Nature ' lays her beams
in music,' and it is the function of science to purify our
organs, so as to enable us to hear the strain.

And now I have to draw your attention to two points
connected with this experiment, of great minuteness, but of
great interest.  You see these flowers by transmitted light,
— by the light which has passed through both the flowers
and the ice.  But when you examine them, by allowing a
beam to fall upon them and to be reflected from them to
your eye, you find in the centre of each flower a spot
which shines with the lustre of burnished silver.  You
might be disposed to think this spot a bubble of air; but
you can, by immersing it in hot water, melt away the ice all
round the spot; and the moment the spot is thus laid bare,
it collapses, and no trace of a bubble of air is to be seen.  *The
spot is a vacuum.*  Observe how truly Nature works; observe
how rigidly she carries her laws into all her operations.
We learned in the last lecture, that ice in melting contracted, and here we find the fact turning up.  The water
of these flowers cannot fill the space occupied by the
ice by whose fusion they are produced, hence the production of a vacuum necessarily accompanies the formation of
every liquid flower.

When I first observed these beautiful figures, I thought at
the moment when the central spot appeared, like a point of
light suddenly formed within the ice, that I heard a

clink, as if the ice had split asunder when the bright spot
was formed.  At first I suspected that it was my imagination which associated sound with the appearance of the
spot, as it is said that people who see meteors often imagine
a rushing noise when they really hear none.  The clink,
however, was a reality; and if you will allow me, I will
now make this trivial fact a starting point from which
I will conduct you through a series of interesting phenomena, to a far-distant question of practical science.

All water holds a large quantity of air within it in a
state of solution; by boiling you may liberate this imprisoned air.  On heating a flask of water you see air
bubbles crowding on its sides long before it boils, and
you see the bubbles rising through the liquid without
condensation, and often floating on the top.  One of the
most remarkable effects of this air in the water is, that it
promotes the ebullition of the liquid.  It acts as a kind of
elastic spring, pushing the atoms of the water apart, and
thus helping them to take the gaseous form.

Now suppose this air removed; having lost the cushion
which separated them, the atoms lock themselves together
in a far tighter embrace.  The cohesion of the water is
vastly augmented by the removal of the air.  Here is a
glass vessel, the so-called water-hammer, which contains
water purged of air.  One effect of the withdrawal of the
elastic buffer is, that the water here falls with the sound of
a solid body.  You hear how the liquid rings against the
end of the tube when I turn it upside down.  Here is another
tube, A B C (fig. 34), bent into the form of a V, and intended
to show how the cohesion of the water is affected by long
boiling.  I bring this water into one arm of the V; by
tilting the tube it flows, as you see, freely into the other
arm.  I restore it to the first arm, and now tap the end of
this arm against the table.  You hear, at first, a loose and
jingling sound.  As long as you hear it the water is not in

true contact with the surface of the tube.   I continue my
tapping :  you mark an alteration in the sound ;  the jingling

FIG. 34.

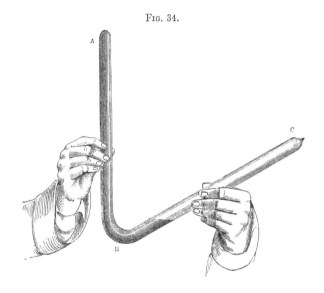

has  disappeared,  and  the  sound  is  now  hard,  like that of
solid  against  solid.   I  now  raise  my  tube.   Observe  what
occurs.   I  turn  the  column  of  water  upside  down, but there
it  stands  in A B.   Its  particles  cling  so  tenaciously  to  the
sides  of  the  tube,  and  lock  themselves  so  firmly  together,
that  it  refuses  to  behave  like  a  liquid  body ;  it  declines  to
obey  the  law  of  gravity.

   So much for the augmentation of cohesion ;  but this very
cohesion enables the liquid  to  resist  ebullition.   Water  thus
freed  of  its  air  can  be  raised  to  a  temperature  100°  and
more  above  its  ordinary  boiling  point,  without  ebullition.
But  mark  what  takes  place  when  the  liquid does boil.   It
has  an  enormous  excess of heat stored up , the locked atoms
finally  part  company,  but  they  do  so  with  the  violence  of  a
spring  which  suddenly  breaks  under  strong  tension,  and

ebullition is converted into explosion.  For the discovery
of this interesting property of water we are indebted to
M. Donny of Ghent.

Turn we now to our ice:—Water, in freezing, completely
excludes the air from its crystalline architecture.  All
foreign bodies are squeezed out in the act of freezing, and
ice holds no air in solution.  Supposing then that we melt
a piece of pure ice under conditions where air cannot
approach it, we have water in its most highly cohesive
condition; and such water ought, if heated, to show the
effects to which I have referred.  That it does so has been
proved by Mr. Faraday.  He melted pure ice under spirit
of turpentine, and found that the liquid thus formed could
be heated far beyond its boiling point, and that the rupture
of the liquid, by the act of ebullition, took place with almost
explosive violence.  And now let us apply these facts to
the six-petaled ice-flowers and their little central star.
They are formed in a place where no air can come.
Imagine the flower forming and gradually augmenting in size.  The cohesion of the liquid is so great, that
it will pull the walls of its chamber together, or even
expand its own volume, sooner than give way.  But as its
size augments, the space which it tries to occupy becomes
too large for it, until finally the liquid snaps, a vacuum is
formed, and a clink is heard.

Let us now take our final glance at this web of relations.
It is very remarkable, that a great number of locomotives
have exploded on quitting the shed where they had remained for a time quiescent.  The number of explosions
which have occurred just as the engineer turned on the steam
is quite surprising.  Now supposing that a locomotive had
been boiling sufficiently long to expel the air contained in
its water; *that* liquid would possess, in a greater or less
degree, the high cohesive quality to which I have drawn
your attention.  It is at least conceivable that while

I

resting previous to starting on its journey, an excess of
heat might be thus stored up in the boiler, and if stored
up, the certain result would be, that the engineer on turning on the steam would, by a mechanical act, produce the
rupture of the cohesion, and steam of explosive force
would instantly be generated.  I do not say that this *is*
the case; but who can say that it is *not* the case.  We
have been dealing throughout with a real agency, which
is certainly competent, if its power be invoked, to produce the most terrible effects.

We have here touched on the subject of steam; let us
bestow a few minutes' further consideration on its formation and action.  As you add heat, or in other words,
motion, to water, the particles from its free surface fly off in
augmented numbers.  We at length approach what is
called the *boiling point* of the liquid, where the conversion
into vapour is not confined to the free surface, but is
most copious at the bottom of the vessel to which the
heat is applied.    When water boils in a glass beaker,
the steam is seen rising in spheres from the bottom to the
top, where it often swims for a time, enclosed above by a
dome-shaped liquid film.  Now, to produce these bubbles
certain resistances must be overcome.  First, we have the
adhesion of the water to the vessel which contains it, and
this force varies with the substance of the vessel.   In the
case of a glass vessel, for example, the boiling point may
be raised two or three degrees by adhesion; while in metal
vessels this is impossible.  The adhesion is overcome by
fits and starts, which may be so augmented by the introduction of salts into the liquid, that a loud bumping sound
accompanies the ebullition; the detachment is in some
cases so sudden and violent as to cause the liquid to jump
bodily out of the vessel.

A second antagonism to the boiling of the liquid is the
attraction of the liquid particles for each other, a force

which, as we have seen, may become very powerful when the liquid is purged of air. This is not only true of water, but of other liquids — of all the ethers and alcohols, for example. If we connect a small flask containing ether or alcohol with an air pump, a violent ebullition occurs in the liquid when the pump is first worked; but after all the air has been removed, we may, in many cases, continue to work the pump, without producing any sensible ebullition; the free surface alone of the liquid yielding vapour.

But that steam should exist in bubbles, in the interior of a mass of liquid, it must be able to resist two other things, the weight of the water above it, and the weight of the atmosphere above the water. What the atmosphere is competent to do may be thus illustrated. I have here a tin vessel containing a little water, which is kept boiling by this small lamp. At the present moment all the space above the water is filled with steam, which issues from this stop-cock. I shut off the cock, withdraw the lamp, and pour cold water upon the tin vessel. The steam within it is condensed, the elastic cushion which pushed the sides outwards in opposition to the pressure of the atmosphere is withdrawn, and observe the consequence. The sides of the vessel are crushed and crumpled up by the atmospheric pressure. This pressure amounts to 15lb. on every square inch: how then can a thing so frail as a bubble of steam exist on the surface of boiling water? simply because the elastic force of the steam within is exactly equal to that of the atmosphere without; the liquid film is pressed between two elastic cushions which exactly neutralise each other. If the steam were predominant, the bubble would burst from within outwards; if the air were predominant, the bubble would be crushed inwards. Here, then, we have the true definition of the boiling point of a liquid. It is that temperature at which the tension of its vapour exactly balances the pressure of the atmosphere.

As we ascend a mountain the pressure of the atmosphere above us diminishes, and the boiling point is correspondingly lowered. On an August morning in 1859 I found the temperature of boiling water on the summit of Mont Blanc to be 184·95° Fahr.; that is, about 27° lower than the boiling point at the sea level. On August 3, 1858, the temperature of boiling water on the summit of the Finsteraarhorn was 187° Fahr. On August 10, 1858, the boiling point on the summit of Monte Rosa was 184·92° Fahr. The boiling point on Monte Rosa is shown by these observations to be almost the same as it was found to be on Mont Blanc, though the latter exceeds the former in height by 500 feet. The fluctuations of the barometer are however quite sufficient to account for this anomaly. The lowering of the boiling point is about 1° Fahr. for every 590 feet that we ascend; and from the temperature at which water boils we may approximately infer the elevation. It is said that to make good tea in London, boiling water is essential; if this be so it is evident that the beverage cannot be procured, in all its excellence, at the higher stations in the Alps.

Let us now make an experiment to illustrate the dependence of the boiling point on external pressure. Here is a flask, F (fig. 35), containing water; here is another and a much larger one, G, from which I have had the air removed by an air pump. The two flasks are connected together by a system of cocks, which enables me to establish a communication between them. The water in the small flask has been kept boiling for some time, the steam generated escaping from the cock y. I now remove the spirit lamp and turn this cock so as to shut out the air. The water ceases to boil, and pure steam now fills the flask above it. Give the water time to cool a little. At intervals you see a bubble of steam rising, because the pressure of the vapour above is gradually becoming less through its

slow condensation. I hasten the condensation by pouring
cold water on the flask, the bubbles are more copiously
generated. By plunging the flask bodily into cold water
we might cause it to boil violently. The water is now at
rest and some degrees below its ordinary boiling point. I
turn this cock *c*, which opens a way for the escape of the
vapour into the exhausted vessel G; the moment the pressure is diminished ebullition sets in in F; and observe how

FIG. 35.

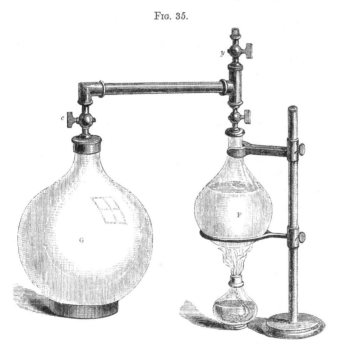

the condensed steam showers in a kind of rain against the
sides of the exhausted vessel. By intentionally promoting
this condensation, and thereby preventing the vapour in
the large flask from reacting upon the surface of the water,
we can keep the small flask bubbling and boiling for a
considerable length of time.

By high heating, the elastic force of steam becomes enormous. The Marquis of Worcester burst cannon with it, and our calamitous boiler explosions are so many illustrations of its power. By the skill of man this mighty agent has been controlled: by it Denis Papin raised a piston, which was pressed down again by the atmosphere, when the steam was condensed; Savery and Newcomen turned it to practical account, and James Watt completed the grand application of the moving power of heat. Pushing the piston up by steam, while the space above the piston is in communication with a condenser or with the free air, and again pushing down the piston, while the space below it is in communication with a condenser or with the air, we obtain a simple to and fro motion, which, by mechanical arrangements, may be made to take any form we please.

But the grand principle of the conservation of force is illustrated here as elsewhere. For every stroke of work done by the steam-engine, for every pound that it lifts, and for every wheel that it sets in motion, an equivalent of heat disappears. A ton of coal furnishes by its combustion a certain definite amount of heat. Let this quantity of coal be applied to work a steam engine; and let all the heat communicated to the machine and the condenser, and all the heat lost by radiation and by contact with the air be collected; it would fall short of the amount produced by the simple combustion of the ton of coal, and it would fall short of it by an amount exactly equivalent to the quantity of work performed. Suppose that work to consist in lifting a weight of 7,720 lbs. a foot high; the heat produced by the coal would fall short of its maximum, by a quantity just sufficient to warm a pound of water 10°.

But my object in these lectures is to deal with nature rather than art, and the limits of our time compel me to pass quickly over the triumphs of man's skill in the application of steam to the purposes of life. Those who have walked through the workshops of Woolwich, or through any of our great factories where machinery is extensively employed, will have been sufficiently impressed with the aid which this great power renders to man. And be it remembered, every wheel which revolves, every chisel, and plane, and saw, and punch, which forces its way through solid iron as if it were so much cheese, derives its moving energy from the clashing atoms in the furnace. The motion of these atoms is communicated to the boiler, thence to the water, whose particles are shaken asunder, and fly from each other with a repellent energy commensurate with the heat communicated. The steam is simply the apparatus through the intermediation of which the atomic motion is converted into the mechanical. And the motion thus generated can reproduce its parent. Look at the planing tools; look at the boring instruments,— streams of water gush over them to keep them cool. Take up the curled iron shavings which the planing tool has pared off: you cannot hold them in your hand they are so hot. Here the moving force is restored to its first form; the energy of the machine has been consumed in reproducing the power from which that energy was derived.

I must now direct your attention to a natural steam-engine which long held a place among the wonders of the world. I allude to the Great Geyser of Iceland. The surface of Iceland gradually slopes from the coast towards the centre, where the general level is about 2,000 feet above the sea. On this, as a pedestal, are planted the Jökull or icy mountains, which extend both ways in a north-easterly direction. Along this chain occur the active volcanos of the island, and the thermal springs follow the same general direction. From the ridges and chasms which diverge from the mountains enormous masses of steam issue at intervals hissing and roaring; and when the escape occurs at the mouth of a cavern, the resonance of the cave often raises the sound to the loudness of thunder. Lower down in the more porous strata we have smoking mud pools, where a repulsive blue-black aluminous paste is boiled, rising at times in huge bubbles, which, on bursting, scatter their slimy spray to a height of fifteen or twenty feet. From the bases of the hills upwards extend the glaciers, and above these are the snow-fields which crown the summits. From the arches aud fissures of the glaciers vast masses of water issue, falling at times in cascades over walls of ice, and spreading for miles over the country before they find definite outlet. Extensive morasses are thus formed, which add their comfortless monotony to the dismal scene already before the traveller's eye. Intercepted by the cracks and fissures of the land, a portion of this water finds its way to the heated rocks underneath; and here, meeting with the volcanic gases which traverse these underground regions, both travel together, to issue, at the first convenient opportunity, either as an eruption of steam or a boiling spring.

The most famous of these springs is the Great Geyser. It consists of a tube 74 feet deep and 10 feet in diameter. The tube is surmounted by a basin, which measures from north to south 52 feet across, and from east to west 60 feet. The interior of the tube and basin is coated with a beautiful smooth siliceous plaster, so hard as to resist the blows of a hammer, and the first question is, how was this wonderful tube constructed — how was this perfect plaster laid on? Chemical analysis shows that the water holds silica in solution, and the conjecture might therefore arise that the water had deposited the silica against the sides of the tube and basin. But this is not the case: the water deposits no sediment; no matter how long it may be kept, no solid substance is separated from it. It may be bottled up and preserved for years as clear as crystal, without showing the slightest tendency to form a precipitate. To answer the

question in this way would moreover assume that the shaft
was formed by some foreign agency, and that the water
merely lined it.   The geyser basin, however, rests upon
the summit of a mound about 40 feet high, and it is evident,
from mere inspection, that the mound has been deposited
by the geyser.   But in building up this mound the spring
must have formed the tube which perforates the mound,
and hence the conclusion that the geyser is the architect
of its own tube.

If we place a quantity of the geyser water in an evaporating basin the following takes place: In the centre
of the basin the liquid deposits nothing, but at the
sides, where it is drawn up by capillary attraction,
and thus subjected to speedy evaporation, we find silica
deposited.   Round the edge a ring of silica is laid
on, and not until the evaporation has continued a considerable time do we find the slightest turbidity in the
middle of the water.   This experiment is the microscopic
representant of what occurs in Iceland.   Imagine the case
of a simple thermal siliceous spring, whose waters trickle
down a gentle incline ; the water thus exposed evaporates
speedily, and silica is deposited.   This deposit gradually
elevates the side over which the water passes until finally
the latter has to take another course.   The same takes
place here, the ground is elevated as before and the spring
has to move forward.   Thus it is compelled to travel round
and round, discharging its silica and deepening the shaft
in which it dwells, until finally, in the course of ages, the
simple spring has produced that wonderful apparatus which
has so long puzzled and astonished both the traveller and
the philosopher.

Previous to an eruption, both the tube and basin
are filled with hot water ; detonations which shake the
ground, are heard at intervals, and each is succeeded by a
violent agitation of the water in the basin.   The water in

the pipe is lifted up so as to form an eminence in the middle of the basin, and an overflow is the consequence. These detonations are evidently due to the production of steam in the ducts which feed the geyser tube, which steam escaping into the cooler water of the tube is there suddenly condensed, and produces the explosions. Professor Bunsen succeeded in determining the temperature of the geyser tube, from top to bottom, a few minutes before a great eruption; and these observations revealed the extraordinary fact, that at no part of the tube did the water reach its boiling point. In the annexed sketch (fig. 36) I have given, on one side, the temperatures actually observed, and on the other side the temperatures at which water would boil, taking into account both the pressure of the atmosphere and the pressure of the superincumbent column of water. The nearest approach to the boiling point is at A, a height of 30 feet from the bottom: but even here the water is 2° Centigrade, or more than $3\frac{1}{2}$° Fahr. below the temperature at which it could boil. How then is it possible that an eruption could occur under such circumstances?

Fix your attention upon the water at the point A; where the temperature is within 2° C. of the boiling point. Call to mind the lifting of the column when the detonations are heard. Let us suppose that by the entrance of steam from the ducts near the bottom of the tube, the geyser column is elevated 6 feet, a height quite within the limits of actual observation; the water at A is thereby transferred to B. Its boiling point at A is 123·8°, and its actual temperature 121·8°; but at B its boiling point is only 120·8°, hence, when transferred from A to B the heat which it possesses is in excess of that necessary to make it boil. This excess of heat is instantly applied to the generation of steam: the column is thus lifted higher, and the water below is further relieved. More

steam is generated; from the middle downwards the mass
suddenly bursts into ebullition, the water above, mixed
with steam clouds, is projected into the atmosphere, and
we have the geyser eruption in all its grandeur.

By its contact with the air the water is cooled, falls back
into the basin, partially refills the tube, in which it gradually

FIG. 36.

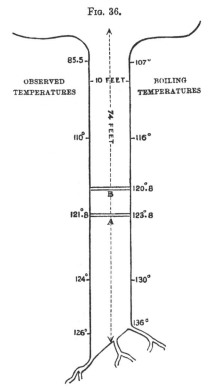

rises, and finally fills the basin as before.   Detonations
are heard at intervals, and risings of the water in the basin.
These are so many futile attempts at an eruption, for not
until the water in the tube comes sufficiently near its boiling temperature, to make the lifting of the column effective,
can we have a true eruption.

Fig. 37.

To Bunsen we owe this beautiful theory, and now let us try to justify it by experiment.  Here is a tube of galvanised iron, 6 feet long, A B (fig. 37), and surmounted by this basin C D.  It is heated by a fire underneath; and to imitate as far as possible the condition of the geyser, I have encircled the tube by a second fire, F, at a height of 2 feet from the bottom.  Doubtless the high temperature of the water at the corresponding part of the geyser tube is due to a local action of the heated rocks.   I fill the tube with water, which gradually becomes heated; and regularly, every five minutes, the water is ejected from the tube into the atmosphere.

Fɪɢ. 38.

But there is another famous spring in Iceland, called the Strokkur, which is usually forced to explode by stopping its mouth with clods.  We can imitate the action of this spring by stopping the mouth of our tube A B with a cork.  I do so: and now the heating progresses.  The steam below will finally attain sufficient tension to eject the cork, and the water, suddenly relieved from the pressure, will burst forth in the atmosphere.  There it goes!  The ceiling of this room is nearly 30 feet from the floor, but the eruption has reached the ceiling, from which the water now drips plentifully.   In fig. 38, I have given a section of the Strokkur.

By stopping the tube with corks, through which tubes of various lengths and widths pass, the action of many of the other eruptive springs may be accurately imitated.   Here, for example, I have an intermittent action; discharges of water and impetuous steam gushes follow each other

in quick succession, the water being squirted in jets 15
or 20 feet high.   Thus, it is proved experimentally, that
the geyser tube itself is the sufficient cause of the eruptions, and we are relieved from the necessity of imagining underground caverns filled with water and steam,
which were formerly regarded as necessary to the production of these wonderful phenomena.

A moment's reflection will suggest to us that there
must be a limit to the operations of the geyser.   When
the tube has reached such an altitude that the water in the
depths below, owing to the increased pressure, cannot attain
its boiling point, the eruptions of necessity cease.  The spring,
however, continues to deposit its silica, and often forms a
*Laug* or cistern.   Some of those in Iceland are 40 feet
deep.   Their beauty, according to Bunsen, is indescribable ;
over the surface curls a light vapour, the water is of the
purest azure, and tints with its lovely hue the fantastic
incrustations on the cistern walls ; while, at the bottom, is
often seen the mouth of the once mighty geyser.   There
are in Iceland traces of vast, but now extinct, geyser
operations.   Mounds are observed whose shafts are filled
with rubbish, the water having forced a passage underneath
and retired to other scenes of action.   We have in fact
the geyser in its youth, manhood, old age, and death, here
presented to us.   In its youth, as a simple thermal spring ;
in its manhood, as the eruptive column ; in its old age, as
the tranquil *Laug* ; while its death is recorded by the
ruined shaft and mound which testify the fact of its once
active existence.

# APPENDIX TO LECTURE IV.

——•◇•——

ABSTRACT OF A LECTURE ON THE VIBRATION AND TONES
PRODUCED BY THE CONTACT OF BODIES OF DIFFERENT
TEMPERATURES.

[*Given at the Royal Institution on Friday, January* 27, 1854.]

In the year 1805, M. Schwartz, an inspector of one of the smelting
works of Saxony, placed a cup-shaped mass of hot silver upon a
cold anvil, and was surprised to find that musical tones proceeded
from the mass. In the autumn of the same year, Professor
Gilbert of Berlin visited the smelting works and repeated the experiment. He observed, that the sounds were accompanied by a
quivering of the hot silver, and that when the vibrations ceased,
the sound ceased also. Professor Gilbert merely stated the facts,
and made no attempt to explain them.

In the year 1829, Mr. Arthur Trevelyan, being engaged in
spreading pitch with a hot plastering iron, and once observing
that the iron was too hot for his purpose, he laid it slantingly
against a block of lead which chanced to be at hand ; a shrill note,
which he compared to that of the chanter of the small Northumber-
land pipes, proceeded from the mass, and, on nearer inspection,
he observed that the heated iron was in a state of vibration. He
was induced by Dr. Reid of Edinburgh to pursue the subject,
and the results of his numerous experiments were subsequently
printed in the Transactions of the Royal Society of Edinburgh.

On April 1, 1831, these singular sounds and vibrations formed
the subject of a Friday evening discourse by Professor Faraday, at
the Royal Institution. Professor Faraday expanded and further

established the explanation of the sounds given by Mr. Trevelyan
and Sir John Leslie.  He referred them to the tapping of the
hot mass against the cold one underneath it, the taps being in
many cases sufficiently quick to produce a high musical note.
The alternate expansion and contraction of the cold mass at the
points where the hot rocker descends upon it, he regarded as the
sustaining power of the vibrations.  The superiority of lead he
ascribed to its great expansibility, combined with its feeble power
of conduction, which latter prevented the heat from being quickly
diffused through the mass.

Professor J. D. Forbes of Edinburgh was present at this lecture,
and not feeling satisfied with the explanation, undertook the
farther examination of the subject; his results are described in a
highly ingenious paper communicated to the Royal Society of
Edinburgh in 1833.  He rejects the explanation supported by
Professor Faraday, and refers the vibrations to ' a new species of
mechanical agency in heat ' — a repulsion exercised by the heat
itself on passing from a good conductor to a bad one.  This conclusion is based upon a number of general laws established by
Professor Forbes.  If these laws be correct, then indeed a great
step has been taken towards a knowledge of the intimate nature
of heat itself, and this consideration was the lecturer's principal
stimulus in resuming the examination of the subject.

He had already made some experiments, ignorant that the
subject had been farther treated by Seebeck, until informed of
the fact by Professor Magnus of Berlin.  On reading Seebeck's
interesting paper, he found that many of the results which it was
his intention to seek had been already obtained.  The portion of
the subject which remained untouched was, however, of sufficient
interest to induce him to prosecute his original intention.

The general laws of Professor Forbes were submitted in succession to an experimental examination.  The first of these laws
affirms that ' the vibrations never take place between substances of
the same nature.'  This the lecturer found to be generally the
case when the hot rocker rested upon a *block*, or on the edge of a
thick plate of the same metal; but the case was quite altered
when a thin plate of metal was used.  Thus a copper rocker laid
upon the edge of a penny-piece did not vibrate permanently;
but when the coin was beaten out by a hammer, so as to present
a thin sharp edge, constant vibrations were obtained.  A silver

rocker resting on the edge of a half-crown refused to vibrate permanently; but on the edge of a sixpence continuous vibrations were obtained.  An iron rocker on the edge of a dinner knife gave continuous vibrations,  A flat brass rocker placed upon the points of two common brass pins, and having its handle suitably supported, gave distinct vibrations.  In these experiments the plates and pins were fixed in a vice, and it was found that the thinner the plate, within its limits of rigidity, the more certain and striking was the effect.  Vibrations were thus obtained with iron on iron, copper on copper, brass on brass, zinc on zinc, silver on silver, tin on tin.  The list might be extended, but the cases cited are sufficient to show that the proposition above cited cannot be regarded as expressing a ' general law.'

The second general law enunciated by Professor Forbes is, that ' *both substances must be metallic.*'  This is the law which first attracted the lecturer's attention.  During the progress of a kindred enquiry, he had discovered that certain non-metallic bodies are endowed with powers of conduction far higher than has been hitherto supposed, and the thought occurred to him that such bodies might, by suitable treatment, be made to supply the place of metals in the production of vibrations.  This anticipation was realised.  Rockers of silver, copper, and brass, placed upon the natural edge of a prism of rock crystal, gave distinct tones; on the clean edge of a cube of fluor spar, the tones were still more musical; on a mass of rock-salt the vibrations were very forcible. There is scarcely a substance, metallic or non-metallic, on which vibrations can be obtained with greater ease and certainty than on rock-salt.  In most cases a high temperature is necessary to the production of the tones, but in the case of rock-salt the temperature need not exceed that of the blood.  A new and singular property is thus found to belong to this already remarkable substance.  It is needless to enter into a full statement regarding the various minerals submitted to experiment.  Upwards of twenty non-metallic substances had been examined by the lecturer, and distinct vibrations obtained with every one of them.

The number of exceptions here exhibited far exceeds that of the substances which are mentioned in the paper of Professor Forbes, and are, it was imagined, sufficient to show that the second general law is untenable.

K

The third general law states, that ' *the vibrations take place with an intensity proportional* (within certain limits) *to the difference of the conducting powers of the metals for heat, the metal having the least conducting power, being necessarily the coldest.*' The evidence adduced against the first law appears to destroy this one also; for if the intensity of the vibrations be proportional to the difference of the conducting powers, then, where there is no such difference, there ought to be no vibrations. But it has been proved in half a dozen cases, that vibrations occur between different pieces of the same metal. The condition stated by Professor Forbes was, however, reversed. Silver stands at the head of conductors; a strip of the metal was fixed in a vice, and hot rockers of brass, copper, and iron, were successively laid upon its edge: distinct vibrations were obtained with all of them. Vibrations were also obtained with a brass rocker which rested on the edge of a half-sovereign. These and other experiments show that it is not necessary that the worst conductor should be the cold metal, as affirmed in the third general law above quoted. Among the metals, antimony and bismuth were found perfectly inert by Professor Forbes; the lecturer however had obtained musical tones from both of these substances.

The superiority of lead as a cold block, Professor Faraday, as already stated, referred to its high expansibility, combined with its deficient conducting power. Against this notion, which he considers to be 'an obvious oversight,' Professor Forbes contends in an ingenious and apparently unanswerable manner. The vibrations, he urges, depend upon the difference of temperature existing between the rocker and the block; if the latter be a bad conductor and retain the heat at its surface, the tendency is to bring both the surfaces in contact to the same temperature, and thus to stop the vibration instead of exalting it. Farther: the greater the quantity of heat transmitted from the rocker to the block during contact, the greater must be the expansion; and hence, if the vibrations be due to this cause, the effect must be a maximum when the block is the best conductor possible. But Professor Forbes, in this argument, seems to have used the term expansion in two different senses. The expansion which produces the vibration is the sudden upheaval of the point where the hot rocker comes in contact with the cold mass underneath: but the expansion due to good conduction would be an expansion of the general mass. Imagine the conductive power of the block to be infinite — that is to say, that the heat imparted by the rocker is instantly diffused equally throughout the block ; then, though the general expansion might be very great, the local expansion at the point of contact would be wanting, and no vibrations would be possible. The inevitable consequence of good conduction is, to cause a sudden abstraction of the heat from the point of contact of the rocker with the substance underneath, and this the lecturer conceived to be the precise reason why Professor Forbes had failed to obtain vibrations when the cold metal was a good conductor. He made use of *blocks*, and the abstraction of heat from the place of contact by the circumjacent mass of metal, was so sudden as to extinguish the local elevation on which the vibrations depend. In the experiments described by the lecturer, this abstraction was to a great extent avoided, by reducing the metallic masses to thin laminæ; and thus the very experiments adduced by Professor Forbes against the theory supported by Professor Faraday, appear, when duly considered, to be converted into strong corroborative proofs of the correctness of the views of the philosopher last mentioned.

EXTRACT FROM A PAPER ON SOME PHYSICAL PROPERTIES
OF ICE.*

In a very interesting paper communicated to the British Association during its last meeting, Mr. James Thomson has explained
the freezing together of two pieces of ice at 32° Fahr., in the following manner : — 'The two pieces of ice, on being pressed together
at their point of contact, will at that place, in virtue of the pressure, be in part liquefied and reduced in temperature, and the cold
evolved in their liquefaction will cause some of the liquid film
intervening between the two masses to freeze.'
    I am far from denying the operation under proper circumstances
of the *vera causa* to which Mr. Thomson refers, but I do not think
it explains the facts.   For freezing takes place without the intervention of any pressure by which Mr. Thomson's effect could
sensibly come into play.
    It is not necessary to squeeze the pieces of ice together; one
bit may be simply laid upon the other, and they will still freeze.
Other substances besides ice are also capable of being frozen to the
ice.   If a towel be folded round a piece of ice at 32° the towel and
ice will freeze together.   Flannel is still better ; a piece of flannel
wrapped round a piece of ice, freezes to it sometimes so firmly
that a strong tearing force is necessary to separate both.   Cotton,
wool, and hair may also be frozen to ice, without the intervention
of any pressure which would render Mr. Thomson's cause sensibly active.
    But there is a class of effects to the explanation of which the
lowering of the freezing point of water, by pressure, may, I think,
be properly applied.   The following statement is true of fifty experiments, or more, made with ice from various quarters.   A
cylinder of ice, two inches high and an inch in diameter, was
placed between two slabs of box-wood, and submitted to a gradually increasing pressure.   Looked at perpendicularly to the axis,
cloudy lines were seen drawing themselves across the cylinder ;
and when the latter was looked at obliquely, these lines were
found to be sections of dim hazy surfaces which traversed the

* Phil. Trans. 1858, p. 225.

cylinder, and gave it an appearance closely resembling that of a crystal of gypsum whose planes of cleavage had been forced out of optical contact by some external force.

Fig. 39.                    Fig. 40.

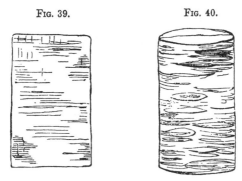

Fig. 39 represents the cylinder looked at perpendicular to its axis, and fig. 40 the same cylinder when looked at obliquely.

To ascertain whether the rupture of optical contact which these experiments disclosed was due to the intrusion of air between two separated surfaces of ice, a cylinder of ice, two inches long and one inch wide, was placed in a copper vessel containing ice-cold water. The ice cylinder projected half an inch above the surface of the water. Placing the copper vessel on a slab of wood, and a second slab of wood upon the cylinder of ice, the whole was subjected to pressure. When the hazy surfaces were well developed in the portion of ice above the water, the cylinder was removed and examined. The planes of rupture extended throughout the entire length of the cylinder, just the same as it had been squeezed in free air.

Still the removal of the cylinder from its vessel might be attended with the intrusion of air into the fissures. I therefore placed a cylinder of ice, two inches long and one inch wide, in a stout vessel of glass, which was filled with ice-cold water. Squeezing the whole, as in the last experiment, the surfaces of discontinuity were seen forming *under the liquid* quite as distinctly as in air.

The surfaces are due to compression, and not to any tearing asunder of the mass by tension, and they are best developed where the pressure, within the limits of fracture, is a maximum.

A cylindrical piece of ice, one of whose ends was not parallel to the other, was placed between slabs of wood and subjected to pressure. Fig. 41 shows the disposition of the experiment. The

FIG. 41.                      FIG. 42.

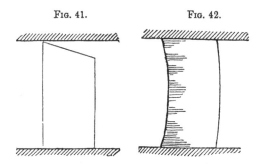

effect upon the ice-cylinder was that shown in fig. 42, the surfaces being developed along that side which had suffered the pressure.

Sometimes the surfaces commence at the centre of the cylinder. A dim small spot is first observed, which, as the pressure continues, expands until it sometimes embraces the entire transverse section of the cylinder.

On examining these surfaces with a pocket lens, they appeared to me to be composed of very minute water-parcels, like what is produced upon a smooth cold surface by the act of breathing. Were they either vacuous plates, or plates filled with air, their aspect would, on optical grounds, be far more vivid than it really was.

A concave mirror was so disposed, that the diffused light of day was thrown full upon the cylinder while under pressure. Observing the expanding surfaces through a lens, they appeared in a state of intense commotion; this was probably due to the molecular tensions of the little water-parcels. This motion followed closely on the edge of the surface as it advanced through the solid ice. Once or twice I observed the hazy surfaces pioneered through the mass by dim offshoots apparently liquid. They constituted a kind of negative crystallisation, having the exact form of the crystalline spines and spurs produced by the congelation of water upon a surface of glass. I have no doubt, then, that these surfaces are produced by the liquefaction of the solid in planes perpendicular to the direction of pressure.

The surfaces are developed with great facility when they correspond to the surfaces of freezing. By care I succeeded in some cases in producing similar effects in surfaces at right angles to the planes of freezing, but this was difficult and uncertain. Wherever the liquid disks before described were observed, the surfaces were always easily developed in the planes of the disks.

# LECTURE V.

[February 20, 1862.]

APPLICATION OF THE DYNAMICAL THEORY TO THE PHENOMENA OF
SPECIFIC AND LATENT HEAT—DEFINITION OF ENERGY: POTENTIAL
AND DYNAMIC ENERGY—ENERGY OF MOLECULAR FORCES—EXPERI-
MENTAL ILLUSTRATIONS OF SPECIFIC AND LATENT HEAT—MECHANICAL
VALUES OF THE ACTS OF COMBINATION, CONDENSATION, AND CON-
GELATION IN THE CASE OF WATER—SOLID CARBONIC ACID—THE
SPHEROIDAL STATE OF LIQUIDS—FLOATING OF SPHEROID ON ITS OWN
VAPOUR—FREEZING OF WATER AND MERCURY IN A RED-HOT CRUCIBLE.

WHENEVER a difficult expedition is undertaken in the Alps, the experienced mountaineer commences the day at a slow pace, so that when the real hour of trial arrives, he may find himself hardened instead of exhausted by his previous work. We, to-day, are about to enter on a difficult ascent, and I propose that we commence it in the same spirit; not with a flush of enthusiasm which the necessity of labour extinguishes, but with patient and determined hearts which will not recoil should a difficulty arise.

I have here a lead weight attached to a string which passes over a pulley at the top of the room. We know that the earth and the weight are mutually attractive; the weight now rests upon the earth and exerts a certain pressure upon its surface. The earth and the weight here *touch each other*; their mutual attractions are as far as possible satisfied, and *motion* by their mutual approach is no longer possible. As far as the attraction of gravity is concerned, the possibility of producing motion ceases as soon as the two attracting bodies are actually in contact.

I draw up this weight. It is now suspended at a height of sixteen feet above the floor; it is just as motionless as when it rested on the floor; but by introducing a space between the floor and it, I entirely change the condition of the weight. By raising it I have conferred upon it a motion-producing power. There is now an action possible to it, which was not possible when it rested upon the earth; *it can fall*, and in its descent can turn a machine or perform other work. It has no energy as it hangs there dead and motionless; but energy is possible to it, and we might fairly use the term *possible energy*, to express this power of motion which the weight possesses, but which has not yet been exercised by falling; or we might call it ' potential energy,' as some eminent men have already done. This potential energy is derived, in the case before us, from the pull of gravity, which pull, however, has not yet eventuated in motion. But I now let the string go; the weight falls, and reaches the earth's surface with a velocity of thirty-two feet a second. At every moment of descent it was pulled down by gravity, and its final moving force is the summation of the pulls. While in the act of falling, the energy of the weight is active. It may be called *actual* energy, in antithesis to *possible*; or it may be called *dynamic* energy in antithesis to *potential*, or we might call the energy with which the weight descends *moving force.* Do not be inattentive to these points; we must be able promptly to distinguish between energy *in store* and energy *in action.* Once for all then, let us take the terms of Mr. Rankine, and call the energy in store ' potential,' and the energy in action ' actual.'* If, after

* Helmholtz in his admirable memoir on ' Die Erhaltung der Kraft' (1847), divided all energy into *Tension* and *vis viva.* (Spannkräfte und Lebendige Kräfte.)

this, I should use the terms 'possible energy,' or ' dynamic
energy,' or 'moving force,' you will have no difficulty
in affixing the exact idea to these terms.    And remember
exactness is here essential.    We must not now tolerate
vagueness in our conceptions.

Our weight started from a height of sixteen feet ; let
us fix our attention upon it after it has accomplished
the first foot of its fall.    The total pull, if I may use the
term, to be expended on it has been then diminished by
the amount expended in its passing through the first foot.
At the height of fifteen feet it has one foot less of potential
energy than it possessed at the height of sixteen feet, but
at the height of fifteen feet it has got an equivalent
amount of dynamic or actual energy which, if reversed in
direction, would raise it again to its primitive height.
Hence as potential energy disappears, dynamic energy
comes into play.    *Throughout the universe the sum of
these two energies is constant.*

It is as yet too early to refer to organic processes, but
could we observe the molecular condition of my arm as I
drew up that weight, it would be seen that in accomplishing
this mechanical act, an equivalent amount of some other
form of motion was consumed.    If the weight were raised by
common heat, a portion of heat would disappear exactly
equivalent to the work done.    The weight is about one
pound, and to raise it sixteen feet would consume as
much heat as would raise the temperature of a cubic foot
of air about 1° F.    Conversely, this quantity of heat
would be generated by the falling of the weight from a
height of sixteen feet.    It is easy to see that, if the force
of gravity were immensely greater than it is, an immensely
greater amount of heat would have to be expended to raise
the weight.    The greater the attraction, the greater would
be the amount of heat necessary to overcome it ; but conversely, the greater would be the amount of heat which a

falling body would then develope by its collision with the earth.

Having made our minds clear that heat is consumed when a weight is forcibly separated from the earth by this agent, and that the amount of heat consumed depends on the energy of the attracting force overcome, we must turn these conceptions, regarding sensible masses, to account, in forming conceptions regarding insensible masses. As an intellectual act it is quite as easy to conceive of the separation of two mutually attracting *atoms*, as to conceive of the separation of the earth and weight. I have already had occcasion to refer more than once to the energy of molecular forces, and here I have to return to the subject. Closely locked together as they are, the atoms of bodies, though we cannot suppose them to be in contact, exert enormous attractions. It would require an almost incredible amount of ordinary mechanical force to widen the distances intervening between the atoms of any solid or liquid, so as to increase the volume of the solid or liquid in any considerable degree. It would also require a force of great magnitude to squeeze the particles of a liquid or solid together, so as to make the body less in size. I have vainly tried to augment the density of a soft metal by pressure. Water, for example, which yields so freely to the hand plunged in it, was for a long time regarded as absolutely incompressible. Great force was brought to bear upon it ; but sooner than shrink sensibly, it oozed through the pores of the metal vessel which contained it, and spread like a dew on the surface* By refined and powerful

* I have to thank my friend, Mr. Spedding, for the following extract in reference to this experiment:—

' Now it is certain that rarer bodies (such as air) allow a considerable degree of contraction, as has been stated ; not that tangible bodies (such as water) suffer compression with much greater difficulty and to a less extent. How far they do suffer it, I have investigated in the following experiment: I had a hollow globe of lead made capable of holding about two pints, and

means we can now compress water, but the force necessary
to accomplish this is very great.

When we wish to overcome molecular forces we must
attack them by their peers.    Heat accomplishes what
mechanical energy, as generally wielded, is incompetent to
perform.    Bodies when heated expand, and to effect this
expansion their molecular attractions must be overcome.
In masses equally large this is a work, in comparison with
which the erection of the Egyptian pyramids dwindles to
the labour of mites ; and where the attractions to be overcome are so vast, we may infer that the quantity of heat
necessary to overcome them will be commensurate.

And now I must ask your entire attention.    I hold in
my hand a lump of lead ; suppose I communicate a certain
amount of heat to the lead, how is that heat disposed of

sufficiently thick to bear considerable force; having made a hole in it, I
filled it with water, and then stopped up the hole with melted lead, so that
the globe became quite solid.    I then flattened the two opposite sides of the
globe with a heavy hammer, by which the water was necessarily contracted
into less space, a sphere being the figure of largest capacity ; and when
the hammering had no more effect in making the water shrink, I made use
of a mill or press; till the water, impatient of farther pressure, exuded
through the solid lead like a fine dew.    I then computed the space lost by
the compression, and concluded that this was the extent of compression
which the water had suffered, but only when constrained by great violence.'
(Bacon's *Novum Organum* published in 1620 : vol. iv. 209 of the translation.)    Note by R. Leslie Ellis, vol. i. p. 324.—'This is perhaps the most
remarkable of Bacon's experiments, and it is singular that it was so little
spoken of by subsequent writers.    Nearly fifty years after the production of
the " Novum Organum," an account of a similar experiment was published by
Megalotti, who was secretary of the Academia del Cimento at Florence ;
and it has since been familiarly known as the Florentine experiment.    I
quote his account of it, " Facemmo lavorar," ' &c.

The writer goes on to remark that the absolute incompressibility of water
is not proved by this experiment, but merely that it is not to be compressed in
the manner described; but the experiment is on other grounds inconclusive.

It is to be remembered that Leibnitz (' Nouveaux Essais ') in mentioning
the Florentine experiment, says that the globe was of gold (p. 229 Erd-
mann), whereas the Florentine academicians expressly say why they preferred
silver to either gold or lead.

within the substance? It is applied to two distinct purposes—it performs two different kinds of work. One portion of it imparts that species of motion which raises the temperature of the lead, and which is sensible to the thermometer; but another portion of it goes to force the atoms of the lead into new positions, and this portion *is lost as heat.* The pushing asunder of the atoms of the lead in this case, in opposition to their mutual attractions, is exactly analogous to the raising of our weight in opposition to the force of gravity. Let me try to make the comparison between the two actions still more strict; suppose that I have a definite amount of force to be expended on our weight, and that I divide this force into two portions, one of which I devote to the actual raising of the weight, while I employ the other to cause the weight, as it ascends, to oscillate, or revolve, like a pendulum or governor, and to oscillate, moreover, with gradually augmented energy; we have, then, the analogue of that which occurs when heat is imparted to the lead. The atoms are pushed apart, but during their recession they vibrate, or revolve, with gradually augmented intensity. Thus the heat communicated to the lead resolves itself, in part, into atomic potential energy, and in part into a kind of atomic music, the musical part alone being competent to act upon our thermometers or to affect our nerves.

In this case, then, the heat accomplishes what we may call *interior work*;* it performs work within the body heated, by forcing its particles to take up new positions. When the body cools, the forces which were overcome in the process of heating come into play, and the heat which was consumed by the forcing asunder of the atoms is now restored by the drawing together of the atoms.

Chemists have determined the relative weights of the

---

* See the excellent memoirs of Clausius in the Philosophical Magazine.

atoms of different substances. Calling the weight of a
hydrogen atom 1, the weight of an oxygen atom, you
know, is 16. Hence to make up a pound weight of
hydrogen, sixteen times the number of atoms contained in
a pound of oxygen would be necessary. The number of
atoms required to make up a pound is evidently inversely
proportional to the atomic weight. We here approach a very
delicate and important point. The experiments of Dulong
and Pétit, and of MM. Regnault and Neumann, render it
extremely probable that all elementary atoms, great and
small, light and heavy, when at the same temperature,
possess the same amount of the energy which we call
heat, the lighter atoms making good by velocity what
they want in mass. Thus, each of the atoms of hydrogen
has the same moving energy as an atom of oxygen at the
same temperature. But, inasmuch as a pound weight of
hydrogen contains sixteen times the number of atoms, it
must also contain sixteen times the amount of heat possessed by a pound of oxygen, at the same temperature.

From this it follows that to raise a pound of hydrogen,
a certain number of degrees in temperature — say from 50°
to 60°— would require sixteen times the amount of heat
needed by a pound of oxygen under the same circumstances. Conversely, a pound of hydrogen, in falling
through 10°, would yield sixteen times the amount of
heat yielded by a pound of oxygen, in falling through the
same number of degrees.

In oxygen and hydrogen we have no sensible amount of
'interior work,' to be performed ; there are no molecular
attractions of sensible magnitude to be overcome. But in
solid and liquid bodies, besides the differences due to the
number of atoms present in the unit of weight, we have
also differences due to the consumption of heat in interior
work. Hence it is clear that the amount of heat which
different bodies contain is not at all declared by their

temperature.   To raise a pound of water, for example, 1°,
would require thirty times the amount of heat necessary
to raise a pound of mercury 1°.   Conversely, the pound
of water, in falling through 1°, would yield up thirty
times the amount of heat yielded up by the pound of
mercury.

Let me illustrate, by a simple experiment, the differences
which exist between bodies, as to the quantity of heat
which they contain.   I have here a cake of beeswax
six inches in diameter and half an inch thick.   Here I
have a vessel containing oil, which is now at a temperature
of 180° C.   In the hot oil I have immersed a number of
balls of different metals—of iron, lead, bismuth, tin and
copper.   At present they all possess the same temperature,
namely, that of the oil.   Well, I lift them out of the oil,
and place them upon this cake of wax c d (fig. 43), which is
supported by the ring of a retort-
stand ; they melt the wax underneath, and sink in it.   But I see
that they are sinking with different velocities.   The iron and
the copper are working themselves much more vigorously
into the fusible mass than the
others ;   the tin   comes   next,
while the lead and the bismuth
lag entirely behind.   There goes
the iron clean through, the copper follows ;   I can see the bot-

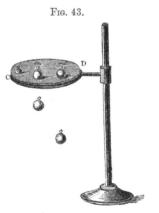

FIG. 43.

tom of the tin ball just peeping through the lower surface
of the cake, but it cannot go farther ; while the lead
and bismuth have made but little way, being unable to
sink to much more than half the depth of the cake.

Supposing, then, I take equal weights of different substances, heat them all ( say to 100°) and then determine

the exact amount of heat which each of them gives out in
cooling from 100° to 0°, I should find very different·
amounts of heat for the different substances.  How could
this problem be solved?  It has been solved by eminent men by observing the *time* which a body requires to
cool.  Of course the greater the amount of heat possessed
and generated by its atoms, the longer would the body
take to cool.  The relative quantities of heat yielded
up by different bodies have also been determined by
plunging them, when heated, into cold water, and observing the gain on the one hand and the loss on the
other.  The problem has also been solved by observing the
quantities of ice which different bodies can liquefy, in falling from 212° Fahr. to 32°, or from 100° C. to 0°.  These
different methods have given concordant results.  Calling
the amount of heat given out by a pound of water, in sinking through one degree of temperature, unity, the following
numbers express the amount of heat given out by a pound
weight of each of the substances whose names are annexed.

| | |
|---|---|
| Water      .   .   .   . | 1·0000 |
| Sulphur    .   .   . | 0·2026 |
| Arsenic    .   .   . | 0·0814 |
| Antimony .   .   . | 0·0508 |
| Bismuth  .   .   . | 0.0308 |
| Zinc       .   .   . | 0·0955 |
| Cadmium .   .   . | 0·0567 |
| Tin  .   .   .   . | 0·0562 |
| Lead .     .   .   . | 0·0314 |
| Iron .     .   .   . | 0·1138 |
| Cobalt     .   .   . | 0·1070 |
| Nickel     .   .   . | 0·1086 |
| Copper     .   .   . | 0·0951 |
| Mercury  .   .   . | 0·0333 |
| Silver     .   .   . | 0·0570 |
| Gold .     .   .   . | 0·0324 |
| Platinum .   .   . | 0·0324 |

A moment's inspection of this table explains why it is
that, in the case of iron and copper, our balls melted

through the wax, while the lead and bismuth balls were incompetent to do so; it will also be seen that tin here occupies the position which we should assign to it from the experiment with the cake of wax; water, we see, stands at the head of all.

Each of these numbers denotes what has been hitherto called, the 'specific heat' or the 'capacity for heat' of the substance to which it is attached. As I stated in a former lecture, those who hold that heat is a fluid, explained these differences by saying that some substances had a greater store of this fluid than others. We may, without harm, continue to use the term 'specific heat' or 'capacity for heat;' now that we know the true nature of the actions covered by the term.

The energy of the forces engaged in this atomic motion and interior work, as measured by any ordinary mechanical standard, is enormous. I have here a pound of iron, which on being heated from 32° to 212° F. expands by about $\frac{1}{800}$th of the volume which it possesses at 32°. Its augmentation of volume would certainly escape the most acute eye; still to give its atoms the motion corresponding to this augmentation of temperature, and to shift them through the small space indicated, an amount of heat is requisite which would raise about eight tons one foot high. Gravity almost vanishes in comparison with these molecular forces; the pull of the earth upon the pound weight, as a mass, is as nothing compared with the mutual pull of its own molecules. Water furnishes a still subtler example. Water expands on both sides of 4° C. or 39° F.; at 4° C. it has its maximum density. Suppose a pound of water heated from $3\frac{1}{2}$° C. to $4\frac{1}{2}$° C.— that is, 1°—its volume at both temperatures is the same; there has been no forcing asunder whatever of the atomic centres, and still, though the volume is unchanged, an amount of heat has been imparted to the water, sufficient,

if mechanically applied, to raise a weight of 1390 lbs. a
foot high. The interior work done here by the heat can
be nothing more than the turning round of the atoms of
water. It separates the attracting poles of the atoms by a
tangential movement, but leaves their centres at the same
distance asunder first and last. The conceptions with
which I here deal, may not be easy to those unaccustomed
to such studies, but they are capable of perfect clearness of
realization to all who have the patience to dwell upon
them sufficiently long.

This is the place to note further, that there are descriptions of interior work different from that of pushing the
atoms more widely apart. Enormous interior work may be
accomplished while the atoms, instead of being pushed
apart, as a whole, approach each other. Polar forces—forces
emanating from distinct points, and acting in distinct
directions, give to crystals their symmetry, and the overcoming of these forces, while it necessitates a consumption
of heat, may also be accompanied by a diminution of
volume. This is illustrated by the deportment of both ice
and bismuth on liquefying. I could readily sketch a
system of atoms illustrative of this position, but every
instructed mind will be able to imagine such combinations
for itself.

The high specific heat of water has one important
bearing which I do not wish to pass over here. Comparing
*equal weights*, the specific heat of water being 1°, that of
air is about 0·25. Hence a pound of water, in losing 1° of
temperature, would warm 4 lbs. of air 1°. But water is
770 times heavier than air; hence, comparing *equal
volumes*, a cubic foot of water in losing 1° of temperature
would raise $770 \times 4 = 3080$ cubic feet of air 1°.

The vast influence which the ocean must exert as a
moderator of climate here suggests itself. The heat of
summer is stored up in the ocean, and slowly given out

during the winter.   Hence one cause of the absence of
extremes in an island climate.   The summers of the island
can never attain the fervid heat of the continental summer,
nor can the winter of the island be so severe as the continental winter.   In various parts of the continent fruits grow
which our summers cannot ripen ; but in these same parts
our evergreens are unknown ; for they cannot live through
the winters.   The winter of Iceland is, as a general rule,
milder than that of Lombardy.

We have hitherto confined our attention to the heat
consumed in the molecular changes of solid and liquid
bodies, while these bodies continue solid and liquid.   We
shall now direct our attention to the phenomena which accompany *changes of the state of aggregation.*   When sufficiently heated, a solid melts, and when sufficiently heated,
a liquid assumes the form of gas.   Let us take the case of
ice, and trace it through the entire cycle.   This block of
ice has now a temperature of 20° F.   I warm it ; a
thermometer fixed in it rises to 32°, and at this point the
ice begins to melt ; the thermometric column, which rose
previously, is now arrested in its march, and becomes perfectly stationary.   I continue to apply warmth, but there
is no augmentation of temperature ; and not till all
the solid has been reduced to liquid does the thermometer
resume its motion.   It is now again ascending ; it reaches
100°, 200°, 212° : here steam-bubbles show themselves in
the liquid ; it boils, and from this point onwards the
thermometer remains stationary at 212°.

But during the melting of the ice and during the
evaporation of the water, heat is incessantly communicated : to simply liquefy the ice, as much heat has been
imparted to it as would raise the same weight of water
143° Fahr., or as would raise 143 times the weight 1° F.
in temperature ; and to convert a pound of water at 212°
into a pound of steam at the same temperature, 967

times as much heat is required as would raise a pound of
water 1° in temperature.   The former number, 143°,
represents what has been hitherto called the *latent heat* of
water : and the latter number, 967°, represents the latent
heat of steam.   It was manifest to those who first used
these terms, that, throughout the entire time of melting,
and throughout the entire time of boiling, heat was
communicated ; but inasmuch as this heat was not revealed by the thermometer, the fiction was invented that it
was rendered latent.   The fluid of heat hid itself in some
unknown way in the interstitial spaces of the water and of
the steam.   According to our present theory, the heat expended in melting is consumed in conferring potential
energy upon the atoms.   It is virtually the lifting of a
weight.   So likewise, as regards the steam, the heat is consumed in pulling the liquid molecules asunder, conferring
upon them a still greater amount of potential energy ; and
when the heat is withdrawn, the vapour condenses and the
molecules again clash with a dynamic energy equal to
that which was employed to separate them, and the precise
quantity of heat then consumed now reappears.

The act of liquefaction consists of interior work—work
expended in moving the atoms into new positions.   The act
of vaporisation is also, for the most part, interior work ; to
which however must be added the external work performed
in the expansion of the vapour, which makes place for
itself by forcing back the atmosphere.

We are indebted to the eminent man, to whom I have
referred so often, for the first accurate determinations of
the calorific power of fuel.   'Rumford estimated the calorific power of a body by the number of parts, by weight, of
water, which one part, by weight, of the body would, on perfect combustion, raise 1° in temperature.   Thus one part, by
weight, of charcoal, in combining with 2⅔ parts of oxygen
to form carbonic acid, will evolve heat sufficient to raise

the temperature of about 8,000 parts by weight of water 1° C. Similarly, one pound of hydrogen, in combining with eight pounds of oxygen to form water, will raise 34,000 lbs. of water 1° C. The relative calorific powers, therefore, of carbon and hydrogen are as 8 : 34.'* The recent refined researches of Favre and Silbermann entirely confirm the determinations of Rumford.

Let us, then, fix our attention upon this wonderful substance, water, and trace it through the various stages of its existence. First we have its constituents as free atoms, which attract each other, fall, and clash together. The mechanical value of this atomic act is easily determined; knowing the number of foot-pounds corresponding to the heating of 1lb. of water 1° C., we can readily calculate the number of foot-pounds equivalent to the heating of 34,000 lbs. of water 1° C. Multiplying the latter number by 1,390,† we find that the concussion of our 1 lb. of hydrogen with 8 lbs. of oxygen is equal, in mechanical value, to the raising of forty-seven million pounds one foot high! I think I did not overrate matters when I said that the force of gravity, as exerted near the earth, was almost a vanishing quantity, in comparison with these molecular forces; and bear in mind the distances which separate the atoms before combination—distances so small as to be utterly immeasurable; still it is in passing over these distances that the atoms acquire a velocity sufficient to cause them to clash with the tremendous energy indicated by the above numbers.

After combination the substance is in a state of vapour, which sinks to 212°, and afterwards condenses to water. In the first instance the atoms fell together to form the compound; in the next instance the molecules of the compound fall together to form a liquid. The

---

* Percy's Metallurgy, p. 53.

† 772 foot-pounds being the mechanical equivalent for 1° F., 1,390 foot-pounds is the equivalent for 1° C.

mechanical value of this act is also easily calculated : 9 lbs. of steam in falling to water, generate an amount of heat sufficient to raise $967 \times 9 = 8,703$ lbs. of water 1° F. Multiplying this number by 772, we have a product of 6,718,716 foot-pounds as the mechanical value of the mere act of condensation.*  The next great fall of our 9 lbs. of water is from the state of liquid to that of ice, and the mechanical value of this act is equal to 993,564 foot-pounds. Thus our 9 lbs. of water, in its origin and progress, falls down three great precipices : the first fall is equivalent to the descent of a ton weight urged by gravity down a precipice 22,320 feet high; the second fall is equal to that of a ton down a precipice 2,900 feet high ; and the third is equal to the descent of a ton down a precipice 433 feet high. I have seen the wild stone-avalanches of the Alps, which smoke and thunder down the declivities with a vehemence almost sufficient to stun the observer.  I have also seen snow-flakes descending so softly as not to hurt the fragile spangles of which they were composed; yet to produce, from aqueous vapour, a quantity of that tender material which a child could carry, demands an exertion of energy competent to gather up the shattered blocks of the largest stone-avalanche I have ever seen, and pitch them to twice the height from which they fell.

I will now relieve the strain which I have hitherto put upon your attention, by introducing a few experimental illustrations of the calorific effects which accompany the change of aggregation.  I place my thermo-electric pile thus upon its back on the table, and on its naked face I

---

* In Rumford's experiments the heat of condensation was included in his estimate of calorific power; deducting the above number from that found for the chemical union of the hydrogen and oxygen, forty millions of foot-pounds would still remain as the mechanical value of the act of combination.

place this thin silver basin, B (fig. 44), into which I pour a quantity of water slightly warmed, the needle of the galvanometer moves to 90°, and remains permanently deflected to 70°. I now place a little powdered nitre, not more than can fit upon a threepenny-piece, in the basin, and allow it to dissolve. I had placed the nitre previously before the fire, so that not only was the liquid warm, but the solid powder was also warm. Observe the effect of

FIG. 44.

their mixture! The nitre dissolves in the water; and to produce this change, all the heat which both the water and the nitre possess, in excess of the temperature of this room, is consumed, and, indeed, a great deal more. The needle, you see, sinks not only to zero, but goes strongly up at the other side, showing that now the face of the pile is powerfully chilled.

I remove the basin, pour the liquid out, and resupply it with warm water, into which I introduce a pinch of common salt. The needle was at 70° when the salt was introduced: it is now sinking, reaches zero, and goes up on the side which indicates cold. But the action is not at all so strong as in the case of saltpetre. The reason is that the amount of interior work required by the salt, and which necessitates the consumption of heat, is much less than that demanded by the nitre. As regards latent heat, then, we have differences similar to those which we have already illustrated as regards specific heat. Again, I cleanse the basin, put fresh water in it, and put a little

sugar in the water; the amount of heat absorbed in the
solution of the sugar is sensible, the liquid is chilled, but
the amount of chilling is much less than in either of the
former cases.  Thus, when you sweeten your hot tea, you
cool it in the most philosophical manner; when you put
salt in your soup, you do the same; and if you were
concerned with the act of cooling alone, and careless of the
flavour of your soup, you might hasten its refrigeration
by adding saltpetre.

In a former lecture I made use of a mixture of pounded
ice and salt to obtain great cold.   Both the salt and the ice
when they are thus mixed together, change their state of
aggregation; the amount of interior work is here so great,
that during its performance the temperature of the mixture sinks 30° Fahr., and more, below the freezing point of
water.   Here is a nest of watch-glasses which I have
wrapped in tinfoil, and immersed in a mixture of ice and

Fig. 45.

salt.    Into each watch-glass I had
poured a little water, in which
the next glass rested.   They are
now all frozen together to a solid
cylinder, by the cold of this mixture of ice and salt.

I will now reverse the process,
and endeavour to show you the
heat developed in  passing from
the liquid to the solid state.   But
first let me show you that heat
is rendered latent when sulphate
of soda is dissolved.   I experiment with the substance exactly
as I experimented with the nitre,
and you see, that as the crystals
melt in the water the pile is chilled.   And now for the
complementary experiment.  This large glass bolt-head,

B (fig. 45), with this long neck, is now filled with a
solution of sulphate of soda. Yesterday Mr. Anderson
dissolved the substance in a pan over our laboratory fire,
and filled this bolt-head with the solution. He then covered
the top carefully with a piece of bladder, and placed the
bottle behind this table, where it has remained undisturbed throughout the night.

The liquid is, at the present moment, supersaturated with
sulphate of soda. When the water was hot, it melted more
than it could melt when cold. But now the temperature has sunk much lower than that which corresponds
to the point of saturation. This state of things is secured by keeping the solution perfectly still, and permitting nothing to fall into it. Water, kept thus still, may
be cooled many degrees below its freezing point. Some
of you may have noticed the water in your jugs, after
a cold winter-night, suddenly freeze on being poured
out in the morning. In cold climates this is not uncommon. Well, the particles of sulphate of soda in this
solution are on the brink of a precipice, and I can push
them over it, by simply dropping a small crystal of the substance, not larger than a grain of sand, into the solution.
Observe what takes place; the bottle now contains a clear
liquid; I drop the bit of crystal in, it does not sink; the
molecules have closed round it to form a solid in which it
is now embedded. The passage of the atoms from a state
of freedom to a state of bondage goes on quite gradually;
you see the solidification extending down the neck of the
bottle. Observe where I have placed my thermo-electric
pile P. Its naked face rests against the convex surface of
the bottle, and the needle of the galvanometer points to
zero. The process of crystallisation has not yet reached
the liquid in front of the pile, but you see it approaching. It is now solidified opposite the pile, and
mark the effect. The atoms, in falling to the solid form,

develope heat ; this heat communicates itself to the glass
envelope, the glass envelope warms the pile, and the
needle, as you see, flies to 90°. The quantity of heat
thus rendered sensible by solidification is exactly equal to
that which was rendered latent by liquefaction.

We have, in these experiments, dealt with the latent
heat of liquids ; let me now direct your attention to a few
experiments illustrative of what has been called the latent
heat of vapours—in other words, the heat consumed in
conferring potential energy, when a body passes from the
liquid to the gaseous state.   As before, I turn my pile upon
its back with its naked face upwards, and on this face I
place the silver basin already used, into which I have
poured a small quantity of a volatile liquid, which I have
purposely warmed.  The needle now moves, indicating
heat.   But scarcely has it attained 90° when it turns
promptly, descends to 0°, and flies with violence up on the
side of cold.   The liquid here used is sulphuric ether; it is
very volatile, and the speed of its evaporation is such that
it consumes, rapidly, the heat at first communicated to it,
and then abstracts heat from the face of the pile.   I remove the ether, and supply its place by alcohol, slightly
warm ; the needle, as before, goes up on the side of heat.
But wait a moment; I will use these small bellows to
promote the evaporation of the alcohol ; now you see the
needle descending, and now it is up at 90° on the side of
cold.   Water is not nearly so volatile as alcohol, still I can
show the absorption of heat by the evaporation of water
also.   We use a kind of pottery for holding water,
which admits of a slight percolation of the liquid, so
as to cause a kind of dewiness on the external surface.
Evaporation goes on from that surface, and the heat necessary to this work, being drawn in great part from the
water within, keeps it cool.   Butter-coolers are made on
the same principle.

To show you the extent to which refrigeration may be carried by the evaporation of water, I have here an instrument (fig. 46), by which water is frozen, through the simple abstraction of its heat by its own vapour. The instrument is called the *cryophorus*, or ice-carrier, and it was invented by Dr. Wollaston. It is made in this way— a little water is put into one of these bulbs; the other bulb, B, when softened by heat, had a tube drawn out from it with a minute aperture at the end. Well, the water was boiled in A, and steam was produced, until it had chased all the air away through the small aperture in the distant bulb. When the bulbs and connecting tube were filled with pure steam, the small orifice was sealed with a blow-

FIG. 46.

pipe. Here, then, we have water and its vapour, with scarcely a trace of air. You hear how the liquid rings, exactly as it does in the case of the water-hammer.

I turn all the liquid into one bulb, A, which I dip into an empty glass to protect it from currents of air. The *empty* bulb, B, I plunge into a freezing mixture; thus, the vapour which escapes from the liquid in the bulb, A, is condensed by the cold, to water, in B. This condensation permits of the formation of new quantities of vapour. As the evaporation continues, the water which supplies the vapour becomes more and more chilled. In a quarter of an hour, or twenty minutes, it will be converted into a cake of ice. Here is the opalescent solid formed in a second instrument, which you saw me arranging before the commencement of the lecture. The whole process consists in the uncompensated transfer of motion from the one bulb to the other.

But the most striking example of the consumption of heat in changing the state of aggregation is furnished by the substance which I have imprisoned in this strong iron bottle. This bottle contains carbonic acid, liquefied by enormous pressure. The substance you know is a gas under ordinary circumstances; here is a jar full of it, which, though it manifests its nature by extinguishing a taper, is not to be distinguished, by the eye, from common air. When the cock attached to the iron bottle is turned, the pressure which acts upon the gas is relieved, the liquid boils—flashes, as it were, suddenly into gas, which rushes from the orifice with impetuous force. But you can see this current of gas; mixed up with it you see a white substance, which is now blown against me, to a distance of eight or ten feet, through the air. What is this white substance? It is carbonic acid snow. The cold produced in passing from the liquid to the gaseous state is so intense that a portion of the carbonic acid is actually frozen to form this snow, and mingles in small flakes with the issuing stream of gas. I can collect this snow in a suitable vessel. Here is a cylindrical box with two hollow handles, through which I will allow the gas to pass. Right and left you see the streams, but a large portion of the frozen mass is retained in the box. I open it, and you see it filled with this perfectly white carbonic acid snow.

The solid very gradually disappears; its conversion into vapour is slow, because it can only slowly collect from surrounding substances the heat necessary to vaporise it. You can handle it freely, but not press it too much, lest it should burn you. It is cold enough to burn the hand. I plunge a piece of it into water, and hold it there: you see bubbles rising through the water—these are pure carbonic acid gas. I collect this gas, and show you that it possesses all the properties of the gas as commonly prepared. The solid acid does not melt in the water; when I release it, it rises to the surface, and floats upon it. I put a bit of the acid into my mouth, taking care not to inhale while it is there. I breathe against this candle; my breath extinguishes the flame. Before the conclusion of the lecture, I will show you how it is possible to preserve so cold a body in the mouth without injury. A piece of iron of equal coldness would do serious damage.

Here, then, we have a solid body intensely cold, which, however, does not chill bodies in contact with it, as it might be expected to do. In fact, no real contact has been established with the acid. Water, we see, will not dissolve it, but sulphuric ether will; and by pouring a quantity of this ether on the snow, I obtain a pasty mass, which has an enormous power of refrigeration. Here I have some thick and irregular masses of glass—the feet, in fact, of drinking-glasses. I place a portion of the solid acid on them, and wet it with ether; you hear the glass crack; it has been shattered by the contraction produced by the intense cold.

In this basin I spread a little paper, and over the paper I pour a pound or two of mercury; on the mercury I place some solid carbonic acid, and over the acid I pour a little ether. Mercury, you know, requires a very low temperature to freeze it. Well, here it is frozen; I turn it out before you, a solid mass; I can hammer the solid; I can also cut it with a knife. To enable me to lift the mercury out of the basin, I have dipped this wire into it; by this I raise it, and plunge it into a glass jar containing water. It liquefies, and showers downwards through the water; but every fillet of mercury freezes the water with which

it comes into contact, and thus round each fillet is formed
a tube of ice, through which you can see the liquid metal
descending. These experiments might be multiplied almost
indefinitely; but enough, I trust, has been shown to illustrate our present subject.

I have now to direct your attention to another, and very
singular class of phenomena, connected with the production
of vapour. Here is a broad porcelain basin, B (fig. 47), filled
with hot water. Here is a silver basin, s, which I now heat to
redness.    If I place the silver basin in the hot water,
what will occur?    You might naturally reply, that

Fig. 47.

the basin will impart its
excess of heat instantly to
the water, and be cooled
down to the temperature of
the latter.   But nothing of
this kind occurs.   The basin
for a time developes a sufficient amount of vapour underneath it, to lift it entirely
out of contact with the water;
or, in the language of the hypothesis, developed in our
third lecture, it is lifted by the discharge of molecular
projectiles against its under surface.    This will go on until
the temperature of the basin sinks, and it is no longer able
to produce vapour of sufficient tension to support it.   Then
it comes into contact with the water, and the ordinary
hissing of a hot metal, together with the cloud which forms
overhead, declares the fact.

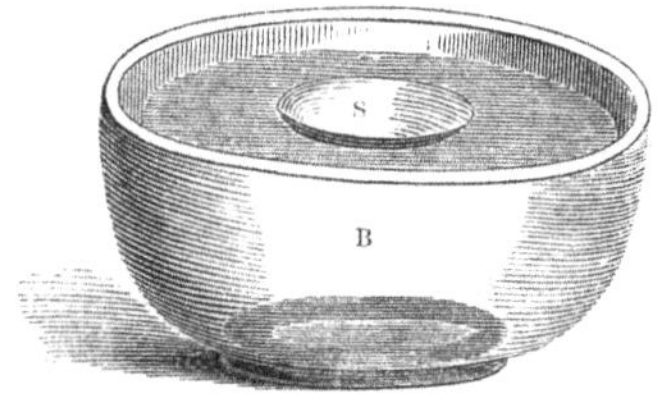

I now reverse the experiment, and instead of placing
the basin in the water, I place the water in the basin —
first of all, however, heating the latter to redness by a
lamp.   You hear no noise of ebullition, no hissing of the
water as I pour it into the hot basin; the drop rolls
about on its own vapour—that is to say, it is sustained

by the recoil of the molecular projectiles discharged from
its under surface. I withdraw the lamp, and allow the
basin to cool, until it is no longer able to produce vapour
strong enough to support the drop. The liquid then
touches the metal; the instant it does so, violent ebullition
sets in, and the cloud which you now observe forms above
the basin.

You cannot, from your present position, see this flattened
spheroid rolling about in the hot basin, but I can show
it to you, and, if I am fortunate, I shall show you something very beautiful. You will bear in mind that there
is an incessant developement of vapour underneath the
drop, which, as incessantly, escapes from it laterally. If
the drop rest upon a flattish surface, so that the lateral

Fig. 48.

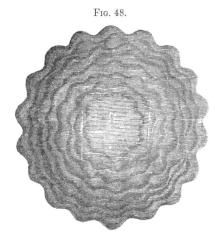

escape is very difficult, the vapour will burst up through the
middle of the drop. But I have here arranged matters,
so that the vapour shall issue laterally; and it sometimes happens that the escape of the vapour is rythmic;
it issues in regular pulses, and then we have our drop
of water moulded to a most beautiful rosette. I have it

now,—a round mass of liquid, two inches in diameter,
with a beautifully crimped border.  I will throw the
beam of the electric lamp upon this drop so as to illuminate it, and holding this lens over it, I hope to cast
its image on the ceiling, or on the screen.  There it is,
(fig. 48), a figure eighteen inches in diameter, and the
vapour breaking, as if in music, from its edge.  If I add a
little ink, so as to darken the liquid, the definition of its
outline is augmented, but the pearly lustre of its surface
is lost.  I withdraw the heat; the undulation continues for some time : the border finally becomes unin-
dented.  The drop is now perfectly motionless—a liquid
spheroid—and now it suddenly spreads upon the surface,

FIG. 49.

contact has been established, and the spheroidal condition
ends.

I dry the silver basin and place it, with its bottom upwards, in front of the electric lamp, and with a lens in
front I bring the rounded outline of the basin to a focus

on the screen; I dip this bit of sponge in alcohol and
squeeze it over the cold basin, so that the drops fall upon
the surface of the metal: you see their magnified images
upon the screen, and you observe that when they strike
the surface they spread out and trickle down along it. Now
I will heat this basin by placing a lamp underneath.
Observe what occurs: when I squeeze the sponge the drops
descend as before, but when they come in contact with the
basin they no longer spread, but roll over the surface as
liquid spheres (fig. 49). See how they bound and dance as if
they had fallen upon elastic springs; and so in fact they
have. Every drop as it strikes the hot surface, and as it rolls
along the surface, developes vapour which lifts it out of contact, thus destroying all cohesion between the surface and
the drop, and enabling the latter to preserve its spherical
or spheroidal form.

I have here an arrangement suggested by Professor
Poggendorf, which shows, in a very beautiful manner, the
interruption of contact between the spheroidal drop and
its supporting surface. From this silver basin, B (fig. 50),

Fig. 50.

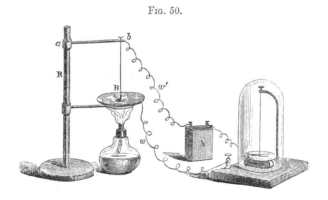

intended to hold the drop, I carry a wire, $w$, round yonder
magnetic needle; the other end of the galvanometer wire

M

I attach to one end of this battery, A. From the opposite pole of the little battery I carry a wire, $w'$, and so
attach it to the arm, $a\,b$, of this retort-stand, $k$, that I can
readily lower it. I heat the basin, pour in the water, and
lower my wire till the end of it dips into the spheroidal
mass : you see no motion of the galvanometer needle;
the only gap in the entire circuit is that which now exists
underneath the drop. If the drop were in contact the
current would pass. I prove this thus: I withdraw the
lamp; the spheroidal state will soon end; the liquid will
touch the bottom. It now does so, and the needle instantly
flies aside.

You can actually see the interval between the drop and
the hot surface upon which it rests. A private experiment
may be made in this way : Let a flattish basin, B (fig. 51),
be turned upside down, and let the bottom of it be slightly
indented so as to be able to bear a drop; heat the basin
by a spirit lamp, and place upon it a drop of ink $d$ with
which a little alcohol has been mixed. Stretch a platinum

FIG. 51.

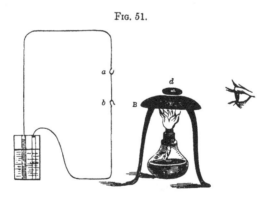

wire, $a\,b$, vertically behind the drop, and render the wire
incandescent by sending a current of electricity through it.
Bring your eye to a level with the bottom of the drop,

and you will be able to see the red-hot wire through the interval between the drop and the surface which supports it. Let me show you this interval. I place my basin, B (fig. 52), as before, with its bottom upward in front of the lamp; I heat the basin and bring carefully down upon it a drop, d, dependent from a pipette. When it rests upon the proper part of the surface, and the lens in front is brought to its proper position, you see a line of bright light between the drop and the silver, indicating that the beam of the lamp has passed underneath the drop to the screen.

The spheroidal condition was first observed by Leidenfrost, and I might give you fifty other illustrations of it. Liquids

FIG. 52.

can be made to roll on liquids. If, moreover, I take this red-hot copper ball and plunge it into a vessel of hot water, a loud sputtering is produced, due to the escape of the vapour generated; still the contact of the liquid and solid is only very partial: let the ball cool, the liquid at length touches it, and then the ebullition is so violent as to project the water from the vessel on all sides.

M. Boutigny has of late lent new interest to this subject by expanding the field of illustration, and applying it to

the explanation of many extraordinary effects.    If the
hand be wet, it may be passed through a stream of molten
metal without injury.    I have seen M. Boutigny myself
pass his wet hand through a stream of molten iron, and
toss with his fingers the fused metal from a crucible : a
blacksmith will lick a white hot iron without fear of burning his tongue.    The tongue is effectually preserved from
contact with the iron, by the vapour developed ; and it was
to the vapour of the carbonic acid, which shielded me from
its contact, that I owed my safety when I put the substance
into my mouth.    To the same protective influence many
escapes from the fiery ordeal of ancient times have been
attributed by M. Boutigny.    I may add, that the explanation of the spheroidal condition given by M. Boutigny has
not been accepted by scientific men.

Boiler explosions have also been ascribed to the water
in the boiler assuming the spheroidal state ; the sudden
developement of steam, by subsequent contact with the
heated metal, causing the explosion.    We are more ignorant of these things than we ought to be.    Experimental

Fig. 53.

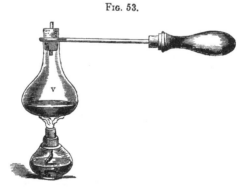

science has brought a series of true causes to light, which
may produce these terrible catastrophes, but practical
science has not yet determined the extent to which they

actually come into operation.   The effect of a sudden
generation of steam has been illustrated by an experiment
which I will now make in your presence.   Here is a copper
vessel, v (fig. 53), with a neck which I can stop with this
cork, through which half an inch of fine glass tubing passes.
I heat the copper vessel, and pour into it a little water.
The liquid is now in the spheroidal state.   I cork the vessel,
and the small quantity of steam developed, while the water
remains spheroidal, escapes through the glass tube.   I now
remove the vessel from the lamp, and wait for a minute or
two :   very soon the water will come into contact with the
copper ;   it now does so, and you observe the result : the
cork is driven, as if by the explosion of gunpowder, to a
considerable height in the atmosphere.

I have reserved what you will probably think the most interesting experiment in connection with this subject, for the
conclusion of to-day's lecture.   M. Boutigny, by means of
sulphurous acid, first froze water in a red-hot crucible;
and Mr. Faraday subsequently froze mercury, by means
of solid carbonic acid.   I will try and reproduce this latter
result ;   but first let me operate with water.   I have here a
hollow sphere of brass about two inches in diameter,
now accurately filled with water ;   into the sphere I have
had this wire screwed, which is to serve as a handle.   I heat
this platinum crucible to glowing redness, and place within it some lumps of solid carbonic acid.   I pour some ether
on the acid—neither of them comes into contact with the
hot crucible—they are protected from contact by the
elastic cushion of vapour which surrounds them ;   I lower
my sphere of water down upon the mass, and carefully
pile fragments of carbonic acid over it, adding also a little
ether.   The pasty mass within the red-hot crucible remains
intensely cold ;   and now you hear a crack !   I am thereby
assured that the experiment will succeed.   The freezing
water has burst the brass sphere, as it burst the iron

bottles in a former experiment.  Round the sphere I have
wound a bit of wire to prevent the ice from falling out.  I
now raise the sphere, peel off the shattered brass shell, and
there you have a solid sphere of ice, extracted from the red-
hot crucible.

I place a quantity of mercury in a conical copper spoon,
and dip it into the crucible.  The ether in the crucible has
taken fire, which I did not intend it to do.  The experiment ought to be so made, that the carbonic acid gas—
the choke-damp of mines—ought to keep the ether from
ignition.  But the mercury will freeze notwithstanding.
Out of the fire, and through the flame, I draw the spoon,
and there is the frozen mass turned out before you on
the table.

horizon appeared clear, while the rest of the firmament was covered by a black cloud, which soon extended to the east, quenched the light there, and at length produced a darkness so dense that the windows in the rooms could not be discerned.  A shower of ashes descended, under which the tree branches bent and broke.  Whence came these ashes?  From the direction of the wind, we should infer that they came from the Peak  of  the  Azores: they came, however, from the volcano Morne Garou in St. Vincent, which lies about 100 miles west of Barbadoes. The ashes had been cast into the current of the upper trade.  A second example of the same kind occurred on January 20, 1835.  On the 24th and 25th the sun was darkened in Jamaica by a shower of fine ashes, which had been discharged from the mountain Coseguina, distant 800 miles.  The people learned in this way that the explosions previously heard were not those of artillery. These ashes could only have been carried by the upper current, as Jamaica lies north-east from the mountain. The same eruption gives also a beautiful proof that the ascending air-current divides itself above, for ashes fell upon the ship Conway in the Pacific, at a distance of 700 miles south-west of Coseguina.

'Even on the highest summits of the Andes no traveller has as yet reached the upper trade.  From this some notion may be formed of the force of the explosions; they were indeed tremendous in both instances. The roaring of Coseguina was heard at San Salvador, a distance of 1,000 miles.  Union, a seaport on the west coast of Conchagua, was in absolute darkness for forty-three hours; as light began to dawn it was observed that the sea-shore had advanced 800 feet upon the ocean, through the mass of ashes which had fallen.  The eruption of Morne Garou forms the last link of a chain of vast volcanic actions.  In June and July 1811, near St.

Miguel, one of the Azores, the island Sabrina rose, accompanied by smoke and flame, from the bottom of a
sea 150 feet deep, attained a height of 300 feet, and a
circumference of a mile. The small Antilles were afterwards
shaken, and subsequently the valleys of the Mississippi,
Arkansas, and Ohio. But the elastic forces found no vent;
they sought one, then, on the north coast of Columbia.
March 26 began as a day of extraordinary heat in Car-
accas; the air was clear and the firmament cloudless.
It was Green Thursday, and a regiment of troops of the line
stood under arms in the barracks of the quarter San Carlos
ready to join in the procession. The people streamed to
the churches. A loud subterranean thunder was heard,
and immediately afterwards followed an earthquake shock
so violent, that the church of Alta Gracia, 150 feet in
height, borne by pillars fifteen feet thick, formed a heap of
crushed rubbish not more than six feet high. In the
evening the almost full moon looked down with mild lustre
upon the ruins of the town, under which lay the crushed
bodies of upwards of 10,000 of its inhabitants. But even
here there was no exit granted to the elastic forces underneath. Finally, on April 27, they succeeded in opening
once more the crater of Morne Garou, which had been
closed for a century; and the earth, for a distance equal to
that from Vesuvius to Paris, rung with the thunder-
shout of the liberated prisoner.'

I have here a terrestrial globe; on which I now trace with
my hand two meridians; they start from the equator of the
globe a foot apart, which would correspond to about 1,000
miles on the earth's surface. But these meridians, as they
proceed northward, gradually approach each other, and
meet at the north pole. It is manifest that the air which
rises between these meridians in the equatorial regions must,
if it went direct to the pole, squeeze itself into an ever-
narrowing bed. Were the earth a cylinder instead of a

sphere, we might have a circulation from the middle of the cylinder quite to each end, and a return current from each end to the middle.   But this, in the case of the earth, is impossible, simply because the space around the poles is unable to embrace the air from the equator.   The cooled equatorial air sinks, and the return current sets in before the poles are attained, and this occurs more or less irregularly.   The two currents, moreover, instead of flowing one over the other, often flow beside each other.   They constitute rivers of air, with incessantly shifting beds.

These are the great winds of our atmosphere which, however, are materially modified by the irregular distribution of land and water.   Winds of minor importance also occur, through the local action of heat, cold, and evaporation. There are winds produced by the heating of the air in Alpine valleys, and which sometimes rush with sudden and destructive violence down the gulleys of the mountains: gentler down-flows of air are produced by the presence of glaciers upon the heights.   There are the land breeze and the sea breeze, due to the varying temperature of the sea-board soil, by day and night.   The morning sun heating the land, produces vertical displacement, and the air from the sea moves landward.   In the evening the land is more chilled by radiation than the sea, and the conditions are reversed; the heavy air of the land now flows seaward.

Thus, then, a portion of the heat of the tropics is sent by an aërial messenger towards the poles, a more equable distribution of terrestrial warmth being thus secured.   But in its flight northward the air is accompanied by another substance—by the vapour of water, which, you know, is perfectly transparent.   Imagine the ocean of the tropics, giving forth its vapour, which promotes by its lightness the ascent of the associated air.   They expand as they ascend : at a height of 16,000 feet the air and vapour

occupy twice the volume which they embraced at the
sea level. To secure this space they must, by their
elastic force, push away the air in all directions round
them; they must perform work; and this work cannot
be performed, save at the expense of the warmth with
which they were in the first instance charged.

The vapour thus chilled is no longer competent to
retain the gaseous form. It is precipitated as cloud : the
cloud descends as rain; and in the region of calms, or
directly under the sun, where the air is first drained of
its aqueous load, the descent of rain is enormous. The
sun does not remain always vertically over the same
parallel of latitude—he is sometimes north of the equator,
sometimes south of the equator, the two tropics limit-
ing his excursion. When he is south of the equator,
the earth's surface north of it is no longer in the region
of calms, but in a region across which the aërial current
from the north flows towards the region of calms. The
moving air is but slightly charged with vapour, and
as it travels from north to south it becomes ever warmer ;
it constitutes a dry wind, and its capacity to retain vapour
is continually augmenting. It is plain, from these considerations, that each place between the tropics must have
its dry season and rainy season ; dry when the sun is at
the opposite side of the equator, and wet when the sun is
overhead.

Gradually, however, as the upper stream, which rises
from the equator, and flows towards the poles, becomes
chilled and dense, it sinks towards the earth ; at the Peak
of Teneriffe it has already sunk below the summit of the
mountain. With the contrary wind blowing at the base,
the traveller finds the stream from the equator blowing
strong over the top. Farther north the equatorial wind
sinks lower still, and finally quite reaches the surface of
the earth. Europe, for the most part, is overflowed by this

equatorial current. Here in London, for eight or nine months in the year, south-westerly winds prevail. But mark what an influence this must have upon our climate. The moisture of the equatorial ocean comes to us endowed with potential energy; with its molecules separate, and therefore competent to clash and develope heat by their collision; it comes, if you will, charged with latent heat. In our northern atmosphere the collision takes place, and the heat generated is a main source of warmth to our climate. Were it not for the rotation of the earth, we should have over us the hot dry blasts of Africa; but owing to this rotation, the wind which starts northward from the Gulf of Mexico is deflected to Europe. Europe is, therefore, the recipient of those stores of latent heat which were amassed in the western Atlantic. The British Isles come in for the greatest share of this moisture and heat, and this circumstance adds itself to that already dwelt upon—the high specific heat of water—to preserve our climate from extremes. It is this condition of things which makes our fields so green, and which gives the blossom to our maidens' cheeks. A German writer, Moritz, expresses himself on these points in the following ardent words:—' Ye blooming youthful faces, ye green meadows and streams of this happy land, how have ye enchanted me! O Richmond, Richmond! never can I forget the evening when, full of delight, I wandered near you, up and down along the flowery banks of the Thames. This, however, must not detain me from that dry and sand-strewn soil on which fate has appointed me my sphere of action.' All this poetry and enchantment are derived directly from aqueous vapour.*

As we travel eastward in Europe, the amount of aqueous precipitation grows less and less; the air becomes more and more drained of its moisture. Even between the east and

* Its relation to Radiant Heat is developed in Lecture XI.

west coasts of our own islands the difference is sensible, and
local circumstances also have a powerful influence on the
amount of precipitation.    Dr. Lloyd finds the mean
yearly temperature of the western coast of Ireland about
two degrees higher than that of the eastern coast, at the
same height, and in the same parallel of latitude.    The
total amount of rain which fell in the year 1851, at various
stations in the island, is given in the following table —

| Station | | | | | Rain in inches |
|---|---|---|---|---|---|
| Portarlington | . | . | . | . | 21·2 |
| Killough | . | . | . | . | 23·2 |
| Dublin | . | . | . | . | 26·4 |
| Athy . | . | . | . | . | 26·7 |
| Donaghadee | . | . | . | . | 27·9 |
| Courtown | . | . | . | . | 29·6 |
| Kilrush | . | . | . | . | 32·6 |
| Armagh | . | . | . | . | 33·1 |
| Killybegs | . | . | . | . | 33·2 |
| Dunmore | . | . | . | . | 33·5 |
| Portrush | . | . | . | . | 37 2 |
| Burincrana . | . | . | . | . | 39·3 |
| Markree | . | . | . | . | 40·3 |
| Castletownsend . | . | . | . | 42·5 |
| Westport | . | . | . | . | 45·9 |
| Cahirciveen . | . | . | . | 59·4 |

With reference to this table, Dr. Lloyd remarks —

' 1. That there is great diversity in the yearly amount
of rain at the different stations, all of which (excepting
four) are but a few feet above the sea level; the greatest
rain (at Cahirciveen) being nearly three times as great as
the least (at Portarlington).

' 2. That the stations of least rain are either inland or on
the eastern coast, while those of the greatest rains are at
or near the western coast.

' 3. That the amount of rain is greatly dependent on the
proximity of a mountain chain or group, being always
considerable in such neighbourhood, unless the station
lie to the north-east of the same.

' Thus, Portarlington lies to the north-east of Slievebloom; Killough to the north-east of the Mourne range; Dublin, north-east of the Wicklow range, and so on. On the other hand, the stations of greatest rain, Cahirciveen, Castletownsend, Westport, &c., are in the vicinity of high mountains, but on a different side.' *

This distribution of heat by the transfer of masses of heated air from place to place, has been called ' *convection,*' in contradistinction to the process of conduction, which will be treated in our next lecture. Heat is distributed in a similar manner through liquids. I have here a glass cell, c (fig. 54), containing warm water; I place it in front of

Fig. 54.

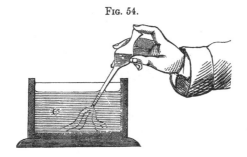

the electric lamp, and by means of a converging lens, throw a magnified image of the cell upon the screen. I now introduce the end of this pipette into the water of the cell, and allow a little cold water to gently enter the hot. The difference of refraction between both enables you to see the heavy cold water falling through the lighter warm water. The experiment succeeds still better when I allow a fragment of ice to float upon the surface of the water. As the ice melts, it sends long heavy striæ downwards

* The greatest rainfall recorded by Sir John Herschel in his table (Meteorology, 110, &c.) occurs at Cherra Pungee, where the annual fall is 592 inches. It is not my object to enter far into the subject of meteorology; for the fullest and most accurate information the reader will refer to the excellent works of Sir John Herschel and Professor Dove.

to the bottom of the cell. You observe, as I cause the
ice to move along the top, how these streams of cold
water descend through the hot. I now reverse the experiment, placing cold water in the cell, and hot water in the
pipette. Care is here necessary to allow the warm water to
enter without any momentum, which would carry it mechanically down. You notice the effect. The point of the
pipette is in the middle of the cell, and you see, as the

FIG. 55.

warm water enters, it speedily turns upwards (fig. 55) and
overflows the top, almost as oil would do under the same
circumstances.

When a vessel containing water is heated at the bottom,
the warmth communicated is thus diffused. You may see
the direction of the ascending warm currents by means of
the electric lamp, and also that of the currents which descend to occupy the place of the lighter water. Here is a
vessel containing cochineal, the fragments of which, being
not much heavier than the water, freely follow the direction
of its currents. You see the pieces of cochineal breaking
loose from the heated bottom; ascending along the middle
of the jar, and descending again by the sides. In the Geyser
of Iceland this convection occurs on a grand scale. A
fragment of paper thrown upon the centre of the water
which fills the pipe is instantly drawn towards the side, and
there sucked down by the descending current.

Partly to this cause, but mainly, perhaps, to the action of winds, currents establish themselves in the ocean, and powerfully influence climate by the heat which they distribute. The most remarkable of these currents, and by far the most important for us, is the so-called Gulf-stream, which sweeps across the Atlantic from the equatorial regions through the Gulf of Mexico, whence it derives its name. As it quits the Straits of Florida it has a temperature of 83° Fahr., thence it follows the coast of America as far as Cape Fear, whence it starts across the Atlantic, taking a north-easterly course, and finally washing the coast of Ireland, and the north-western shores of Europe generally. As might be expected, the influence of this body of warm water makes itself most evident in our winter. It then entirely abolishes the difference of temperature due to the difference of latitude of north and south Britain; if we walk from the Channel to the Shetland Isles, in January, we encounter everywhere the same temperature. The Isothermal line runs north and south. The presence of this water renders the climate of western Europe totally different from that of the opposite coast of America. The river Hudson, for example, in the latitude of Rome, is frozen over for three months in the year. Starting from Boston in January, and proceeding round St. John's, and thence to Iceland, we meet everywhere the same temperature. The harbour of Hammerfest derives great value from the fact that it is clear of ice all the year round. This is due to the Gulf-stream which sweeps round the North Cape, and so modifies the climate there, that at some places, by proceeding northward, you enter a warmer region. The contrast between northern Europe and the east coast of America caused Halley to surmise that the north pole of the earth had shifted; that it was formerly situate somewhere near Behring's Straits, and that the intense cold observed in these regions is really the cold of the ancient pole, which had not been entirely subdued since the axis changed its direction. But now we know that the Gulf-stream and the diffusion of heat by winds and vapours are the real causes of European mildness. On the western coast of America, between the Rocky mountains and the ocean, we find a European climate.

Europe, then, is the condenser of the Atlantic; and the mountains are the chief condensers in Europe. On them, moreover, when they are sufficiently high, the condensed vapour descends, not in a liquid, but a solid form. Let us look to this water in its birthplace, and follow it through its subsequent course. Clouds float in the air, and hence the surmise that they are composed of vesicles or bladders of water, thus forming *shells* instead of *spheres*. Eminent travellers say that they have seen these bubbles, and their statements are entitled to all respect. It is certain, however, that the water-particles at high elevations possess, on or after precipitation, the powers of building themselves into crystalline forms; they thus bring forces into play which we have hitherto been accustomed to regard as molecular, and which could not be ascribed to the aggregates necessary to form vesicles.

Snow, perfectly formed, is not an irregular aggregate of ice-particles; in a calm atmosphere, the aqueous atoms arrange themselves so as to form the most exquisite figures. You have seen those six-petalled flowers which form themselves within a block of ice when a beam of heat is sent through it. The snow-crystals, formed in a calm atmosphere, are built upon the same type: the molecules arrange themselves to form hexagonal stars. From a central nucleus shoot six spiculæ, every two of which are separated by an angle of 60°. From these central ribs smaller spiculæ shoot right and left with unerring fidelity to the angle 60°, and from these again other smaller ones diverge at the same angle. The six-leaved blossoms assume the most wonderful variety of form; their tracery is of the finest frozen gauze; and round about their corners other rosettes of smaller dimensions often cling. Beauty is superposed upon beauty, as if Nature, once committed to her task, took delight in showing, even within the narrowest limits, the wealth of her resources.*

These frozen blossoms constitute our mountain snows; they load the Alpine heights, where their frail architecture is soon destroyed by the accidents of the weather. Every winter they fall, and every summer they disappear, but this rythmic action does not perfectly compensate itself. Below a certain line warmth is predominant, and the quantity which falls every winter is entirely swept away; above this line cold is predominant, the quantity which falls is in excess of the quantity melted, and an annual residue remains. In winter the snows reach to the plains; in summer they retreat to the *snow-line*,—to that particular line where the snow-fall of every year is exactly balanced by the consumption, and above which is the region of eternal snows. But if a residue remains annually above the snow line, the mountains must be loaded with a burden which increases every year. Supposing at a particular point above the line referred to, a layer of three feet a year is added to the mass; this deposit, accumulating even through the brief period of the Christian era, would produce an elevation of 5,580 feet. And did such accumulations continue throughout geologic instead of historic ages, there is no knowing the height to which the snows would pile themselves. It is manifest no accumulation of this kind takes place; the quantity of snow on the mountains is not augmenting in this way; for some reason or other the sun is not permitted to lift the ocean out of its basins and pile its waters permanently upon the hills.

But how is this annually augmenting load taken off the

---

* See fig. 56, in which are copied some of the beautiful drawings of Mr. Glaisher.

Fig. 56.

shoulders of the mountains ? The snows sometimes detach themselves and rush down the slopes in avalanches, melting to water in the warmer air below. But the violent rush of the avalanche is not their only motion; they also creep by almost insensible degrees down the slopes. As layer, moreover, heaps itself upon layer, the deeper portions of the mass become squeezed and consolidated; the air first entrapped in the meshes of the snow is squeezed out, and the compressed mass approximates more and more to the character of ice. You know how the granules of a snow-ball will adhere; you know how hard you can make it if mischievously inclined : the snowball is incipient ice; augment your pressure, and you actually convert it into ice. But even after it has attained a compactness which would entitle it to be called ice, it is still capable of yielding more or less, as the snow yields, to pressure. When, therefore, a sufficient depth of the substance collects upon the earth's surface, the lower portions are squeezed out by the pressure of the upper ones, and if the snow rests upon a slope, it will yield principally in the direction of the slope, and move downwards.

This motion is incessantly going on along the slopes of every snow-laden mountain; in the Himalayas, in the Andes, in the Alps; but in addition to this motion, which depends upon the power of the substance itself to yield to pressure, there is also a sliding motion over the inclined bed. The consolidated snow moves bodily over the mountain slope, grinding off the asperities of the rocks, and polishing their hard surfaces. The under surface of the mighty polisher is also scarred and furrowed by the rocks over which it has passed; but as the compacted snow descends, it enters a warmer region, is more copiously melted and sometimes, before the base of its slope is reached, it is wholly cut off by fusion. Sometimes, however, large and deep valleys receive the gelid masses thus sent down; in these valleys it is further consolidated, and through them

it moves, at a slow but measurable pace, imitating in all its
motions those of a river. The ice is thus carried far beyond
the limits of perpetual snow, until, at length, the consumption below equals the supply above, and at this point the
glacier ceases. From the snow-line downwards in summer,
we have *ice*; above the snow-line, both summer and winter,
we have, on the surface, *snow*. The portion below the
snow-line is called a *glacier*, that above the snow-line is
called the *névé*. The névé, then, is the feeder of the glacier.

Several valleys thus filled may unite in a single valley,
the tributary glaciers welding themselves together to form
a trunk glacier. Both the main valley and its tributaries
are often sinuous, and the tributaries must change their
direction to form the trunk. The width of the valley, also,
often changes : the glacier is forced through narrow
gorges, widening after it has passed them; the centre of
the glacier moves more quickly than the sides, and the
surface more quickly than the bottom. The point of
swiftest motion follows the same law as that observed in
the flow of rivers, changing from one side of the centre to
the other, as the flexure of the valley changes.* Most of
the great glaciers in the Alps have, in summer, a central
velocity of two feet a day. There are points on the Mer-
de-Glace, opposite the Montenvert, which have a daily
motion of thirty inches in summer, and in winter have
been found to move at half this rate.

The power of accommodating itself to the channel
through which it moves has led eminent men to assume
that ice is viscous; and the phenomena at first sight
seem to enforce this assumption. The glacier widens,
bends, and narrows, and its centre moves more quickly
than its sides; a viscous mass would undoubtedly do the
same. But the most delicate experiments on the capacity
of ice to yield to strain, to stretch out like treacle, honey,

---

* For the data on which this law is founded see Appendix to this Lecture.

or tar, have failed to detect this stretching power. Is there, then, any other physical quality to which the power of accommodation possessed by glacier ice, may be referred?

Let us approach this subject gradually. We know that vapour is continually escaping from the free surface of a liquid; that the particles at the surface attain their gaseous liberty sooner than the particles within the liquid; it is natural to expect a similar state of things with regard to ice; that when the temperature of a mass of ice is uniformly augmented, the first particles to attain liquid liberty are those at the surface; for here they are entirely free, on one side, from the controlling action of the surrounding particles. Supposing, then, two pieces of ice raised throughout to 32°, and melting at this temperature at their surfaces; what may be expected to take place if we place the liquefying surfaces close together? We thereby virtually transfer these surfaces to the centre of the ice, where the motion of each molecule is controlled all round by its neighbours. As might reasonably be expected, the liberty of liquidity at each point where the surfaces touch each other, is arrested, and the two pieces freeze together at these points. Let us make the experiment: Here are two masses which I have just cut asunder by a saw; I place their flat surfaces together; half a minute's contact will suffice; they are now frozen together, and by taking hold of one of them I thus lift them both.

This is the effect to which attention was first directed by Mr. Faraday in June 1850, and which is now known under the name of *Regelation*. On a hot summer's day, I have gone into a shop in the Strand where fragments of ice were exposed in a basin in the window; and with the shopman's permission have laid hold of the topmost piece of ice, and by means of it have lifted the whole of the pieces bodily out of the dish. Though the thermometer at the time stood at 80°, the pieces of ice had

frozen together at their points of junction. Even under hot
water this effect takes place; I have here a basin of water
as hot as my hand can bear; I plunge into it these two
pieces of ice, and hold them together for a moment: they
are now frozen together, notwithstanding the presence of
the heated liquid.   A pretty experiment of Mr. Faraday's
is to place a number of small fragments of ice in a dish of
water deep enough to float them.   When one piece
touches the other, if only at a single point, regelation
instantly sets in.   Thus a train of pieces may be caused to
touch each other, and, after they have once so touched,
you may take the terminal piece of the train and, by means
of it, draw all the others after it.   When we seek to bend
two pieces thus united at their point of junction, the
frozen points suddenly separate by fracture, but at the
same moment other points come into contact, and regelation sets in between them.   Thus a wheel of ice might be
caused to roll on an ice surface, the contacts being incessantly ruptured, with a crackling noise, and others as
quickly established by regelation.   In virtue of this property of regelation, ice is able to reproduce many of the
phenomena which are usually ascribed to viscous bodies.*

Here, for example, is a straight bar of ice: I can by
passing it successively through a series of moulds, each
more curved than the last, finally turn it out as a
semi-ring.   The straight bar in being squeezed into the
curved mould breaks, but by continuing the pressure new
surfaces come into contact, and the continuity of the mass
is restored.   I take a handful of those small ice fragments
and squeeze them together, they freeze at their points of
contact and now the mass is one aggregate.   The making
of a snow-ball, as remarked by Mr. Faraday, illustrates the
same principle.   In order that this freezing shall take
place, the snow ought to be at 32° and moist.   When below

* See note on the Regelation of Snow Granules in the Appendix to this Lecture.

32° and dry, on being squeezed it behaves like salt. The crossing of snow-bridges in the upper regions of the Swiss glaciers is often rendered possible solely by the regelation of the snow granules. The climber treads the mass carefully, and causes its granules to regelate: he thus obtains an amount of rigidity which, without the act of regelation, would be quite unattainable. To those un-accustomed to such work, the crossing of snow bridges, spanning, as they often do, fissures 100 feet and more in depth, must appear quite appalling.

If I still further squeeze this mass of ice fragments, I bring them into still closer proximity. My hand, however, is incompetent to squeeze them very closely together. I place them in this boxwood mould, which is a shallow cylinder, and placing a flat piece of boxwood overhead, I introduce both between the plates of a small hydraulic press, and squeeze the mass forcibly into the mould. I now relieve the pressure and turn the substance out before you: it is converted into a coherent cake of ice. I place it in this lenticular cavity and again squeeze it. It is crushed by the pressure, of course, but new contacts es tablish themselves, and there you have the mass a lens of ice. I now transfer my lens to this hemispherical cavity, H (fig. 57), and bring down upon it a hemispherical protu-berance, P, which is not quite able to fill the cavity. I

Fig. 57.

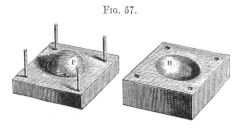

squeeze the mass: the ice, which a moment ago was a lens, is now squeezed into the space between the two spherical

surfaces: I remove the protuberance, and here I have the
interior surface of a cup of glassy ice. By care I release
it from the mould, and there it is, a hemispherical cup,
which I can fill with cold sherry, without the escape of a
drop. I scrape with a chisel a quantity of ice from this
block, and placing the spongy mass within this spherical
cavity, c (fig. 58), I squeeze it and add to it, till finally I can
bring down another spherical cavity, D, upon it, enclosing
it as a sphere between both. As I work the press the mass
becomes more and more compacted. I add more material,
and again squeeze ; by every such act the mass is made
harder, and there you have a snow-ball before you such as
you never saw before. It is a sphere of hard translucent
ice, B. Thus, you see, broken ice can be compacted together

FIG. 58.

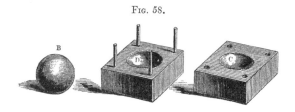

by pressure, and in virtue of the property of regelation,
which cements its touching surfaces, the substance may
be made to take any shape we please. Were the experiment worth the trouble, I feel satisfied that I could form
*a rope of ice* from this block, and afterwards coil the rope
into a knot. Nothing of course can be easier than to
produce statuettes of the substance from suitable moulds.

It is easy to understand how a substance so endowed
can be squeezed through the gorges of the Alps—can
bend so as to accommodate itself to the flexures of the
Alpine valleys, and can permit of a differential motion of its
parts, without at the same time possessing a sensible trace
of viscosity. The hypothesis of viscosity, first started by

Rendu, and worked out with such ability by Prof. Forbes, accounts, certainly, for half the facts. Where pressure comes into play, the deportment of ice is apparently that of a viscous body; where tension comes into play, the analogy with a viscous body ceases.

I have thus briefly sketched the phenomena of existing glaciers, as far as they are related to our present subject; but the scientific explorer of mountain regions soon meets with appearances which carry his mind back to a state of things very different from that which now obtains. The unmistakable traces which they have left behind them show that vast glaciers once existed in places, from which they have for ages disappeared. Go, for example, to the glacier of the Aar in the Bernese Alps, and observe its present performances; look to the rocks upon its flanks as they are at this moment, rounded, polished, and scarred by the moving ice. And having by patient and varied exercise educated your eye and judgement in these matters, walk down the glacier towards its end, keeping always in view the evidences of the glacier's action. After quitting the ice, continue your walk down the valley towards the Grimsel : you see everywhere the same unmistakable record. The rocks which rise from the bed of the valley are rounded like hogs' backs; these are the ' roches mou-tonnés' of Charpentier and Agassiz; you observe upon them the larger flutings of the ice, and also the smaller scars scratched by pebbles, which the glacier held as emery on its under surface. All the rocks of the Grimsel have been thus planed down. Walk down the valley of Hasli and examine the mountain sides right and left; without the key which I now suppose you to possess, you would be in a land of enigmas; but with this key all is plain, you see everywhere the well-known scars and flutings and furrowings. In the bottom of the valley you have the rocks filed down in some places to dome-shaped masses,

and, in others, polished so smooth that to pass over them,
even when the inclination is moderate, steps must be hewn.
All the way down to Meyringen, and beyond it, if you
wish to pursue the enquiry, these evidences abound. For
a preliminary lesson in the recognition of the traces of
ancient glaciers no better ground can be chosen than this.

Similar evidences are found in the valley of the Rhone;
you may track them through the valley for eighty miles,
and lose them at length in the lake of Geneva. But on
the flanks of the Jura, at the opposite side of the Canton
de Vaud, the evidences reappear. All along these lime-
stone slopes you have strewn the granite boulders of Mont
Blanc. Right and left also from the great Rhone valley the
lateral valleys show that they were once held by ice. On
the Italian side of the Alps the remains are, if possible,
more stupendous than on the northern side. Grand as
are the present glaciers to those who explore them in all
their lengths, they are mere pigmies in comparison with
their predecessors.

Not in Switzerland alone — not alone in proximity with
existing glaciers — are these well-known vestiges of the
ancient ice discernible; in the hills of Cumberland they are
almost as clear as in the Alps. Where the bare rock has
been exposed for ages to the action of the weather, the
finer marks have in most cases disappeared; and the
mammillated forms of the rocks are the only evidences.
But the removal of the soil which has protected them,
often discloses rock surfaces which are scarred as sharply,
and polished as cleanly, as those which are now being
scratched and polished by the glaciers of the Alps. Round
about Scawfell the traces of the ancient ice appear, both in
*roches moutonnés* and *blocs perchés*; and there are ample
facts to show that Borrodale was once occupied by glacier
ice. In North Wales, also, the ancient glaciers have placed
their stamp so firmly upon the rocks, that the ages which

have since elapsed have failed to obliterate even their superficial marks. All round Snowdon these evidences abound. On the south-west coast of Ireland also rise the Reeks of Magillicuddy, which tilt upwards, and catch upon their cold crests the moist winds of the Atlantic; precipitation is copious, and rain at Killarney seems the rule of Nature. In this moist region every crag is covered with rich vegetation; but the vapours which now descend as mild and fertilising rain, once fell as snow, which formed the material for noble glaciers. The Black Valley was once filled by ice, which planed down the sides of the Purple Mountain, as it moved towards the Upper Lake. The ground occupied by this lake was entirely held by the ancient ice, and every island that now emerges from its surface is a glacier-dome. The fantastic names which many of the rocks have received are suggested by the shapes into which they have been sculptured by the mighty moulding plane which once passed over them. North America is also thus glaciated. But the most notable observation in connection with this subject is one recently made by Dr. Hooker during a visit to Syria: he has found that the celebrated cedars of Lebanon grow upon ancient glacier moraines.

To determine the condition which permitted of the formation of those vast masses of ice has long been a problem with philosophers, and a consideration of the solutions which have been offered from time to time will not be uninstructive. I have no new hypothesis, but it seems possible to give a truer direction and more definite aim to our enquiries. The aim of all the writers on this subject, with whom I am acquainted, has been directed to the attainment of *cold*. Some eminent men have thought, and some still think, that the reduction of temperature during the glacier epoch was due to a temporary diminution of solar radiation; others have thought that, in its motion through space, our system may have traversed regions of low temperature, and that during its passage through these regions, the ancient glaciers were produced. Others, with greater correctness, have sought to lower the temperature by a re-distribution of land and water. If I understand the writings of the eminent men who have propounded and advocated the above hypotheses, many of them seem to have overlooked the fact, that the enormous extension of glaciers in bygone ages, demonstrates, just as rigidly, the operation of heat as the action of cold.

Cold will not produce glaciers. You may have the bitterest north-east winds here in London throughout the winter without a single flake of snow. Cold must have the fitting object to operate upon, and this object—the aqueous vapour of the air—is the direct product of heat. Let us put this glacier question in another form: the latent heat of aqueous vapour, at the temperature of its production in the tropics, is about 1,000° Fahr., for the latent heat grows larger as the temperature of evaporation descends. A pound of water then vaporised at the equator, has absorbed 1,000 times the quantity of heat which would raise a pound of the liquid one degree in temperature. But the quantity of heat which would raise a pound of water one degree would raise a pound of cast-iron ten degrees: hence, simply to convert a pound of the water of the equatorial ocean into vapour, would require a quantity of heat sufficient to impart to a pound of cast-iron 10,000 degrees of temperature. But the fusing point of cast-iron is 2,000° Fahr.; therefore, for every pound of vapour produced, a quantity of heat has been expended by the sun sufficient to raise 5 lbs. of cast-iron to its melting point. Imagine, then, every one of those ancient glaciers with its mass of ice quintupled; and let the place of the mass, so augmented, be taken by an equal mass of cast-iron raised to the white heat of fusion, and we have the exact expression of the solar action involved in the production of the ancient glaciers. Substitute the hot iron for the cold ice—our speculations would instantly be directed to account for the high temperature of the glacial epoch, and a complete reversal of some of the hypotheses above quoted would probably ensue.

It is perfectly manifest that by weakening the sun's action, either through a defect of emission, or by the steeping of the entire solar system in space of a low temperature, we should be cutting off the glaciers at their source. Vast masses of mountain ice indicate, infallibly, commensurate masses of atmospheric vapour, and a proportionately vast action on the part of the sun. In a distilling apparatus, if you required to augment the quantity distilled, you would not surely attempt to obtain the low temperature necessary to distillation, by taking the fire from under your boiler; but this, if I understand them aright, is what has been done by those philosophers who have sought to produce the ancient glaciers by diminishing the sun's heat. It is quite manifest that the thing most needed to produce the glaciers is an *improved condenser* ; we cannot afford to lose an iota of solar action ; we need, if anything, more vapour, but we' need a condenser so powerful that this vapour, instead of falling in liquid showers to the earth, shall be so far reduced in temperature as to descend in snow. The problem, I think, is thus narrowed to the precise issue on which its solution depends.

# APPENDIX TO LECTURE VI.

————◦◦——

ABSTRACT OF A DISCOURSE ON THE MER-DE-GLACE.*

A PORTION of a series of observations made upon the Mer-de-Glace of Chamouni during the months of July and August last year, formed the basis of this discourse.

The law first established by [M. Agassiz and] Prof. J. D. Forbes, that the central portions of a glacier moved faster than the sides, was amply illustrated by the deportment of lines of stakes placed across the Mer-de-Glace at several places, and across the tributaries of the glacier. The portions of the Mer-de-Glace derived from these tributaries were easily traceable throughout the glacier by means of the *moraines*. Thus, for example, that portion of the trunk stream derived from the Glacier du Géant, might be distinguished, in a moment, from the portion derived from the other tributaries, by the absence of the débris of the moraines upon the surface of the former. The commencement of the dirt formed a distinct junction between both portions. Attention has been drawn by Prof. Forbes to the fact, that the eastern side of the glacier in particular is ' excessively crevassed; ' and he accounts for this crevassing by supposing that the Glacier du Géant moves most swiftly, and in its effort to drag its more sluggish companions along with it, tears them asunder, and thus produces the fissures and dislocation for which the eastern side of the glacier is remarkable. The speaker said that too much weight must not be attached to this explanation. It was one of those suggestions which are perpetually thrown out by men of science during the course of an investigation, and the fulfillment or non-fulfillment of

* Given at the Royal Institution of Great Britain, on Friday, June 4, 1858. By John Tyndall, F.R.S.

which cannot materially affect the merits of the investigator. Indeed, the merits of Forbes must be judged on far broader grounds ; and the more his labours are compared with those of other observers, the more prominently does his comparative intellectual magnitude come forward. The speaker would not content himself with saying that the book of Prof. Forbes was the best book which had been written upon the subject. The qualities of mind, and the physical culture invested in that excellent work, were such as to make it, in the estimation of the physical investigator at least, outweigh all other books upon the subject taken together.* While thus acknowledging its merits, let a free and frank comparison of its statements with facts be instituted. To test whether the Glacier du Géant moved quicker than its fellows, five different lines were set out across the Mer-de-Glace, in the vicinity of the Montenvert, and in each of these it was found that the point of swiftest motion did not lie upon the Glacier du Géant at all ; but was displaced so as to bring it comparatively close to the eastern side of the glacier. These measurements prove that the statement referred to is untenable ; but the deviation of the point of swiftest motion from the centre of the glacier will doubtless be regarded by Prof. Forbes as of far greater importance to his theory. At the place where these measurements were made, the glacier turns its convex curvature to the eastern side of the valley, being concave towards the Montenvert. Let us take a bolder analogy than even that suggested in the explanation of Forbes, where he compares the Glacier du Géant to a strong and swiftly-flowing river. Let us enquire how a river would behave in sweeping round a curve similar to that here existing. The point of swiftest motion would undoubtedly lie on that side of the centre of the stream towards which it turns its convex curvature. Can this be the case with

---

* Since the above was written, my 'Glaciers of the Alps' has been published, and, soon after its appearance, a 'Reply' to those portions of the book which referred to the labours of M. Rendu was extensively circulated by Principal Forbes. For more than two years I have abstained from answering my distinguished censor; not from inability to do so, but because I thought, and think, that, within the limits of the case, it is better to submit to misconception, than to make science the arena of a purely personal controversy.

the ice? If so, then we ought to have a shifting of the point
of maximum motion towards the western side of the valley, when
the curvature of the glacier so changes as to turn its convexity
to the western side.   Such a change of flexure occurs opposite the
passages called *Les Ponts*, and at this place the view just enunciated was tested.   It was soon ascertained that the point of
swiftest motion here lay at a different side of the axis from that
observed lower down.   But to confer strict numerical accuracy
upon the result, stakes were fixed at certain distances from the
western side of the glacier, and others *at equal distances* from
the eastern side.   The velocities of these stakes were compared
with each other, two by two ; a stake on the western side being
always compared with a second one, which stood at the same
distance from the eastern side.   The results of this measurement
are given in the following table, the numbers denoting inches :—

| 1st pair | 2nd pair | 3rd pair | 4th pair | 5th pair |
|---|---|---|---|---|
| West 15 | West $17\frac{1}{4}$ | West $22\frac{1}{4}$ | West $23\frac{3}{4}$ | West $23\frac{3}{4}$ |
| East $12\frac{1}{2}$ | East $15\frac{1}{2}$ | East $15\frac{1}{2}$ | East $18\frac{1}{4}$ | East $19\frac{1}{2}$ |

It is here seen that in each case the western stake moved more
rapidly than its eastern fellow stake ; thus proving, beyond a
doubt, that opposite the Ponts the western side of the Mer-de-
Glace moves quickest—a result precisely the reverse of that
observed where the curvature of the valley was different.

But another test of the explanation is possible.   Between the
Ponts and the promontory of Trélaporte, the glacier passes a
point of contrary flexure, its convex curvature opposite to Trélaporte being turned towards the base of the Aiguille du Moine,
which stands on the eastern side of the valley.   A series of stakes
was placed across the glacier here ; and the velocities of those
placed at certain distances from the western side were compared,
as before, with those of stakes placed at the same distances from
the eastern side.   The following table shows the result of these
measurements ; the numbers, as before, denote inches :—

| 1st pair | 2nd pair | 3rd pair |
|---|---|---|
| West . . $12\frac{3}{4}$ | West . . 15 | West $17\frac{1}{4}$ |
| East . . $14\frac{3}{4}$ | East . . $17\frac{1}{2}$ | East 19 |

Here we find that in each case the *eastern* stake moved faster
than its fellow.   The point of maximum motion has therefore

once more crossed the axis of the glacier, being now upon its eastern side.

Determining the points of maximum motion for a great number of transverse sections of the Mer-de-Glace, and uniting these points, we have the *locus* of the curve described by the point referred to. Fig. 59 represents a sketch of the Mer-de-Glace. The dotted line is drawn along the centre of the glacier; the defined line, which crosses the axis of the glacier at the points A A, is then the locus of the point of swiftest motion. It is a curve more deeply sinuous than the valley itself, and crosses the central line of the valley at each point of contrary flexure. The speaker drew attention to the fact that the position of towns upon the banks of rivers is usually on the convex side of the stream, where the rush of the water renders silting-up impossible : the Thames was a case in point; and the same law which regulated its flow and determined the position of the adjacent towns, is at this moment operating, with silent energy, among the Alpine glaciers.

Fig. 59.

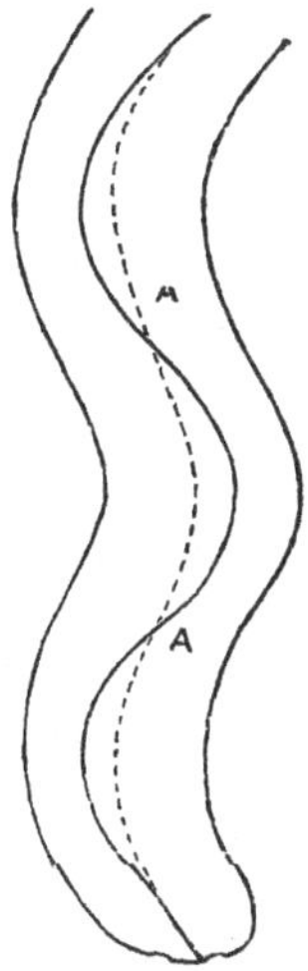

Another peculiarity of glacier motion is now to be noticed.

Before any observations had been made upon the subject, it was surmised by Prof. Forbes that the portions of a glacier near its bed were retarded by friction against the latter. This view was afterwards confirmed by his own observations, and by those of M. Martins. Nevertheless the state of our knowledge upon the subject, rendered further confirmation of the fact highly desirable. A rare opportunity for testing the question was furnished by an almost vertical precipice of ice, constituting the side of the Glacier du Géant, which was exposed near the Tacul. The precipice was about 140 feet in height. At the top and near the bottom stakes were fixed, and by hewing steps in the ice, the speaker succeeded in fixing a stake in the face of the precipice, at a point about 40 feet above the base. After the lapse of a sufficient number of days, the progress of the three stakes was measured : reduced to the diurnal rate, the motion was as follows :—

Top stake  .    .  6·00 inches
Middle stake  .    .  4·59  ,,
Bottom stake  .    .  2·56  ,,

We thus see that the top stake moved with more than twice
the velocity of the bottom one; while the velocity of the middle
stake lies between the two.  But it also appears that the augmentation of velocity upwards is not proportional to the distance
from the bottom, but increases in a quicker ratio.  At a height
of 100 feet from the bottom, the velocity would undoubtedly be
practically the same as at the surface.  Measurements made upon
an adjacent ice cliff proved this.  We thus see the perfect
validity of the reason assigned by Forbes for the continued
verticality of the walls of transverse crevasses.  Indeed a comparison of the result with his anticipations and reasonings will
prove alike their sagacity and their truth.

The most commanding view of the Mer-de-Glace and its
tributaries is obtained from a point above the remarkable cleft in
the mountain range underneath the Aiguille de Charmoz, which
is sure to attract the attention of an observer standing at the
Montenvert.  This point, which is marked G on the map of
Forbes, the speaker succeeded in attaining.  A Tübingen professor once visited the glaciers of Switzerland, and seeing these
apparently rigid masses enclosed in sinuous valleys, went home
and wrote a book, flatly denying the possibility of their motion.
An inspection from the point now referred to would have doubtless confirmed him in his opinion; and indeed nothing can be
more calculated to impress the mind with the magnitude of the
forces brought into play than the squeezing of the three tributaries of the Mer-de-Glace through the neck of the valley at
Trélaporte.  But let us state numerical results.  Previous to its
junction with its fellows, the Glacier du Géant measures 1,134
yards across.  Before it is influenced by the thrust of the Talèfre,
the Glacier de Léchaud has a width of 825 yards; while the
width of the Talèfre branch across the base of the cascade, before
it joins the Léchaud, is approximately 638 yards.  The sum of
these widths is 2,597 yards.  At Trélaporte those three branches
are forced through a gorge 893 yards wide, with a central
velocity of 20 inches a day !  The result is still more astonishing,
if we confine our attention to one of the tributaries — that of the
Léchaud.  Before its junction with the Talèfre, the glacier has a

width of 37½ English chains.  At Trélaporte this broad ice river is squeezed to a driblet of less than 4 chains in width—that is to say, to about one-tenth of its previous horizontal transverse dimension.  Whence is the force derived which drives the glacier through the gorge?  The speaker believed that it must be a pressure from behind.  Other facts also suggest that the Glacier du Géant is throughout its length in a state of forcible longitudinal compression.  Taking a series of points along the axis of this glacier—if these points, during the descent of the glacier, preserved their distances asunder perfectly constant—there could be no longitudinal compression.  The mechanical meaning of this term, as applied to a substance capable of yielding like ice, must be that the hinder points are incessantly advancing upon the forward ones.  The speaker was particularly anxious to test this view, which first occurred to him from à priori considerations. Three points, A B C, were therefore fixed upon the axis of the Glacier du Géant, A being the highest up the glacier.  The distance between A and B was 545 yards, and that between B and C was 487 yards.  The daily velocities of these three points, determined by the theodolite, were as follows:—

A   .   .  20·55 inches  
B   .   .  15·43  „  
C   .   .  12·75  „

The result completely corroborates the foregoing anticipation. The hinder points are incessantly advancing upon those in front, and that to an extent sufficient to shorten a segment of this glacier, measuring 1,000 yards in length, at the rate of 8 inches a day.  Were this rate uniform at all seasons, the shortening would amount to 240 feet in a year.  When we consider the compactness of this glacier, and the uniformity in the width of the valley which it fills, this result cannot fail to excite surprise; and the exhibition of force thus rendered manifest must, in the speaker's opinion, be mainly instrumental in driving the glacier through the jaws of the granite vice at Trélaporte.

In virtue of what quality, then, can ice be bent and squeezed, and change its form in the manner indicated in the foregoing observations?  The only theory worthy of serious consideration at the present day is that of Prof. Forbes, which attributes these

effects to the viscosity of the ice. The speaker did not agree
with this theory; as the term viscosity appeared to him to be
wholly inapplicable as expressive of the physical constitution
of the glacier ice. He had already moulded ice into cups, bent
it into rings, changed its form in a variety of ways by artificial
pressure, and he had no doubt of his ability to mould a compact
mass of Norway ice which stood upon the table into a statuette;
but would viscosity be the proper term to apply to the process of
bruising and regelation by which this result could be attained?
He thought not. A mass of ice at 32° is very easily crushed,
but it has as sharp and definite a fracture as a mass of glass.
There is no sensible evidence of viscosity.

The very essence of viscosity is the ability of yielding to a
force of *tension*, the texture of the substance, after yielding, being
in a state of equilibrium, so that it has no strain to recover from;
and the substances chosen by Prof. Forbes, as illustrative of the
physical condition of a glacier, possess this power of being drawn
out in a very eminent degree. But it has been urged, and justly
urged, that we ought not to conclude that viscosity is absent
because hand specimens do not show it, any more than we ought
to conclude that ice is not blue because small fragments of the
substance do not exhibit this colour. To test the question of
viscosity, then, we must appeal to the glacier itself. Let us do so.
First, an analogy between the motion of a glacier through a
sinuous valley, and of a river in a sinuous channel, has been
already pointed out. But the analogy fails in one important
particular: the river, and much more so a mass of flowing treacle,
honey, tar, or melted caoutchouc, sweeps round its curves without rupture of continuity. The viscous mass *stretches*, but the
icy mass *breaks*, and the 'excessive crevassing' pointed out by
Prof. Forbes himself is the consequence. Secondly, the inclina-
tions of the Mer-de-Glace and its three tributaries were taken,
and the association of transverse crevasses with the changes of
inclination was accurately noted. Every Alpine traveller knows
the utter dislocation and confusion produced by the descent of the
Mer-de-Glace from the Chapeau downwards. A similar state of
things exists in the ice-cascade of the Talèfre. Descending from
the Jardin, as the ice approaches the fall, great transverse chasms
are formed, which at length follow each other so speedily as to
reduce the ice masses between them to mere plates and wedges,

along which the explorer has to creep cautiously. These plates and wedges are in some cases bent and crumpled by the lateral pressure, and on some masses vortical forces appeared to have acted, turning large pyramids 90° round, so as to set their structure at right angles to its normal position. The ice afterwards descends the fall, the portions exposed to view being a fantastic assemblage of frozen boulders, pinnacles, and towers, some erect, some leaning, falling at intervals with a sound like thunder, and crushing the ice crags on which they fall to powder. The descent of the ice through this outlet has been referred to as a proof of its viscosity ; but the description just given does not, it was believed, harmonise with our ideas of a viscous substance.

But the proof of the non-viscosity of the substance must be sought at places where the change of inclination is very small. Nearly opposite l'Angle there is a change from 4 to 9 degrees, and the consequence is a system of transverse fissures, which renders the glacier here perfectly impassable. Further up the glacier, transverse crevasses are produced by a change of inclination from 3 to 5 degrees. This change of inclination is accurately protracted in fig. 60 ; the bend occurs at the point B; it is

FIG. 60.

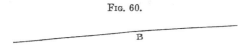

scarcely perceptible, and still the glacier is unable to pass over it without breaking across. Thirdly, the crevasses are due to a state of strain, from which the ice relieves itself by breaking : the rate at which they widen may be taken as a measure of the amount of relief demanded by the ice. Both the suddenness of their formation, and the slowness with which they widen, are demonstrative of the non-viscosity of the ice. For were the substance capable of stretching even at the small rate at which they widen, there would be no necessity for their formation.

Further, the marginal crevasses of a glacier are known to be a consequence of the swifter flow of its central portions, which throws the sides into a state of strain, from which they relieve themselves by breaking. Now it is easy to calculate the amount of stretching demanded of the ice in order to accommodate itself to the speedier central flow. Take the case of a glacier, half a mile wide. A straight transverse element, or slice, of such a glacier, is bent in twenty-four hours to a curve. The ends of the slice move a little, but the centre moves more : let us suppose the versed side of the curve formed by the slice in twenty-four hours to be a foot, which is a fair average. Having the chord of this arch, and its versed side, we can calculate its length. In the case of the Mer-de-Glace, which is about half-a-mile wide, the amount of stretching demanded would be about the eightieth of an inch in twenty-four hours. Surely, if the glacier possessed a property which could with any propriety be called viscosity, it ought to be able to respond to this moderate demand ; but it is not able to do so : instead of stretching as a viscous body, in obedience to this slow strain, it breaks as an eminently fragile one, and marginal crevasses are the consequence. It may be urged that it is not fair to distribute the strain over the entire length of the curve : but reduce the distance as we may, a residue must remain which is demonstrative of the non-viscosity of the ice.

To sum up, then, two classes of facts present themselves to the glacier investigator—one class in harmony with the idea of viscosity, and another as distinctly opposed to it. Where *pressure* comes into play we have the former, where *tension* comes into play we have the latter. Both classes of facts are reconciled by the assumption, or rather the experimental verity, that the fragility of ice and its power of regelation render it possible for it to change its form without prejudice to its continuity.

---

## NOTE ON THE REGELATION OF SNOW-GRANULES.*

I this morning (March 21, 1862) noticed an extremely interesting case of regelation. A layer of snow, between one and two inches thick, had fallen on the glass roof of a small green-house, into which a door opened from the mansion to which the green-house was attached. Air, slightly warmed, acting on the glass surface underneath, melted the snow in immediate contact

* Phil. Mag. 1862, vol. xxiii. p. 312.

with the glass, and the layer in consequence slid slowly down the glass roof. The inclination of the roof was very gentle, and the motion correspondingly gradual. When the layer overshot the edge of the roof, it did not drop off, but bent like a flexible body and hung down over the edge for several inches. The continuity of the layer was broken into rectangular spaces by the inclined longitudinal sashes of the roof, and from local circumstances one side of the roof was warmed a little more than the other : hence the subdivisions of the layer moved with different velocities, and overhung the edge to different depths. The bent and down-hanging layer of snow in some cases actually curved up inwards.

Faraday has shown that when small fragments of ice float on water, if two of them touch each other, they instantly cement themselves at the point of contact; and on causing a row of fragments to touch, by laying hold of the terminal piece of the row, you can draw all the others after it. A similar cementing must have taken place among the particles of snow now in question, which were immersed in the water of liquefaction near the surface of the glass. But Faraday has also shown that when two fragments of ice are thus united, a hinge-like motion sets in when you try to separate the one from the other by a lateral push: one fragment might, in fact, be caused to roll round another, like a wheel, by the incessant rupture, and re-establishment of regelation.

The power of motion thus experimentally demonstrated, rendered it an easy possibility for the snow in question to bend itself in the manner observed. The lowermost granules, when the support of the roof had been withdrawn, rolled over each other without a destruction of continuity, and thus enabled the snow-layer to bend as if it were viscous. The curling up was evidently due to a contraction of the inner surface of the layer, produced, no doubt, by the accommodation of the granules to each other, as they slowly diminished in size.

J. T.

# LECTURE VII.

[March 6, 1862.]

CONDUCTION A TRANSMISSION OF MOTION—GOOD CONDUCTORS AND
BAD CONDUCTORS—CONDUCTIVITY OF THE METALS FOR HEAT:
RELATION BETWEEN THE CONDUCTIVITY OF HEAT AND THAT OF
ELECTRICITY—INFLUENCE OF TEMPERATURE ON THE CONDUCTION
OF ELECTRICITY—INFLUENCE OF MOLECULAR CONSTITUTION ON THE
CONDUCTION OF HEAT—RELATION OF SPECIFIC HEAT TO CONDUCTION
—PHILOSOPHY OF CLOTHES : RUMFORD'S EXPERIMENTS—INFLUENCE
OF MECHANICAL TEXTURE ON CONDUCTION—INCRUSTATIONS OF
BOILERS—THE SAFETY LAMP—CONDUCTIVITY OF LIQUIDS AND GASES :
EXPERIMENTS OF RUMFORD AND DESPRETZ—COOLING EFFECT OF
HYDROGEN GAS—EXPERIMENTS OF MAGNUS ON THE CONDUCTIVITY
OF GASES.

I THINK we are now sufficiently conversant with our
subject to distinguish between the sensible motions
produced by heat, and heat itself. Heat is not the clash
of winds; it is not the quiver of a flame, nor the ebullition of water, nor the rising of a thermometric column,
nor the motion which animates steam as it rushes from a
boiler in which it has been compressed. All these are
mechanical motions into which the motion of heat may be
converted; but heat itself is *molecular* motion—it is an
oscillation of ultimate particles. But such particles, when
closely grouped, cannot oscillate without communication
of motion from one to the other. To this propagation of
the motion of heat, through ordinary matter, we must this
day devote our attention.

Here is a poker, the temperature of which I am scarcely
conscious of: I feel it as a hard and heavy body, but it

neither warms me nor chills me; it has been before the
fire, and the motion of its particles at the present moment
chances to be the same as that which actuates my nerves;
there is neither communication nor withdrawal, and hence
the temperature of the poker on the one hand, and my
sensations on the other, remain unchanged. But I thrust
the end of the poker into the fire; it is heated; the particles in contact with the fire are thrown into a state of
more intense oscillation; the swinging atoms strike their
neighbours, these again theirs, and thus the molecular
music rings along the bar. The motion, in this instance,
is communicated from particle to particle of the poker,
and finally appears at its most distant end. If I now lay
hold of the poker, its motion is communicated to my
nerves, and produces pain; the bar is what we call hot,
and my hand, in popular language, is burned. Convection we have already defined to be the transfer of heat, by
sensible masses, from place to place; but this molecular
transfer, which consists in each *atom* taking up the motion
of its neighbours, and sending it on to others, is called the
*conduction* of heat.

Let me exemplify this property of conduction in a
homely way. I have here a basin filled with warm water,
and in the water I place this cylinder of iron, an inch in

FIG. 61.

diameter, and two inches in height; this cylinder is to be
my source of heat. I lay my thermo-electric pile, *o* (fig. 61),
thus flat, with its naked face turned upwards, and on that
face I place a cylinder of copper, *c*, which now possesses the

temperature of this room.   We observe no deflection of the
galvanometer.   I now place my warm cylinder, $i$, having
first dried it, upon the cool cylinder, which is supported by
the pile.   The upper cylinder is not at more than a blood
heat; but you see that I have scarcely time to make this remark before the needle flies aside, indicating that the heat
has reached the face of the pile.   Thus the molecular
motion imparted to the iron cylinder by the warm water has
been communicated to the copper one, through which it has
been transmitted, in a few seconds, to the face of the pile.

Different bodies possess different powers of transmitting
molecular motion; in other words, of conducting heat.  Copper, which we have just used, possesses this power in a
very eminent degree.   I will now remove the copper,
allow the needle to return to 0°, and then lay upon the
face of the pile this cylinder of glass.   On the cylinder of
glass I place my iron cylinder, which has been re-heated
in the warm water.   There is, as yet, no motion of the
needle, and you would have to wait a long time to see it
move.   We have already waited thrice the time which
the copper required to transmit the heat, and you see the
needle continues motionless.   I place cylinders of wood,
chalk, stone, and fireclay, in succession on the pile, and
heat their upper ends in the same manner; but in the
time which we can devote to an experiment, not one of
these substances is competent to transmit the heat to
the pile.   The molecules of these substances are so hampered or entangled, that they are incompetent to pass
the motion freely from one to another.   The bodies are all
*bad conductors* of heat.   On the other hand, I place cylinders of zinc, iron, lead, bismuth, &c., in succession on the
pile; each of them, you see, has the power of transmitting
the motion of heat swiftly through its mass.   In comparison with the wood, stone, chalk, glass, and clay, they
are all *good conductors* of heat.

As a general rule, though it is not without its exceptions, the metals are the best conductors of heat.   But the metals differ notably among themselves as regards their powers of conduction.   In illustration of this I will compare copper and iron.   Here, behind me, are two bars, A B, A C (fig. 62),

placed end to end, with balls of wood attached by wax at equal distances from the place of junction.   Under the junction I place a spirit-lamp, which heats the ends of the bars ; the heat will be propagated right and left through both. This bar is iron, this one is copper ; the heat will travel to the greatest distance along the best conductor, liberating a greater number of its balls.

But for my present purpose I want a quicker experiment.   Here, then, are two plates of metal, the one of copper, the other of iron, which are united together, so as to form a long continuous plate, c I (fig. 63).   To it a handle is attached, which gives the whole instrument the shape of a T.   From c to the middle, the plate is copper, from I to the middle it is iron.   At c I have soldered a small bar of bismuth to the plate ; at I a similar bar ; and from both bars wires, g g, lead to the galvanometer.   I warm the junction I by placing my finger on it; an electric current is there generated, and you observe the deflection.   The red end of the needle moves towards you.   I withdraw my finger, and the needle sinks to 0°.   I now warm, 'n the same manner, the junction c ; the needle is deflected, but in the opposite

direction.   If I place a finger on each end, at the same
time, these currents neutralise each other, and we have no
deflection.    I now place a spirit-lamp, with a very small

FIG. 63.

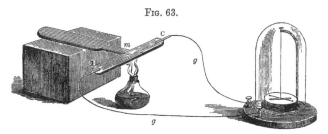

flame, directly under the middle of the compound plate;
the heat will propagate itself from the centre towards the
two ends, passing on one side through copper, and on the
other through iron.    If the heat reach both ends at the
same instant, the one end will neutralise the other, and the
needle will rest quiescent.   But if one end be reached
sooner than the other, we shall obtain a deflection, and
the direction in which the needle moves will declare which
end is heated.   Now for the experiment: I place the
lamp underneath, and in three seconds the needle flies
aside.   The red end moves towards me, which proves that
the end c is heated ; the molecular motion has propagated
itself most swiftly through the copper.   I allow the lamp
to remain until each metal has taken up as much heat as
it can appropriate, until the ends of the plates become
stationary in temperature; that is to say, until the quantity
of heat which they receive from the lamp is exactly equal
to the quantity dissipated in the space around them.   The
copper still asserts its predominance; the needle still indicates that the end c is most heated: and thus we prove
copper to be a better conductor of heat than iron.   This
little experiment illustrates how in natural philosophy we
turn one agent to account in the investigation of another.
Every new discovery is a new instrument: it was once

an end, but it is soon a means; and thus the growth of science is secured.

One of the first attempts to determine with accuracy the conductivity of different bodies for heat, was that suggested by Franklin, and carried out by Ingenhausz. He coated a number of bars of various substances with wax, and, immersing the ends of the bars in hot oil, he observed the distance to which the wax was melted on each of the bars. The good conductors melted the wax to the greatest distance; and the melting distance furnished a measure of the conductivity of the bar.

The second method was that pointed out by Fourier, and followed out experimentally by M. Despretz. A B (fig. 64) represents a bar of metal with holes drilled in it, intended to contain small thermometers. At the end of the bar was

FIG. 64.

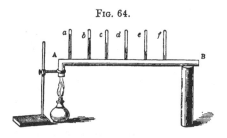

placed a lamp as a source of heat; the heat propagated itself through the bar, reaching the thermometer *a* first, *b* next, *c* next, and so on. For a certain time the thermometers continued to rise, but afterwards the state of the bar became stationary, each thermometer marking a constant temperature. The better the conduction, the smaller is the difference between any two successive thermometers. The decrement, or *fall* of heat, if I may use the term, from the hot end towards the cold, is greater in the bad conductors than in the good ones, and from the decrement of temperature shown by the thermometers we can

P

deduce, and express by a number, the conductivity of the bar.    This same method was followed by MM. Wiedemann and Franz, in a very important investigation, but instead of using thermometers they employed a suitable modification of the thermo-electric pile.   Of the numerous and highly interesting results of these experimenters the following is a résumé :—

| Name of Substance | Conductivity | |
|---|---|---|
|  | For Electricity | For Heat |
| Silver  .    .    .    . | 100 | 100 |
| Copper    .    .    . | 73 | 74 |
| Gold  .    .    .    . | 59 | 53 |
| Brass  .    .    .    . | 22 | 24 |
| Tin   .    .    .    . | 23 | 15 |
| Iron   .    .    .    . | 13 | 12 |
| Lead  .    .    .    . | 11 | 9 |
| Platinum   .    .    . | 10 | 8 |
| German Silver   .    . | 6 | 6 |
| Bismuth    .    .    . | 2 | 2 |

This   table shows, that, as regards their  conductive powers, the metals differ very widely from  each other. Calling, for example, the conductive power of silver 100, that of German silver is only 6.   You may illustrate this difference in a very simple way by plunging two spoons, one of German silver and the other of pure silver, into the same vessel of hot water.   After a little time you find the free end of the silver spoon much hotter than that of its neighbour ; and if bits of phosphorus be placed on the ends of the spoons, that on the silver will fuse and ignite in a very short time, while the heat transmitted through the other spoon will never reach an intensity sufficient to ignite the phosphorus.

Nothing is more interesting to the natural philosopher than the tracing out of connections and relationships between the various agencies of nature.   We know that they are a common brotherhood, we know that they are mutually convertible, but as yet we know very little as to the precise form of the conversion.  We have every reason to conclude that heat and electricity are both modes of motion; we know experimentally that from electricity we can get heat, and from heat, as in the case of our thermo-electric pile, we can get electricity.  But although we have, or think we have, tolerably clear ideas of the character of the motion of heat, our ideas are very unclear as to the precise nature of the change which this motion must undergo, in order to appear as electricity—in fact, we know as yet nothing about it.

Our table, however, exhibits one important connection between heat and electricity.  Beside the numbers expressing conductivity for heat, MM. Wiedemann and Franz have placed the numbers expressing the conductivity of the same metals for electricity.  They run side by side: the good conductor of heat is the good conductor of electricity, and the bad conductor of heat is the bad conductor of electricity.*  Thus we may infer, that the physical quality which interferes with the transmission of heat, interferes, in a proportionate degree, with the transmission of electricity.  This common susceptibility of both forces indicates a relationship which future investigations will no doubt clear up.

Let me point out another evidence of communion between heat and electricity.  I have here a length of wire made up of pieces of two different kinds of wire; there are three pieces of platinum, each four or five inches long, and three pieces of silver of the same length and thickness.  It is a proved fact, that the amount of heat developed in a wire by a current of electricity of a certain strength, is directly proportional to the resistance of the wire.†  We may figure the atoms as throwing themselves

* Professor Forbes had previously noticed this.
† Joule, Phil. Mag. 1841, vol. xix. p. 263.

P 2

as barriers across the track of the electric current—the
current knocking against them, and imparting its motion
to them, and rendering the wire hot. In the case of the
good conductor, on the contrary, the current may be
figured as gliding freely round the atoms without disturbing them in any great degree. I will now send the
self-same current from a battery of twenty of Grove's cells
through this compound wire. You see three spaces white-
hot, and three dark spaces between them. The white-hot
portions of the wire are platinum, and the dark portions
are silver. The electric current breaks impetuously upon
the molecules of the platinum, while it glides with little
resistance among the atoms of silver, thus producing, in
the metals, different calorific effects.*

Now I wish to show you that the motion of heat
interferes with the motion of electricity. You are acquainted with the little platinum lamp which stands in
front of the table. It consists simply of a little coil of
platinum wire suitably attached to a brass stand. I can
send a current through that coil and cause it to glow.
But you see I have introduced into the circuit two
feet additional of thin platinum wire, and on establishing
the connection, the same current passes through this wire
and the coil. Both, you see, are raised to redness—both are
in a state of intense molecular motion. What I wish now
to prove is, that this motion of heat, which the electricity
has generated in these two feet of wire, and in virtue of
which the wire glows, offers a hindrance to the passage of
the current. The electricity has raised up a foe in its own
path. I will cool this wire, and thereby cause the heat to
subside. I shall thus open a wider door for the passage
of the electricity. But if more electricity passes, it will
announce itself at the platinum lamp; it will raise that

* May not the condensed ether which surrounds the atoms be the
vehicle of electric currents?

red heat to whiteness, and the change in the intensity of the light will be visible to you all.

Fig. 65.

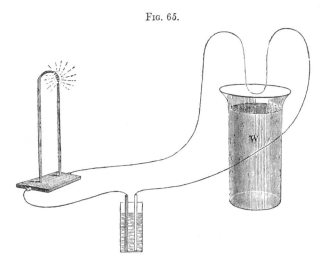

Thus, then, I plunge my red-hot wire into a beaker of water w (fig. 65): observe the lamp, it becomes almost too bright to look at. I raise the wire out of the water and allow the motion of heat once more to develope itself; the motion of electricity is instantly impeded, and the lamp sinks in brightness. I again dip the wire into the cold water, deeper and deeper: observe how the light becomes intensified—deeper still, so as to quench the entire two feet of wire; the augmented current raises the lamp to its maximum brightness, and now it suddenly goes out. The circuit is broken, for the coil has actually been fused by the additional flow of electricity.

Let us now devote a moment's time to the conduction of cold. To all appearance cold may be conducted like heat. Here is a copper cylinder, which I warm a little by holding it for a moment in my hand. I place it on the pile, and the needle goes up to 90°, declaring heat. On this cylinder I place a second one, which, as you observe, I have chilled by sinking it for some time in this mass of ice. We wait a moment, the needle moves: it is now descending to zero, passes it, and goes on to 90° on the side of cold. Analogy might well lead you to suppose that the cold is conducted downwards from the top cylinder to the bottom one, as the heat was conducted in our former experiments. I have no objection to the term 'conduction of cold,' if it be used with a clear knowledge of the real physical process involved. The real process is, that the warm intermediate cylinder first delivers up its motion, or heat, to the cold cylinder overhead, and, having thus lost its own possession of heat, it draws upon that of the pile. In our former experiments we had conduction of motion *to* the pile; in our present one we have conduction of motion *from* the pile. In the former case the pile is heated, in the latter chilled; the heating produces a positive current, the chilling produces a negative current; but it is in both cases the propagation of motion with which we have to do, the heating and the chilling depending solely upon the *direction* of propagation. I place one of these metal cylinders, which I have purposely cooled, on the face of our pile; a violent deflection follows, declaring the chilling of the instrument. Are we to suppose the cold to be an entity communicated to the pile? No. The pile here is the warm body; its molecular motion is in excess of that possessed by the cylinder; and when both come into contact the pile seeks to make good the defect. It imparts a quantity of its own motion to the cylinder, and by its bounty becomes impoverished: it chills itself, and generates the current due to cold.

I remove the cold metal cylinder, and place upon the pile a cylinder of wood, having the same temperature as the metal one. The chill is very feeble, and the consequent deflection very small. Why does not the cold wood produce an action equal to that of the cold metal? Simply because the heat communicated to it by the pile is accumulated at its under surface; it cannot escape through the bad conducting wood as it escapes through the metal, and thus the quantity of heat withdrawn from the pile, by the wood, is less than that withdrawn by the copper. A similar effect is produced when the human nerves are substituted for the pile. Suppose you come into a cold room and lay your hand upon the fire-irons, the chimney-piece, the chairs, the carpet, in succession; they appear to you of different temperatures: the iron chills you more than the marble, the marble more than the wood, and so on. Your hand is affected exactly as the pile was affected in the last experiment. It is needless to say that the reverse takes place when you enter a hot room; that is to say, a room hotter than your own bodies. I should certainly suffer if I were to lie down upon a plate of metal in a Turkish bath; but I do not suffer when I lie down on a bench of wood. By preserving the body from contact with good conductors, very high temperatures may be endured. Eggs may be boiled and beefsteaks cooked, by the heat of an apartment in which the living bodies of men sustain no injury.

The exact philosophy of this last experiment is worthy of a moment's consideration. With it the names of Blagden and Chantrey are associated, those eminent men having exposed themselves, in ovens, to temperatures considerably higher than that of boiling water. Let us compare the condition of the two living human beings, with that of two marble statues placed in the same oven. The statues become gradually hotter, until finally they assume the temperature of the air of the oven; the two sculptors, under the same circumstances, do not similarly rise in temperature. If they did, the tissues of the body would be infallibly destroyed, the temperature which they

endured being more than sufficient to stew the muscles in
their own liquids. But the fact is, that the heat of the blood
is scarcely affected by an augmentation of the external heat.
This heat, instead of being applied to increase the temperature of the body, is applied to the performance of
work, in altering the aggregation of the body; it prepares
the perspiration, forces it through the pores, and in part
vaporises it. Heat is here converted into potential energy;
it is consumed in work. This is the waste-pipe, if I may
use the term, through which the excess of heat overflows;
and hence it is, that under the most varying conditions of
climate the temperature of the human blood is practically
constant. The blood of the Laplander is sensibly as
warm as that of the Hindoo; while an Englishman, in
sailing from the north pole to the south, finds his blood-
temperature hardly heightened by his approach to the
equator, and hardly diminished by his approach to the
antarctic pole.

When the communication of heat is gradual—as it
always is when the body is surrounded by an imperfect
conductor—the heat is consumed in the manner indicated
as fast as it is supplied; but if the supply of heat be so
quick (as it would be in the case of contact with a good
conductor) that the conversion into this harmless potential
energy cannot be executed with sufficient rapidity, the
injury of the tissues is the result. Some people have
professed to see in this power of the living body to resist
a high temperature, a conservative action peculiar to the
vital force. No doubt all the actions of the animal organ-
ism are connected with what we call its vitality; but the
action here referred to is the same in kind as the melting
of ice, or the vaporisation of water. It consists simply in
the diversion of heat from the purposes of temperature to
the performance of work.

Thus far we have compared the conducting power of

different bodies together; but the same substance may possess different powers of conduction in different directions.   Many crystals are so built that the motion of heat runs with greater facility along certain lines of atoms than along others.   Here, for instance, is a large rock-crystal — a crystal of quartz forming an hexagonal pillar, which, if complete, would be terminated by two six-sided pyramids.   Heat travels with greater facility along the axis of this crystal than across it.   This has been proved in a very simple manner by M. de Senarmont.   I have here two plates of quartz, one of which is cut parallel to the axis of the crystal, and the other perpendicular to it. I coat the plates with a layer of white wax, laid on by a camel's-hair pencil.   The plates are pierced at the centre, and into the hole I insert a wire, which I warm by an

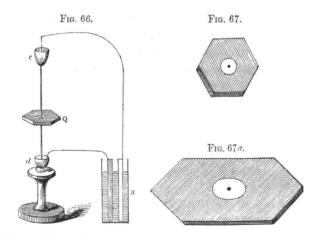

FIG. 66.

FIG. 67.

FIG. 67<i>a</i>.

electric current.  B (fig. 66) is the battery whence the current proceeds; <i>c</i> is a capsule of wood, through the bottom of which a sewing-needle passes; <i>d</i> is a second capsule, into which dips the point of the needle, and Q is the perforated plate of quartz.   Each capsule contains a drop

of mercury. When the current passes from $c$ to $d$, the
needle is heated, and the heat is propagated in all directions. The wax melts around the place where the heat
is applied; and on this plate, which is cut perpendicular to the axis of the quartz, I find the figure of the
melted wax to be a perfect circle (fig. 67). The heat has
travelled with the same rapidity all round, and melted
the wax to the same distance in all directions. I make
a similar experiment with the other plate : the wax is now
melting; but I notice that its figure is no longer a circle.
The heat travels more speedily along the axis than across
it, and hence the wax figure is an ellipse instead of a
circle (fig. 67$a$). When the wax dries, I will project magnified images of these two plates upon the screen, and you will
then see the circular figure of the melted wax on the one,
and the oval figure of the wax on the other. Iceland spar
conducts better along the crystallographic axis than at
right angles to it, while a crystal of tourmaline conducts best at right angles to its axis. The metal bismuth,
with which you are already acquainted, cleaves with great
facility in one direction, and, as has been well shown
by MM. Svanberg and Matteucci, it conducts both heat
and electricity better along the planes of cleavage than
across them.

In wood we have an eminent example of this difference of
conductivity. Upwards of twenty years ago MM. De la Rive
and De Candolle instituted an enquiry into the conductive
power of wood,* and, in the case of five specimens examined,
established the fact that the velocity of transmission was
greater along the fibre than across it. The manner of
experiment was that usually adopted in enquiries of this
nature, and which was applied to metals by M. Despretz.†
A bar of the substance was taken, one end of which was

---

\* Mém. de la Soc. de Genève, vol. iv. p. 70.

† Annales de Chim. et de Phys. December 1827.

brought into contact with a source of heat, and allowed to remain so until a stationary temperature was assumed. The temperatures attained by the bar, at various distances from its heated end, were ascertained by means of thermometers fitting into cavities made to receive them; from these data, with the aid of a well-known formula, the conductivity of the wood was determined.

To determine the velocity of calorific transmission in different directions through wood, the instrument shown in fig. 68 was devised some years ago by myself. Q Q′ R R′ is an oblong piece of mahogany, A is a bar of antimony, B is a bar of bismuth. The united ends of the two bars are kept in close contact by the ivory jaws I I′, and the other ends are let into a second piece of ivory, in which they are firmly fixed. Soldered to these ends are two pieces of platinum wire, which proceed to the little ivory cups M M, enter through the sides of the cups, and communicate with a drop of mercury placed in the interior. The mahogany is cut away, so that the bars A and B are sunk to a depth which places their upper surfaces a little below the general level of the slab of mahogany. The ivory jaws I I′ are sunk similarly. Two small projections are observed in the figure jutting from I I′; across, from one projection to the other, a fine membrane is stretched, thus enclosing a little chamber $m$, in front of the wedge-like end of the bismuth and antimony junction; the chamber has an ivory bottom. S is a wooden slider, which can be moved smoothly back and forward along a bevelled groove, by means of the lever L. This lever turns on a pivot near Q, and fits into a horizontal slit in the slider, to which it is attached by the pin $p'$ passing through both; in the lever an oblong aperture is cut, through which $p'$ passes, and in which it has a certain amount of lateral play, so as to enable it to push the slider forward in a straight line. Two projections are seen at the end of the slider,

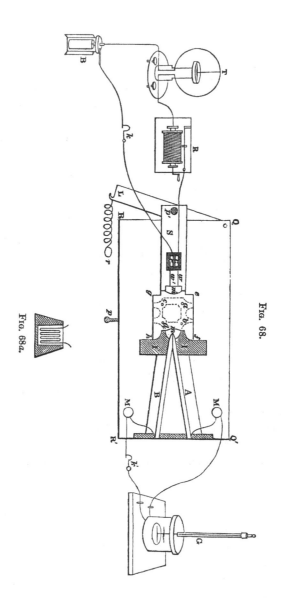

Fig. 68.

Fig. 68a.

and across, from projection to projection, a thin membrane is stretched ; a chamber $m'$ is thus formed, bounded on three sides and the bottom by wood, and in front by the membrane. A thin platinum wire, bent up and down several times, so as to form a kind of grating, is laid against the back of this chamber, and imbedded in the end of the slider by the stroke of a hammer ; the end in which the wire is imbedded is then filed down, until about half the wire is removed, and the whole is reduced to a uniform flat surface. Against the common surface of the slider and wire, an extremely thin plate of mica is glued, sufficient, simply, to interrupt all contact between the bent wire and a quantity of mercury which the chamber $m'$ is destined to contain ; the ends $w\ w'$ of the bent wire proceed to two small cisterns $c\ c'$, hollowed out in a slab of ivory ; the wires enter through the substance into the cisterns, and come thus into contact with mercury, which fills the latter. The end of the slider and its bent wire are shown in fig. 68$a$. The rectangular space $e f g h$ (fig. 68) is cut quite through the slab of mahogany, and a brass plate is screwed to the latter underneath ; from this plate (which, for reasons to be explained presently, is cut away, as shown by the dotted lines in the figure) four conical ivory pillars $a\ b\ c\ d$ project upwards ; though appearing to be upon the same plane as the upper surfaces of the bismuth and antimony bars, the points are in reality 0·3 of an inch below the said surfaces.

The body to be examined is reduced to the shape of a cube, and is placed, by means of a pair of pliers, upon the four supports $a\ b\ c\ d$ ; the slider s is then drawn up against the cube, and the latter becomes firmly clasped between the projections of the piece of ivory $\mathrm{I}\ \mathrm{I}'$ on the one side, and those of the slider s on the other. The chambers $m\ m'$ being filled with mercury, the membrane in front of each is pressed gently against the cube by the interior

fluid mass, and in this way perfect contact, which is absolutely essential, is secured.

The problem which requires solution is the following:—
It is required to apply a source of heat of a strictly
measurable character, and always readily attainable, to
that face of the cube which is in contact with the membrane at the end of the slider, and to determine the
quantity of this heat which crosses the cube to the opposite face, in a minute of time.  For the solution of this
problem, two things are required — first, the source of
heat to be applied to the left hand of the face of the cube,
and secondly, a means of measuring the amount which has
made its appearance at the opposite face at the expiration
of a minute.

To obtain a source of heat of the nature described,
the following method was adopted: — B is a small
galvanic battery, from which a current proceeds to the
tangent galvanometer T; passes round the ring of the
instrument, deflecting in its passage the magnetic needle,
which hangs in the centre of the ring.  From T the
current proceeds to the rheostat R; this instrument consists of a cylinder of serpentine stone, round which a
German silver wire is coiled spirally; by turning the
handle of the instrument, any required quantity of this
powerfully resisting wire is thrown into the circuit, the
current being thus regulated at pleasure.  The sole use of
these two last instruments, in the present series of experiments, is to keep the current perfectly constant from day
to day.  From the rheostat the current proceeds to the
cistern c, thence through the bent wire, and back to the
cistern c', from which it proceeds to the other pole of the
battery.

The bent wire, during the passage of the current,
becomes gently heated; this heat is transmitted through
the mercury in the chamber m' to the membrane in front

of the chamber; this membrane becomes the proximate source of heat which is applied to the left-hand face of the cube. The quantity of heat transmitted from this source, through the mass of the cube, to the opposite face, in any given time, is estimated from the deflection which it is able to produce upon the needle of a galvanometer, connected with the bismuth and antimony pair. G is a galvanometer used for this purpose; from it proceed wires to the mercury cups M M, which, as before remarked, are connected by platinum wires with A and B.

The action of mercury upon bismuth, as a solvent, is well known; an amalgam is speedily formed when the two metals come into contact. To preserve the thermo-electric couple from this action, their ends are protected by a sheathing of the same membrane as that used in front of the chambers $m\,m'$.

Previous to the cube's being placed between the two membranes, the latter, by virtue of the fluid masses behind them, bulge out a little, thus forming a pair of soft and slightly convex cushions. When the cube is placed on its supports, and the slider is brought up against it, both cushions are pressed flat, and thus make the contact perfect. The surface of the cube is larger than the surface of the membrane; * and thus the former is always firmly caught between the opposed rigid projections, the slider being held fast in this position by means of the spring $r$, which is then attached to the pin $p$. The exact manner of experiment is as follows:—Having first seen that the needle of the galvanometer points to zero, when the thermo-circuit is complete, the latter is interrupted by means of the break-circuit key $k'$. At a certain moment, marked by the second-hand of a watch, the voltaic circuit is closed by the key $k$, and the current is permitted to circulate for sixty seconds; at the sixtieth

* The edge of each cube measured 0·3 inch.

second the voltaic circuit is broken by the left hand at $k$,
while, at the same instant, the thermo-circuit is closed
by the right hand at $k'$.  The needle of the galvanometer is instantly deflected, and the limit of the first
impulsion is noted; the amount of this impulsion depends,
of course, upon the quantity of heat which has reached
the bismuth and antimony junction through the mass of
the cube, during the time of action.  The limit of the
first impulsion being noted, the cube is removed, and
the instrument is allowed to cool, until the needle of the
galvanometer returns to zero.  Another cube being introduced, the voltaic circuit is once more closed, the current
permitted to circulate sixty seconds, then interrupted by
the left hand, the thermo-circuit being closed at the same
moment with the right, and the limit of the first swing is
noted as before.

Judging from the description, the mode of experiment
may appear complicated, but in reality it is not so.  A
single experimenter has the most complete command over
the entire arrangement.  The wires from the small galvanic battery (a single cell) remain undisturbed from
day to day; all that is to be done is to connect the
battery with them, and everything is ready for experiment.

There are in wood three lines, at right angles to each
other, which the mere inspection of the substance enables
us to fix upon as the necessary resultants of molecular
action: the first line is parallel to the fibre; the second is
perpendicular to the fibre, and to the ligneous layers which
indicate the annual growth of the tree; while the third is
perpendicular to the fibre, and parallel, or rather tangential,
to the layers.  From each of a number of trees a cube
was cut, two of whose faces were parallel to the ligneous
layers, two perpendicular to them, while the remaining
two were perpendicular to the fibre.  It was proposed to

examine the velocity of calorific transmission through the wood in these three directions. It may be remarked that the cubes were fair average specimens of the woods, and were in all cases well-seasoned and dry.

The cube was first placed upon its four supports $a\,b\,c\,d$, so that the line of flux from $m'$ to $m$ was parallel to the fibre, and the deflection produced by the heat transmitted in sixty seconds was observed. The position of the cube was then changed, so that its fibre stood vertical, the line of flux from $m'$ to $m$ being perpendicular to the fibre, and parallel to the ligneous layers; the deflection produced by a minute's action in this case was also determined. Finally, the cube was turned 90° round, its fibre being still vertical, so that the line of flux was perpendicular to both fibre and layers, and the consequent deflection was observed. In the comparison of these two latter directions the chief delicacy of manipulation is necessary. It requires but a rough experiment to demonstrate the superior velocity of propagation along the fibre, but the velocities in all directions perpendicular to the fibre are so nearly equal that it is only by great care, and, in the majority of cases, by numerous experiments, that a difference of action can be securely established.

The following table contains some of the results of the enquiry; it will explain itself:—

| Description of Wood | DEFLECTIONS | | |
|---|---|---|---|
| | I. Parallel to fibre | II. Perpendicular to fibre and parallel to ligneous layers | III. Perpendicular to fibre and to ligneous layers |
| | ° | ° | ° |
| 1 American Birch . . | 35 | 9·0 | 11.0 |
| 2 Oak . . . . . | 34 | 9·5 | 11·0 |
| 3 Beech . . . . | 33 | 8·8 | 10·8 |
| 4 Coromandel-wood . . | 33 | 9·8 | 12·3 |
| 5 Bird's-eye Maple . . | 31 | 11·0 | 12·0 |
| 6 Lance-wood . . . | 31 | 10·6 | 12·1 |
| 7 Box-wood . . . | 31 | 9·9 | 12·0 |
| 8 Teak-wood . . . | 31 | 9·9 | 12·4 |
| 9 Rose-wood . . . | 31 | 10·4 | 12·6 |
| 10 Peruvian-wood . . . | 30 | 10·7 | 11·7 |
| 11 Green-heart . . . | 29 | 11·4 | 12·6 |
| 12 Walnut . . . . | 28 | 11·0 | 13·0 |
| 13 Drooping Ash . . . | 28 | 11·0 | 12·0 |
| 14 Cocoa-wood . . . | 28 | 11·9 | 13·6 |
| 15 Sandal-wood . . . | 28 | 10·0 | 11·7 |
| 16 Tulip-wood . . . | 28 | 11·0 | 12·1 |
| 17 Camphor-wood . . . | 28 | 8·6 | 10·0 |
| 18 Olive-tree . . . | 28 | 10·5 | 13·2 |
| 19 Ash . . . . . | 27 | 9·5 | 11·5 |
| 20 Black Oak . . . | 27 | 8·0 | 9·4 |
| 21 Apple-tree . . . | 26 | 10·0 | 12·5 |
| 22 Iron-wood . . . | 26 | 10·2 | 12·4 |
| 23 Chestnut . . . . | 26 | 10·1 | 11·5 |
| 24 Sycamore . . . . | 26 | 10·6 | 12·2 |
| 25 Honduras Mahogany . | 25 | 9·0 | 10·0 |
| 26 Brazil-wood . . . | 25 | 11·9 | 13·9 |
| 27 Yew . . . . | 24 | 11·0 | 12·0 |
| 28 Elm . . . . . | 24 | 10·0 | 11·5 |
| 29 Plane-tree . . . | 24 | 10·0 | 12·0 |
| 30 Portugal Laurel . . | 24 | 10·0 | 11·5 |
| 31 Spanish Mahogany . . | 23 | 11·5 | 12·5 |
| 32 Scotch Fir . . . | 22 | 10·0 | 12·0 |

The above table furnishes us with a corroboration of the result arrived at by De la Rive and De Candolle, regarding the superior conductivity of the wood in the direction of the fibre. Evidence is also afforded as to how little mere density affects the velocity of transmission. There appears to be neither law nor general rule here. American Birch, a comparatively light wood, possesses undoubtedly a higher transmissive power than any other in the list. Iron-wood, on the contrary, with a specific gravity of 1·426, stands low. Again, Oak and Coromandel-wood—the latter so hard and dense that it is used for sharp war-instruments by savage tribes—stand near the head of the list, while Scotch Fir and other light woods stand low.

If we cast our eyes along the second and third columns of the table, we shall find that in every instance the velocity of propagation is greatest in a direction perpendicular to the ligneous layers. The law of molecular action, as regards the transmission of heat through wood, may therefore be expressed as follows :—

*At all the points not situate in the centre of the tree, wood possesses three unequal axes of calorific conduction, which are at right angles to each other. The first, and principal axis, is parallel to the fibre of the wood; the second, and intermediate axis, is perpendicular to the fibre and to the ligneous layers; while the third and least axis is perpendicular to the fibre and parallel to the layers.*

MM. De la Rive and De Candolle have remarked upon the influence which its feeble conducting power in a lateral direction must exert in preserving within a tree the warmth which it acquires from the soil. In virtue of this property a tree is able to resist sudden changes of temperature which would probably be prejudicial to it : it resists alike the sudden abstraction of heat from within and the sudden accession of it from without. But Nature has gone further, and clothes the tree with a sheathing of worse-conducting material than the wood itself, even in its worst direction. The following are the deflections obtained by submitting a number of cubes of bark, of the

same size as the cubes of wood, to the same conditions of
experiment:—

|  | Deflection | Corresponding deflection produced by the wood |
|---|---|---|
| Beech-tree Bark . . . . | 7° | 10·8° |
| Oak-tree Bark . . . . | 7 | 11·0 |
| Elm-tree Bark . . . . | 7 | 11·5 |
| Pine-tree Bark . . . . | 7 | 12·0 |

The direction of transmission, in these cases, was from
the interior surface of the bark outwards.

The average deflection produced by a cube of wood,
when the flux is lateral, may be taken at

$$12°;$$

a cube of rock-crystal (pure silica), of the same size, produces a deflection of

$$90°.$$

Two bodies so diverse, where they cover any considerable
portion of the earth's surface, must affect the climate very
differently. There are the strongest experimental grounds
for believing that rock-crystal possesses a higher conductive
power than some of the metals.

The following numbers express the transmissive power
of a few other organic structures: cubes of the substances
were examined in the usual manner:—

| | |
|---|---|
| Tooth of Walrus . . . . | 16 |
| Tusk of East-Indian Elephant . . | 17 |
| Whalebone . . . . . | 9 |
| Rhinoceros'-horn . . . . | 9 |
| Cow's-horn . . . . . | 9 |

Sudden changes of temperature are prejudicial to animal
and vegetable health; the substances used in the construction of organic tissues are exactly such as are best
calculated to resist those changes.

The following results further illustrate this point. Each

of the substances mentioned was reduced to the cubical form, and submitted to an examination similar in every respect to that of wood and quartz.  While, however, a cube of the latter substance produces a deflection of 90°, a cube of

| | |
|---|---|
| Sealing-wax produces a deflection of . . | 0° |
| Sole leather . . . . . . . | 0 |
| Bees'-wax . . . . . . . | 0 |
| Glue . . . . . . . . | 0 |
| Gutta-percha . . . . . . . | 0 |
| India-rubber . . . . . . . | 0 |
| Filbert-kernel . . . . . . | 0 |
| Almond-kernel . . . . . . | 0 |
| Boiled ham-muscle . . . . . | 0 |
| Raw veal-muscle . . . . . . | 0 |

The substances here named are animal and vegetable productions; and the experiments demonstrate the extreme imperviousness of every one of them.  Starting from the principle that sudden accessions or deprivations, of heat are prejudicial to animal and vegetable health, we see that the materials chosen are precisely those which are best calculated to avert such changes.

I wish now to direct your attention to what may, at first sight, appear to you a paradoxical experiment.  Here is a short prism of bismuth, and here another of iron, of the same size.  I coat the ends of both prisms with white wax, and then place them, with their coated surfaces upwards, on the lid of this vessel, which contains hot water.  The motion of heat will propagate itself through the prisms, and you are to observe the melting of the wax.  It is already beginning to yield, but on which ?  On the bismuth.  And now the white has entirely disappeared from the bismuth, the wax overspreads it in a transparent liquid layer, while the wax on the iron is not yet melted.  How is this result to be reconciled with the fact stated in our table (page 211), that, the conduction of iron being 12, the conduction of bismuth is only 2 ?  In this experiment the bismuth

seems to be the best conductor. We solve this enigma by
turning to our table of specific heats (Lecture V.); we there
find that, the specific heat of iron being 1138, that of bismuth is only 308; to raise it, therefore, a certain number
of degrees in temperature, iron requires more than three
times the absolute quantity of heat required by bismuth.
Thus, though the iron is really a much better conductor
than the bismuth, and is at this moment accepting, in
every unit of time, a much greater amount of heat than the
bismuth, still, in consequence of the number of its atoms,
or the magnitude of its interior work, the augmentation
of temperature, in the case of iron, is slow. Bismuth,
on the contrary, can immediately devote a large proportion of the heat imparted to it to the augmentation of
temperature; and thus it apparently outstrips the iron in
the transmission of that motion to which temperature is
due.

You see here very plainly the incorrectness of the statements sometimes made in books, and certainly made
very frequently by candidates in our science examinations,
regarding the experiment of Ingenhausz, to which I have
already referred. It is usually stated, that the greater the
*quickness* with which the wax melts, the better is the
conductor. If the bad conductor and the good conductor
have the same specific heat, this is true, but in other cases,
as proved by our last experiment, it may be entirely incorrect. The proper way of proceeding, as already indicated, is to wait until both the iron and the bismuth have
attained a constant temperature—till each of them, in fact,
has accepted, and is transmitting, all the motion which
it can accept, or transmit, from the source of heat;
when this is done, it is found that the quantity transmitted
by the iron is six times greater than that transmitted by
the bismuth. You remember our experiments with the
Trevelyan instrument, and know the utility of having a

highly expansible body as the bearer of the rocker. Lead is good, because it is thus expansible. But the coefficient of expansion of zinc is slightly higher than that of lead; still zinc does not answer well as a block. The reason is, the specific heat of zinc is more than three times that of lead, so that the heat communicated to the zinc by the contact of the rocker, produces only about one-third the augmentation of temperature, and a correspondingly small amount of local expansion.

These considerations also show that in our experiments on wood the quantity of heat transmitted by our cube in one minute's time, cannot, in strictness, be regarded as the expression of the conductivity of the wood, unless the specific heat of the various woods be the same. On this point no experiments have been made. But as regards the influence of molecular structure, the experiments hold good, for here we compare one direction with another, *in the same cube*. With respect to organic structures, I may add that, even allowing them time to accept all the motion which they are capable of accepting from a source of heat, their power of transmitting that motion is exceedingly low. They are really bad conductors.

It is the imperfect conductibility of woollen textures which renders them so eminently fit for clothing. They preserve the body from sudden accessions or losses of heat. The same quality of non-conductibility manifests itself when we wrap flannel round a block of ice. The ice thus preserved is not easily melted. In the case of a human body on a cold day, the woollen clothing prevents the transmission of motion from within outwards; in the case of the ice on a warm day, the self-same fabric prevents the transmission of motion from without inwards. Animals which inhabit cold climates are furnished by Nature with their necessary clothing. Birds especially need this protection, for they are still more warm-blooded than the mammalia. They are furnished with feathers, and between the feathers the interstices are filled with down, the molecular constitution and mechanical texture of which render it, perhaps, the worst of all conductors. Here we have another example of that harmonious relation of life to the conditions of life, which is incessantly presented to the student of natural science.

The indefatigable Rumford made an elaborate series of experiments on the conductivity of the substances used in clothing.* His method was this :—A mercurial thermometer was suspended in the axis of a cylindrical glass tube ending with a globe, in such a manner that the centre of the bulb of the thermometer occupied the centre of the globe; the space between the internal surface of the globe and the bulb was filled with the substance whose conductive power was to be determined; the instrument was then heated in boiling water, and afterwards, being plunged into a freezing mixture of pounded ice and salt, the times of cooling down 135° Fahr. were noted. They are recorded in the following table :—

| Surrounded with | | | | Seconds |
|---|---|---|---|---|
| Twisted silk | . | . | . | 917 |
| Fine lint . | . | . | . | 1032 |
| Cotton wool | . | . | . | 1046 |
| Sheep's wool | . | . | . | 1118 |
| Taffety | . | . | . | 1169 |
| Raw silk . | . | . | . | 1264 |
| Beavers' fur | . | . | . | 1296 |
| Eider down | . | . | . | 1305 |
| Hares' fur . | . | . | . | 1312 |
| Wood ashes | . | . | . | 927 |
| Charcoal . | . | . | . | 937 |
| Lamp-black. | . | . | . | 1117 |

Among the substances here examined, hares' fur offered the greatest impediment to the transmission of the heat.

* Phil. Trans. 1792, p. 48.

The transmission of heat is powerfully influenced by the mechanical state of the body through which it passes. The raw and twisted silk of Rumford's table illustrate this. Pure silica, in the state of hard rock-crystal, is a better conductor than bismuth or lead; but if the crystal be reduced to powder, the propagation of heat through that powder is exceedingly slow. Through transparent rock-salt heat is copiously conducted, through common table-salt very feebly. I have here some asbestos, which is composed of certain silicates in a fibrous condition; I place it on my hand, and on it I place a red-hot iron ball: you see I can support the ball without inconvenience. The asbestos intercepts the heat. That this division of the substance should interfere with the transmission might reasonably be inferred; for, heat being motion, anything which disturbs the continuity of the molecular chain, along which the motion is conveyed, must affect the transmission. In the case of the asbestos the fibres of the silicates are separated from each other by spaces of air; to propagate itself, therefore, the motion has to pass from the silicate to the air, a very light body, and again from the air to the silicate, a comparatively heavy body; and it is easy to see that the transmission of motion through this composite texture must be very imperfect. In the case of an animal's fur, this is more especially the case; for here not only do spaces of air intervene between the hairs, but the hairs themselves, unlike the fibres of the asbestos, are very bad conductors. Lava has been known to flow over a layer of ashes underneath which was a bed of ice, and the non-conductivity of the ashes has saved the ice from fusion. Red-hot cannon-balls may be wheeled to the gun's mouth in wooden barrows partially filled with sand. Ice is packed in saw-dust to prevent it from melting; powdered charcoal is also an eminently bad conductor. But there are cases where sawdust, chaff, or charcoal could not be used with safety,

on account of their combustible nature. In such cases,
powdered gypsum may be used with advantage; in the
solid crystalline state it is incomparably a worse conductor
than silica, and it may be safely inferred, that in the
powdered state its imperviousness far transcends that of
sand, each grain of which is a good conductor. A jacket
of gypsum powder round a steam boiler would materially
lessen its loss of heat.

Water usually holds certain minerals in solution. In
percolating through the earth, it dissolves more or less of
the substances with which it comes into contact. For
example, in chalk districts the water always contains a
quantity of carbonate of lime; such water is called *hard
water*. Sulphate of lime is also a common ingredient of
water. In evaporating, the water only is driven off, the
mineral is left behind, and often in quantities too great
to be held in solution by the water. Many springs are
strongly impregnated with carbonate of lime, and the
consequence is, that when the waters of such springs reach
the surface and are exposed to the air, where they can
partially evaporate, the mineral is precipitated, and forms
incrustations on the surfaces of plants and stones, over
which the water trickles. In the boiling of water the
same occurs; the minerals are precipitated, and there is
scarcely a kettle in London which is not internally coated
with a mineral incrustation. This is an extremely serious difficulty as regards steam boilers; the crust is a
bad conductor, and it may become so thick as materially
to intercept the passage of heat to the water. I have here
an example of this mischief. This is a portion of a boiler
belonging to a steamer, which was all but lost through the
exhaustion of her coals: to bring this vessel into port her
spars and every piece of available wood were burnt. On
examination this formidable incrustation was found within
the boiler: it is mainly carbonate of lime, which by its

non-conducting power rendered a prodigal expenditure of fuel necessary to generate the required quantity of steam. Doubtless the slowness of many kettles in boiling would be found due to a similar cause.

I wish now to bring before you one or two instances of the action of good conductors in preventing the local accumulation of heat. I have here two spheres of the same size, both covered closely with white paper. One of them is copper, the other is wood. I place a spirit lamp underneath each of them, and after a time we will observe the effect. The motion of heat is, of course, communicating itself to each ball, but in one it is quickly conducted away from the place of contact with the flame, through the entire mass of the ball; in the other this quick conduction does not take place, the motion therefore accumulates at the point where the flame plays upon the ball; and here you have the result. I turn up the wooden ball, the white paper is quite charred; I turn up the other ball,—so far from being charred, it is *wet* at its under surface by the condensation of the aqueous vapour generated by the lamp. Here is a cylinder covered closely with paper; I hold its centre thus over the lamp, turning it so that the flame shall play all round the cylinder: you see a well-defined black mark, on one side of which the paper is charred, on the other side not. The cylinder is half brass and half wood, and this black mark shows their line of junction: where the paper covers the wood, it is charred; where it covers the brass, it is not sensibly affected.

If the entire moving force of a common rifle bullet were communicated to a heavy cannon-ball, it would produce in the latter a very small amount of motion. Supposing the rifle bullet to weigh two ounces, and to have a velocity of 1,600 feet a second, the moving force of this bullet communicated to a 100 lb. cannon-ball would impart to

the latter a velocity of only 32 feet a second.  Thus
with regard to a flame; its molecular motion is very
intense, but its weight is extremely small, and if communicated to a heavy body, the intensity of the motion must
fall.  For example, I have here a sheet of wire gauze, with
meshes wide enough to allow air to pass through them
with the utmost freedom; and here is a jet of gas burning brilliantly.  I bring down the wire gauze upon the
flame; you would imagine that the flame could readily
pass through the meshes of the gauze : but no, not a flicker
gets through (fig. 69).  The combustion is entirely confined

FIG. 69.                              FIG. 70.

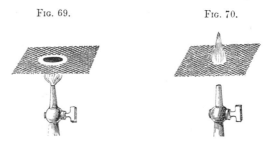

to the space under the gauze.   I extinguish the flame, and
allow the unignited gas to stream from the burner.   I
place the wire gauze thus above the burner: the gas, I
know, is now freely passing through the meshes.   I ignite
the gas above; there you have the flame, but it does not
propagate itself downwards to the burner (fig. 70).   You
see a dark space of four inches between the burner and the
gauze, a space filled with gas in a condition eminently
favourable to ignition, but still it does not ignite.   Thus,
you see, this metallic gauze, which allows the gas to pass
freely through, intercepts the flame.   And why?   A certain heat is necessary to cause the gas to ignite; but by
placing the wire gauze over the flame, or the flame over
the wire gauze, you transfer the motion of that light
and quivering thing to the comparatively heavy gauze.

The intensity of the molecular motion is greatly lowered by being communicated to so great a mass of matter —so much lowered, indeed, that it is incompetent to propagate the combustion to the opposite side of the gauze.

We are all, unhappily, too well acquainted with the terrible accidents that occur through explosions in coal mines. You know that the cause of these explosions is the presence of a certain gas—a compound of carbon and hydrogen—generated in the coal strata. When this gas is mixed with a sufficient quantity of air, it explodes on ignition, the carbon of the gas uniting with the oxygen of the air, to produce carbonic acid; the hydrogen of the gas uniting with the oxygen of the air, to produce water. By the flame of the explosion the miners are burnt; but even should this not destroy life, they are often suffocated afterwards by the carbonic acid produced. The original gas is the miner's ' fire-damp,' the carbonic acid is his ' choke-damp.' Sir Humphry Davy, after having assured himself of the action of wire gauze, which I have just exhibited before you, applied it to the construction of a lamp which should enable the miner to carry his light into an explosive atmosphere. Previous to the introduction of the *safety-lamp*, the miner had to content himself with the light from sparks produced by the collision of flint and steel, for it was found that these sparks were incompetent to ignite the fire-damp.

Davy surrounded a common oil lamp by a cylinder of wire gauze (fig. 71). As long as this lamp is fed by pure air, the flame burns with the ordinary brightness of an oil-flame; but when the miner comes into an atmosphere which contains ' fire-damp,' his flame enlarges, and becomes less luminous ; instead of being fed by the pure oxygen of the air, it is now in part surrounded by inflammable gas. This he ought to take as a warning to retire. Still, though a continuous explosive atmosphere may extend from the air outside, through the meshes of the gauze, to the flame within, the ignition is not propagated across the gauze. The lamp may be filled with an almost lightless flame, and still explosion does not occur. A defect in the gauze, the destruction of the wire at any point by oxidation, hastened by the flame playing against it, would cause an explosion. The motion of the lamp through the air might also force, mechanically, the flame through the meshes. In short, a certain amount of intelligence and caution is necessary in using the lamp. The intelligence, unhappily, is not always possessed, nor the caution always exercised, by the miner; and the consequence is, that, even with the safety-lamp, explosions still occur. Before permitting a man or a boy to enter a mine, would it not be well to place these results, by experiment, visibly before him? Mere advice will not enforce caution; but let the miner have the physical image of what he is to expect, clearly and vividly before his mind, and he will find it a restraining and a monitory influence, long after the effect of cautioning *words* has passed away.

FIG. 71.

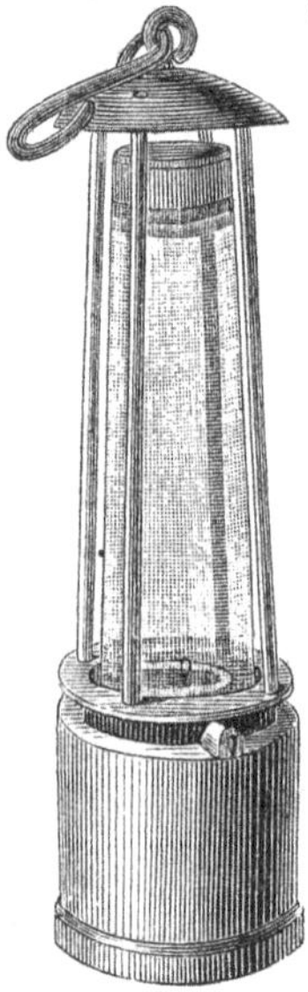

A word or two now on the conductivity of liquids and gases. Rumford made numerous experiments on this subject, showing at once clearness of conception and skill of execution. He supposed liquids to be non-conductors, clearly distinguishing the 'transport' of heat by convection from true conduction; and in order to prevent convection in his liquids, he heated them *at the top*. In this way he found the heat of a warm iron cylinder incompetent to pass downwards through 0·2 of an inch of olive oil; he

also boiled water in a glass tube, over ice, without melting
the substance.  The later experiments of M. Despretz show,
however, that liquids possess true, though extremely feeble,
powers of conduction.  Rumford also denied the conductivity of gases, though he was well acquainted with their
convection.*  The subject of gaseous conduction has been
recently taken up by Professor Magnus, of Berlin, who
considers that his experiments prove that hydrogen gas
conducts heat like a metal.

The cooling action of air may be thus prettily illustrated
—here is a platinum wire formed into a coil; I send a
voltaic current through the coil, till it glows bright red.
I now stretch out the coil so as to form a straight wire ;
the glow instantly sinks—you can now hardly see it.
This effect is due entirely to the freer access of the cold air
to the stretched wire.    Here, again, is a receiver R (fig. 72)
which can be exhausted at pleasure ; attached to the
bottom is a vertical metal rod, $m\,n$, and through the top
another rod, $a\,b$, passes, which can be moved up and down
through an air-tight collar, so as to bring the ends of the
two rods within any required distance of each other.    At
present the rods are united by two inches of platinum wire,
$b\,m$, which I can heat to any required degree of intensity
by a voltaic current.   I have here a small battery, and
now I make my connections; the wire is barely luminous
enough to be seen ; in fact, the current from a single
cell only is now sent through it.   It is surrounded by air,
which, no doubt, is carrying off a portion of its heat.   I
exhaust the receiver—the wire glows more brightly than
before.   I allow air to enter—the wire, for a time, is quite
quenched, rendered perfectly black ; but after the air has
ceased to enter, its first feeble glow is restored.   The current of air here passing over the wire, and destroying its
glow, acts like the current which the wire itself establishes

* Phil. Trans. 1792: Essays, vol. ii. p. 56.

by heating the air in contact with it. The cooling of the wire in both cases is due to convection, and not to true conduction.

Fig. 72.

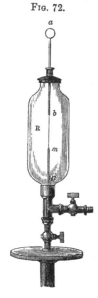

The same effect is obtained in a greatly increased degree, if hydrogen be used instead of air. We owe this interesting observation to Mr. Grove, and it formed the starting-point of M. Magnus's investigation. The receiver is now exhausted, and the wire is almost white-hot. Air cannot do more than reduce that whiteness to bright redness; but observe what hydrogen can do. On the entrance of this gas the wire is totally quenched, and even after the receiver has been filled with the gas, and the inward current has ceased, the glow of the wire is not restored. The electric current now passing through the wire is from two cells; I try three cells, the wire glows feebly; five cause it to glow more brightly, but even with five it is but a bright red. Were the hydrogen not there, the current now passing through the wire would infallibly fuse it. Let us see whether this is not the case. I commence exhaustion,—the first few strokes of the pump produce a scarcely sensible effect; but I continue to work the pump, and now the effect begins to be visible. The wire whitens and appears to thicken. To those at a distance it is now as thick as a goose-quill; and now it glows upon the point of fusion; I continue to work the pump,—the light suddenly vanishes, the wire is fused.

This extraordinary cooling power of hydrogen has been usually ascribed to the mobility of its particles, which enables currents to establish themselves in this gas with greater facility than in any other. But Prof. Magnus

conceives the chilling of the wire to be an effect of conduction. To impede, if not prevent, the formation of
currents, he passes his platinum wire along the axis of a
narrow glass tube, which he fills with hydrogen. Although
in this case the wire is surrounded by a mere film of the
gas, and currents, in the ordinary sense, are scarcely to be
assumed, the film shows itself just as competent to quench
the wire, as when the latter is caused to pass through a
large vessel containing the gas. He also heated the closed
top of a vessel, and found that the heat was conveyed more
quickly from it to a thermometer, placed at some distance
below the source of heat, when the vessel was filled with hydrogen, than when it was filled with air. He found this to be
the case, even when the vessel was loosely filled with cotton
wool or eider down. Here, he contends, currents could not
be formed ; the heat must be conveyed to the thermometer
by the true process of conduction, and not by convection.

Beautiful and ingenious as these experiments are, I do
not think they conclusively establish the conductivity of
hydrogen. Let us suppose the wire in Prof. Magnus's first
experiment to be stretched along the axis of a wide
cylinder containing hydrogen, we should have convection,
in the ordinary sense, on heating the wire. Where does
the heat thus dispersed ultimately go ? It is manifestly
given up to the sides of the cylinder, and if we narrow our
cylinder we simply hasten the transfer. The process of narrowing may continue till a narrow tube is the result,—the
convection between centre and sides will continue and produce the same cooling effect as before. The heat of the
gas being instantly lowered by communication to the
heavy tube, it is prepared to re-abstract the heat from the
wire. With regard also to the vessel heated at the top,
it would require a surface mathematically horizontal, and
a perfectly uniform application of heat to that surface—it
would, moreover, be necessary to cut the heat sharply off

from the sides of the vessel—to prevent convection. Even in the interstices of the eider down and of the cotton wool the convective mobility of hydrogen will make itself felt, and taking everything into account, I think the experimental question of gaseous conduction is still an open one.

# LECTURE VIII.

[March 13, 1862.]

COOLING A LOSS OF MOTION: TO WHAT IS THIS MOTION IMPARTED ?—
EXPERIMENTS ON SOUND BEARING ON THIS QUESTION—EXPERIMENTS
ON LIGHT BEARING ON THIS QUESTION—THE THEORIES OF
EMISSION AND UNDULATION—LENGTH OF WAVES AND NUMBER OF
IMPULSES OF LIGHT—PHYSICAL CAUSE OF COLOUR—INVISIBLE
RAYS OF THE SPECTRUM—THE CALORIFIC RAYS BEYOND THE RED—
THE CHEMICAL RAYS BEYOND THE BLUE—DEFINITION OF RADIANT
HEAT—REFLECTION OF RADIANT HEAT FROM PLANE AND CURVED
SURFACES: LAWS THE SAME AS THOSE OF LIGHT—CONJUGATE
MIRRORS.

APPENDIX:—ON SINGING FLAMES.

WE have this day reached the boundary of one of the two great divisions of our subject; hitherto we have dealt with heat while associated with solid, liquid or gaseous bodies. We have found it competent to produce changes of volume in all these bodies. We have also observed it reducing solids to liquids, and liquids to vapours; we have seen it transmitted through solids by the process of conduction, and distributing itself through liquids and gases by the process of convection. We have now to follow it into conditions of existence, different from any which we have examined hitherto.

I hang this heated copper ball in the air; you see it glow, the glow sinks, the ball becomes obscure; in popular language the ball cools. Bearing in mind what has been said on the nature of heat, we must regard this cooling

as a loss of motion on the part of the ball. But motion
cannot be lost without being imparted to something; to
what then is the molecular motion of this ball transferred?
You would, perhaps, answer to the air, and this is partly
true: over the ball air is passing, and rising in a heated
column, which is quite visible against the screen, when we
allow the electric beam to pass through the warmed air.
But not the whole, nor even the chief part, of the molecular motion of the ball is lost in this way. If the ball
were placed in vacuo it would still cool. Rumford, of whom
we have heard so much, contrived to hang a small thermometer, *by a single fibre of silk*, in the middle of a glass
globe exhausted by means of mercury, and he found that
the calorific rays passed to and fro across the vacuum; thus
proving that the transmission of the heat was independent
of the air. Davy, with an apparatus which I have here
before me, showed that the heat rays from the electric
light passed freely through an air-pump vacuum; and we
can repeat his experiment substantially for ourselves. I
simply take the receiver made use of in our last lecture (fig.
72), and, removing the remains of the platinum wire, then
destroyed, I attach to each end of the two rods, $m\,n$ and $a\,b$,
a bit of retort carbon. I now exhaust the receiver, bring
the coal points together, and send a current from point to
point. The moment I draw the points a little apart,
the electric light blazes forth: and here I have the thermo-electric pile ready to receive a portion of the rays.
The galvanometer needle at once flies aside, and this
has been accomplished by rays which have crossed the
vacuum.

But if not to the air, to what is the motion of our
cooling ball communicated? We must ascend by easy
stages to the answer to this question. It was a very
considerable step in science when men first obtained a
clear conception of the way in which sound is transmitted

through air, and it was a very important experiment which
Hauksbee made before the Royal Society in 1705, by
which he showed that sound could *not* propagate itself
through a vacuum.   Now I wish to make manifest to you
this conveyance of the vibrations of sound through the
air.   I have here a bell turned up-side-down, and supported by a stand.   I draw a fiddle-bow across the edge
of the bell, you hear its tone; the bell is now vibrating,
and if I throw sand upon · its flattish bottom, it would
arrange itself there so as to form a definite figure,
or if I filled it with water I should see the surface
fretted with the most beautiful crispations.   These crispations would show that the bell, when it emits this
note, divides itself into four swinging parts, which are
separated from each other by lines of no swinging.   Here
is a sheet of tracing paper, drawn tightly over this hoop,
so as to form a kind of fragile drum.   I hold it over the
vibrating bell, but not so as to touch the latter ; you hear the
shivering of the membrane. It is a little too slack, so I will
tighten it by warming it before the fire, and repeat the experiment.   You no longer hear a shivering, but a loud musical
tone superadded to that of the bell.   I raise the membrane
and lower it; I move it to and fro, and you hear the rising
and the sinking of the tone. Here is a smaller drum, which
I pass round the bell, holding the membrane vertical ; it
actually bursts into a roar when I bring it within half
an inch of the bell.   The motion of the bell, communicated to the air, has been transmitted by it to the membrane, and the latter is thus converted into a sonorous
body.

   I have here two plates of brass, A B (fig. 73), united together
by this metal rod.   I have darkened the plates by bronzing
them, and on both of them I strew a quantity of white sand.
I now take the connecting brass rod by its centre, between
the finger and thumb of my left hand, and holding it upright

I draw, with my right, a piece of flannel, over which I have
shaken a little powdered resin, along the rod. You hear
the sound; but observe the behaviour of the sand : a single
stroke of my finger, you
see, has caused it to jump
into a series of concentric
rings, which must be quite
visible to you all. I repeat
the experiment, operating
more gently; you hear the
clear, weak, musical sound,
you see the sand shivering,
and creeping, by degrees,
to the lines which it formerly occupied ; and there
are the curves as sharply
drawn upon the surface of
the lower disk as if they
had been arranged with a
camel's-hair pencil. On
the upper disk you see a
series of concentric circles
of the same kind. In fact
the vibrations which I have
imparted to the rod have
communicated themselves
to both the disks, and divided each of them into a
series of vibrating seg-
ments, which are separated
from each other by lines
of no vibration, on which
the sand finds peace.

FIG. 73.

Now let me show you the transmission of these vibrations from the lower disk through the air. On the floor I

place this paper drum, D, strewing dark-coloured sand uniformly over it ; I might stand on the table—I might stand
as high as the ceiling, and produce the effect which I am
now going to show you. Pointing the rod which unites my
plates in the direction of the paper drum, I draw my
resined rubber vigorously over the rod: observe the effect,—
a single stroke has caused that sand to spring into a reti-
culated pattern. A precisely similar effect is produced by
sound on the drum of the ear ; the tympanic membrane is
caused to shudder in the same manner as that drum-head
of paper, and its motion, conveyed to the auditory nerves,
and transmitted thence to the brain, awakes in us the
sensation of sound.

Here is a still more striking example of the conveyance of
the motion of sound through the air. By permitting a jet of
gas to issue through the small orifice of this tube, I obtain
a slender flame, and by turning the cock I reduce the flame
to a height of about half an inch. I introduce the flame
into this glass tube, A B (fig. 74), which is twelve inches long.
Now I must ask your permission to address that flame,
and if I am skilful enough to pitch my voice to the precise
note, I am sure the flame will respond ; it will start suddenly into a melodious song, and continue singing as long
as the gas continues to burn. The burner is now arranged
within the tube, which covers it to a depth of a couple
of inches. If I were to lower it more, the flame would
start into singing on its own account, as in the well-known
case of the hydrogen harmonica ; but, with the present
arrangement, it cannot sing till I tell it to do so. Now I
emit a sound, which you will pardon if it is not musical.
The flame does not respond ; I have not spoken to it in
the proper language. Let me try again ; I pitch my voice
a little higher ; there, the flame stretches its little throat,
and every individual in this large audience hears the sound
of it. I stop the song, and stand at a greater distance

from the flame, and now that I have ascertained the
proper pitch, the experiment is sure to succeed; from a
distance of twenty or thirty
feet I can cause that flame
to sing. I now stop it, turn
my back upon it, and strike
the note as before; you see
how obedient it is to my
voice; when I call, it an-
swers, and with a little
practice I have been able to
command the flame to sing
and to stop, and it has
strictly obeyed the injunc-
tion. Here, then, we have
a striking example of the
conveyance of the vibra-
tions of the organ of voice
through the air, and of their
communication to a body
which is eminently sensitive
to their action.*

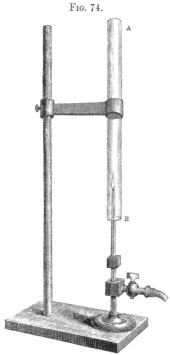

Fig. 74.

Why do I make those ex-
periments on sound? Simply
to give you clear conceptions regarding what takes place
in the case of heat; to lead you up from the tangible to
the intangible; from the region of sense into that of
physical theory.

After philosophers had become aware of the manner in
which sound was produced and transmitted, analogy led
some of them to suppose that light might be produced

* Though not belonging to our present subject, so many persons have
evinced an interest in this experiment that I have been induced to reprint
two short papers in the Appendix to this Lecture, in which the experiment is
more fully described.

and transmitted in a somewhat similar manner.   And
perhaps in the whole history of science there was never a
question more hotly contested than this one.   Sir Isaac
Newton supposed light to consist of minute particles
darted out from luminous bodies: this was the celebrated
Emission Theory.   Huyghens, the contemporary of Newton,
found great difficulty in conceiving of this cannonade of
particles; that they should shoot with inconceivable
velocity through space and not disturb each other.   This
celebrated man entertained the view that light was produced by vibrations similar to those of sound.   Euler
supported Huyghens, and one of his arguments, though
not quite physical, is so quaint and curious that I will
repeat it here.   He looks at our various senses, and at
the manner in which they are affected by external objects.
' With regard to smell,' he says, ' we know that it is produced by material particles which issue from a volatile
body.   In the case of hearing, nothing is detached from
the sounding body, and in the case of feeling we must
touch the body itself.   The distance at which our senses
perceive bodies is, in the case of touch, no distance, in the
case of smell a small distance, in the case of hearing, a
considerable distance, but in the case of sight greatest of
all.   It is therefore more probable that the same mode of
propagation subsists for sound and light, than that odours
and light should be propagated in the same manner;—that
luminous bodies should behave, not as volatile substances,
but as sounding ones.'
   The authority of Newton bore these men down, and not
until a man of genius within these walls took up the
subject, had the Theory of Undulation any chance of
coping with the rival Theory of Emission.   To Dr. Thomas
Young, who was formerly Professor of Natural Philosophy
in this Institution, belongs the immortal honour of stem-
ming this tide of authority, and of establishing on a

safe basis, the theory of undulation. There have been
great things done in this edifice, but hardly a greater than
this. And Young was led to his conclusion regarding
light, by a series of investigations on sound. He, like
ourselves, at the present moment, rose from the known to
the unknown, from the tangible to the intangible. This
subject has been illustrated and enriched by the labours
of genius ever since the time of Young; but one name
only will I here associate with his,—a name which, in connection with this subject, can never be forgotten: that is,
the name of Augustin Fresnel.

According to the notion now universally received, light
consists, first, of a vibratory motion of the particles of the
luminous body; but how is this motion transmitted to our
organs of sight? Sound has the air as its medium, and
long pondering on the phenomena of light, and refined
and conclusive experiments, devised with the express
intention of testing the idea, have led philosophers to the
conclusion, that space is occupied by a substance almost infinitely elastic, through which the pulses of light make
their way. Here your conceptions must be perfectly clear.
The intellect knows no difference between great and small:
it is just as easy, as an intellectual act, to conceive of a
vibrating atom as to conceive of a vibrating cannon ball;
and there is no more difficulty of conceiving of this Ether,
as it is called, which fills space, than in imagining all space
to be filled with jelly. You must imagine the atoms vibrating,
and their vibrations you must figure as communicated to
the ether in which they swing, being propagated through it
in waves; these waves enter the pupil, cross the ball of the
eye, and break upon the retina at the back of the eye.
The act, remember, is as real, and as truly mechanical as the
breaking of the sea waves upon the shore. Their motions
are communicated to the retina, transmitted thence along

the optic nerve to the brain, and there announce themselves to consciousness as light.

I have here an electric lamp, known well to all of you, and on the screen in front of you I project an image of the incandescent coal points which produce the electric light. I will first bring the points together and then separate them. Observe the effect. You have first the place of contact rendered luminous, then you see the glow conducted downwards to a certain distance along the stem of coal. This, as you know, is in reality the conduction of motion. I interrupt the circuit. The points continue to glow for a short time; the light is now subsiding. The coal points are now quite dark, but have they ceased to radiate? By no means. At the present moment there is a copious radiation from these points, which, though incompetent to affect sensibly the nerves of vision, are quite competent to affect other nerves of the human system. To the eye of the philosopher who looks at such matters without reference to sensation, these obscure radiations are precisely the same in kind as those which produce the impression of light. You must therefore figure the particles of the heated body as being in a state of motion ; you must figure the motion communicated to the surrounding ether, and transmitted through the ether with a velocity, which we have the strongest reason for believing is the same as that of light. Thus when you turn towards a fire on a cold day, and expose your chilled hands to its influence, the warmth that you feel is due to the impact of these ethereal billows upon your skin; they throw the nerves into motion, and the consciousness corresponding to this motion is what we popularly call warmth. Our task during the lectures which remain to us is to examine heat under this *radiant* form.

To investigate this subject we possess our invaluable

thermo-electric pile, the face of which is now coated
with lamp-black, a powerful absorber of radiant heat. I
hold the instrument in front of the cheek of Mr. Anderson;
he is a radiant body, and observe the effect produced by
his rays; the pile drinks them in, they generate electricity,
and the needle of the galvanometer moves up to 90°. I
withdraw the pile from the source of heat, and allow the
needle to come to rest, and now I place this slab of ice
in front of the pile. You have a deflection in the opposite
direction, as if rays of cold were striking on the pile. But
you know that in this case the pile is the hot body;
it radiates its heat against the ice; the face of the pile is
thus chilled, and the needle, as you see, moves up to 90° on
the side of cold. Our pile is therefore not only available
for the examination of heat communicated to it by direct
contact, but also for the examination of radiant heat.
Let us apply it at once to a most important investigation,
and examine, by means of it, the distribution of thermal
power in the electric spectrum.

Let me in the first place show you this spectrum. I do so
by sending a slice of pure white light from the orifice, o (fig.
75), through this prism, *a b c*, which is built up of plane glass

Fig. 75.

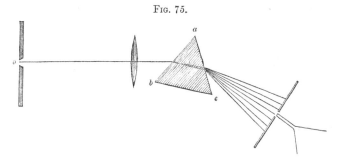

sides, but is filled with the liquid bisulphide of carbon. It
gives a richer display of colour than glass does, and this is
one reason why I use it in preference to glass. Here then you

have the white beam disentangled, and reduced to the colours
which compose it; you have this burning red, this vivid
orange, this dazzling yellow, this brilliant green, and these
various shades of blue; the blue space being usually sub-
divided into blue, indigo and violet. I will now cause a
thermo-electric pile of particular construction to pass gradually through all these colours in succession, so as to test
their heating powers, and I will ask you to observe the
needle of the galvanometer which is to declare the magnitude of that power.

For this purpose I have here (fig. 76) a beautiful piece of
apparatus, designed by Melloni, and executed, with his
accustomed skill, by M. Ruhmkorff.* You observe here a
polished brass plate, A B, attached
to a stem, and this stem is mounted on a horizontal bar, which, by
means of a screw, has motion
imparted to it. By turning this
ivory handle in one direction I
cause the plate of brass to approach; by turning it in the
other, I cause it to recede, and the
motion is so fine and gradual,
that I could, with ease and certainty, push the screen through a
space less than $\frac{1}{2000}$th of an inch.
You observe a narrow vertical slit
in the middle of this plate, and
something dark behind it. That
dark space is the blackened face of a thermo-electric pile,
P, the elements of which are ranged in a single row, and
not in a square, as in our other instrument. I will allow
distinct slices of the spectrum to fall on that slit; each will

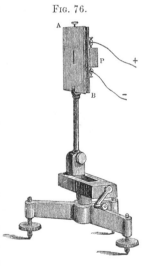

FIG. 76.

* Kindly lent to me by M. Gassiot.

impart whatever heat it possesses to the pile, and the
quantity of the heat will be marked by the needle of our
galvanometer.

At present a small but brilliant spectrum falls upon the
plate, A B, but the slit is quite out of the spectrum. I turn
the handle, and the slit gradually approaches the violet
end of the spectrum; the violet light now falls upon the
slit, but the needle does not move sensibly. I pass on
to the indigo, the needle is still quiescent; the blue
also shows no action. I pass on to the green, the needle
barely stirs: now the yellow falls upon the slit; the motion
of the needle is now perhaps for the first time visible to you;
but the deflection is small, though I now expose the pile
*to the most luminous part of the spectrum.** I will now
pass on to the orange, which is less luminous than the
yellow, but you observe, though the light diminishes the
heat increases; the needle moves still farther. I pass on
to the red, which is still less luminous than the orange,
and you see that I here obtain the greatest thermal power
exhibited by any of the visible portions of the spectrum.

The appearance, however, of this burning red might
lead you to suppose it natural for such a colour to be
hotter than any of the others. But now pay attention. I
will cause my slit to pass entirely out of the spectrum, quite
beyond the extreme red. Look to the galvanometer! The
needle goes promptly up to the stops. So that we have
here a heat-spectrum which we cannot see, and whose
thermal power is far greater than that of any visible part
of the spectrum. In fact, the electric light with which we
deal, emits an infinity of rays which are converged by our
lens, refracted by our prism, which form the prolongation of
our spectrum, but which are utterly incompetent to excite
the optic nerve to vision. It is the same with the sun. Our
orb is rich in these obscure rays; and though they are for

---

* I am here dealing with a large lecture-room galvanometer.

the most part cut off by our atmosphere, multitudes of them still reach us. To the great William Herschel we are indebted for the discovery of them.

Thus we prove that the spectrum extends on the red side much beyond its visible limits: and were I, instead of being compelled to make use of lenses and prisms of glass, fortunate enough to possess lenses and prisms of rock salt, I could show you, as Melloni has done, that those rays extend a great way farther than it is now in my power to prove. In fact, glass, though sensibly transparent to light, is, in a great measure, opaque to these obscure rays; instead of reaching the screen, they are for the most part lodged in the glass.

The visible spectrum, then, simply marks an interval of radiant action, in which the radiations are so related to our organisation that they excite the impression of light; beyond this interval, *in both directions*, radiant power is exerted—obscure rays fall—those falling beyond the red being powerful to produce heat, while those falling beyond the violet are powerful to promote chemical action. These latter rays can actually be rendered visible; or more strictly expressed, the undulations or waves which are now striking here beyond the violet against the screen, and which are scattered from it so as to strike the eyes of every person present, though they are incompetent to excite vision in those eyes; those waves, I say, may be caused to impinge upon another body, and to impart their motion to it, and actually to convert the dark space beyond the violet into a brilliantly illuminated one. I have here the proper substance. The lower half of this sheet of paper has been washed with a solution of sulphate of quinine, while I have left the upper half in its natural state. I will hold the sheet, so that the straight line dividing its prepared from its unprepared half, shall be horizontal and shall cut the spectrum into two equal parts; the upper half will remain unaltered, and you will be able to compare with it the under half, on which I hope to find the spectrum elongated. You see this effect; we have here a splendid fluorescent band, several inches in width, where a moment ago there was nothing but darkness. I remove the prepared paper, and the light disappears. I re-introduce it, and the light flashes out again, showing you, in the most emphatic manner, that the visible limits of the ordinary spectrum by no means mark the limits of radiant action. I dip my brush in this solution of sulphate of quinine, and dab it against the paper; wherever the solution falls light flashes forth. The existence of these extra violet rays has been long known; it was known to Thomas Young, who actually experimented on them; but to Prof. Stokes we are indebted for the complete investigation of this subject. He rendered the rays thus visible.

How then are we to conceive of the rays, visible and invisible, which fill this large space upon the screen? Why are some of them visible and others not? Why are the visible ones distinguished by various colours? Is there anything that we can lay hold of in the undulations which produce these colours, to which, as a physical cause, we must assign the colour? Observe first, that the entire beam of white light is drawn aside, or refracted by the prism, but the violet is pulled aside more than the indigo, the indigo more than the blue, the blue more than the green, the green more than the yellow, the yellow more than the orange, and the orange more than the red. These colours are differently refrangible, and upon this depends the possibility of their separation. To every particular degree of refraction belongs a definite colour and no other. But why should light of one degree of refrangibility produce the sensation of red, and of another degree the sensation of green? This leads us to consider more closely the cause of these sensations.

A reference to the phenomena of sound will materially help our conceptions here. Figure clearly to your minds a harp-string vibrating to and fro; it advances and causes the particles of air in front of it to crowd together; it thus produces a *condensation* of the air. It retreats, and the air particles behind it separate more widely; in other words, a *rarefaction* of the air occurs behind the retreating wire. The string again advances and produces the condensation as before, it again retreats and produces a rarefaction. Thus the condition of the air through which the sound of the string is propagated consists of a regular sequence of condensations and rarefactions, which travel with a velocity of about 1,100 feet a second.

The condensation and rarefaction constitute what is called a sonorous pulse or *wave,* and the length of the wave is the distance from the middle of the condensation to the middle of the rarefaction. Of course these blend gradually into each other. The length of the wave is also measured by the distance from the centre of one condensation to the centre of the next one. Now the quicker a string vibrates the more quickly will these pulses follow each other, and the shorter, at the same time, will be the length of each individual wave. Upon these differences the *pitch* of a note in music depends. If a violin player wishes to produce a higher note, he shortens his string by pressing his finger on it; he thereby augments the rapidity of vibration. If his point of pressure exactly *halves* the length of his string, he obtains the *octave* of the note which the string emits when vibrating as a whole. ' Boys are chosen as choristers to produce the shrill notes, men to produce the bass notes; the reason being, that the boy's organ vibrates more speedily than the man's;' and the hum of a gnat is shriller than that of a beetle, because the smaller insect can send a greater number of impulses per second to the ear.

We have now cleared our way towards the clear com-

prehension of the physical cause of colour. This spectrum is to the eye what the gamut is to the ear; its
different colours represent notes of different pitch. The
vibrations which produce the impression of red are slower,
and the ethereal waves which they generate are longer,
than those which produce the impression of violet, while
the other colours are excited by waves of some intermediate
length. The length of the waves both of sound and light,
and the number of shocks which they respectively impart
to the ear and eye, have been strictly determined. Let us
here go through a simple calculation. Light travels
through space at a velocity of 192,000 miles a second.
Reducing this to inches we find the number to be
12,165,120,000. Now it is found that 39,000 waves of red
light placed end to end would make up an inch; multiply
the number of inches in 192,000 miles by 39,000, we obtain
the number of waves of red light in 192,000 miles: this
number is 474,439,680,000,000. *All these waves enter the
eye in a single second.* To produce the impression of
red in the brain, the retina must be hit at this almost
incredible rate. To produce the impression of violet, a
still greater number of impulses is necessary; it would
take 57,500 waves of violet to fill an inch, and the number
of shocks required to produce the impression of this
colour, amounts to six hundred and ninety-nine millions
of millions per second. The other colours of the spectrum,
as already stated, rise gradually in pitch from the red to
the violet.

But beyond the violet we have rays of too high a pitch
to be visible, and beyond the red we have rays of too low
a pitch to be visible. The phenomena of light are in this
case also paralleled by those of sound. If it did not involve
a contradiction, we might say that there are musical
sounds of too high a pitch to be heard, and also sounds of
too low a pitch to be heard. Speaking strictly, there are

waves transmitted through the air from vibrating bodies, which, though they strike upon the ear in regular recurrence, are incompetent to excite the sensation of a musical note. Probably sounds are heard by insects which entirely escape our perceptions; and, indeed, as regard human beings, the selfsame note may be of piercing shrillness to one person, while it is absolutely unheard by another. Both as regards light and sound, our organs of sight and hearing embrace a certain practical range, beyond which, on both sides, though the objective cause exists, our nerves cease to be influenced by it.

When therefore I place this red-hot copper ball before you, and watch the waning of its light, you will have a perfectly clear conception of what is occurring here. The atoms of the ball oscillate, but they oscillate in a resisting medium on which their moving force is expended, and which transmits it on all sides with inconceivable velocity. The oscillations competent to produce light are now exhausted; the ball is quite dark, still its atoms oscillate, and still their oscillations are taken up and transmitted on all sides by the ether. The ball cools as it thus loses its molecular motion, but no cooling to which it can be practically subjected can entirely deprive it of its motion. That is to say, all bodies, whatever may be their temperature, are radiating heat. From the body of every individual here present, waves are speeding away, some of which strike upon this cooling ball and restore a portion of its lost motion. But the motion thus received by the ball is far less than what it communicates, and the difference between them expresses the ball's loss of motion. As long as this state of things continues the ball will continue to show an ever-lowering temperature: its temperature will sink until the quantity it emits is equal to the quantity which it receives, and at this point its temperature becomes constant. Thus, though you are

conscious of no reception of heat, when you stand before
a body of your own temperature, an interchange of rays
is passing between you. Every superficial atom of each mass
is sending forth its waves, which cross those that move
in the opposite direction, every wave asserting its own
individuality amid the entanglement of its fellows. When
the sum of motion received is greater than that given out,
warming is the consequence; when the sum of motion
given out is greater than that received, chilling takes
place. This is Prevost's Theory of Exchanges, expressed
in the language of the Wave Theory.

Let us occupy the remainder of this lecture by illustrating experimentally the analogy between light and
radiant heat, as regards reflection. You observed when I
placed my thermo-electric pile in front of Mr. Anderson's
face, that I had attached to it an open cone which I did
not use in my former experiments. This cone is silvered
inside, and it is intended to augment the action of feeble
radiations, by converging them upon the face of the
thermo-electric pile. It does this by reflection; instead
of shooting wide of the pile, as they would do if the
reflector were removed, they meet the silvered surface and
glance from it against the pile. The augmentation of
the effect is thus shown. I place the pile at this end of
the table with its reflector off, and at a distance of four or
five feet I place this copper ball, hot—but not red-hot;
you observe scarcely any motion of the needle of the
galvanometer. Disturbing nothing, I now attach the
reflector to the pile; the needle instantly goes up to 90°,
declaring the augmented action.

The law of this reflection is precisely the same as that
of light. Observe this apparently solid luminous cylinder,
issuing from our electric lamp, and marking its track thus
vividly upon the dust of our darkened room. I take a mirror
in my hand, and permit the beam to fall upon it; the beam

rebounds from the mirror; it now strikes the ceiling. This horizontal beam is the incident beam, this vertical one is the reflected beam, and the law of light, as many of you know, is, that the angle of incidence is equal to the angle of reflection. The incident and reflected beams now enclose a right angle, and when this is the case I may be sure that both beams

FIG. 77.

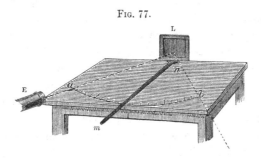

form, with a perpendicular to the surface of the mirror, an angle of 45°.

I place the lamp at this corner, E, of the table (fig. 77); behind the table I place a looking-glass, L, and on the table you observe I have drawn a large arc, *a b*. Attached to the mirror is this long straight lath, *m n*, and the looking-glass, resting upon rollers, can be turned by the lath, which is to serve as an index. I have here drawn a dark central line, and when the mirror exactly faces the middle of the audience, our lath and this line coincide. Those in front may see that the lath itself and its reflection in the mirror form a straight line, which proves that the central dark line is now perpendicular to the mirror. Right and left of this central line I have divided the arc into ten equal parts; commencing at the end E with 0°, I have graduated the arc up to 20°. I first turn the index so that it shall be in the line of the beam emitted by the lamp. The beam now falls upon the mirror, striking it as a perpendicular, and you see it is reflected back along the

line of incidence. I now move my index to 1 ; the reflected
beam, as you observe, draws itself along the table, cutting
the figure 2. I move the index to 2, the beam is now at
4 ; I move the index to 3, the beam is now at 6 ; I move
it to 5, the beam is now at 10 ; I move it to 10, the beam
is now at 20. If I stand midway between the incident
and reflected beams, and stretch out my arms, my finger-
tips touch each of them. One lies as much to the left of
the perpendicular as the other does to the right. The
angle of incidence is equal to the angle of reflection. But
we have also demonstrated that the beam moves twice as
fast as the index ; and this is usually expressed in the
statement, that the angular velocity of a reflected ray is
twice that of the mirror which reflects it.

I have already shown you that these incandescent coal-
points emit an abundance of obscure rays—of rays of
pure heat, which have no illuminating power; my object
now is to show you that those rays of heat emitted by the
lamp, have obeyed precisely the same laws as the rays of
light. I have here a piece of black glass ; so black that
when I look through it at the electric light, or even at
the noonday sun, I see nothing. You observe the disap-
pearance of the beam when I place this glass in front of
the lamp. It cuts off every ray of light ; but, strange as
it may appear to you, it is, in a considerable degree, transparent to the obscure rays of the lamp. I now extinguish
the light by interrupting the current, and I lay my thermo-
electric pile on the table at the number 20, where the
luminous beam fell a moment ago. The pile is connected with the galvanometer, and the needle of the
instrument is now at zero. I ignite the lamp, no light
makes it appearance, but observe the galvanometer; the
needle has already swung to 90°, through the action of
the non-luminous rays upon the pile. If I move the instrument right or left from its present position the needle

immediately sinks; the calorific rays have pursued the precise track of the luminous rays; and for them, also, the angle of incidence is equal to the angle of reflection. Repeating the experiments that I have already executed with light, bringing the index in succession to 1, 2, 3, 5, &c., I prove that in the case of radiant heat also, the angular velocity of the reflected ray is twice that of the mirror.

The heat of the fire obeys the same law. I have here a sheet of tin—a homely reflector, but it will answer my purpose. At this end of the table I place the thermo-electric pile, and at the other end my tin screen. The needle of the galvanometer is now at zero. Well, I turn the reflector so as to cause the heat striking it to rebound towards the pile; it now meets the instrument, and the needle at once declares its arrival. Observe the positions of the fire, of the reflector, and of the pile; you see that they are just in the positions which make the angle of incidence equal to that of reflection.

But in these experiments the heat is, or has been, associated with light. Let me now show that the law holds good for rays emanating from a truly obscure body. Here is a copper ball, c (fig. 78), heated to dull redness; I plunge it in water until its light totally disappears, but I leave it warm. It is still giving out radiant heat of a slightly greater intensity than that emitted by the human body. I place it on this candlestick as a support, and here I place my pile, p, turning its conical reflector away from the ball, so that no *direct* ray from the latter can reach the pile. You see the needle remains at zero. I place here my tin reflector, m n, so that a line drawn to it from the ball, shall make the same angle with a perpendicular to the polished tin reflector, as a line drawn from the pile. The axis of the conical reflector lies in this latter line. True to the law, the heat-rays emanating from the ball rebound from it and strike the pile, and you observe the consequent prompt motion of the needle.

Like the rays of light, the rays of heat emanating from
our ball proceed in straight lines through space, diminish-

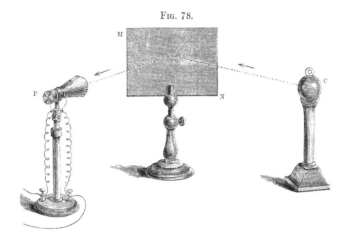

FIG. 78.

ing in intensity exactly as light diminishes.   Thus, this
ball, which when close to the pile causes the needle of the
galvanometer to fly up to 90°, at a distance of 4 feet 6
inches, shows scarcely a sensible action.   Its rays are
squandered on all sides, and comparatively few of them
reach the pile.   But I now introduce between the pile and
the ball this tin tube, A B (fig. 79), 4 feet long.   It is polished

FIG. 79.

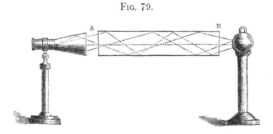

within, and therefore capable of reflection.   The calorific
rays which strike the interior surface obliquely, are reflected

from side to side of the tube, and thus those rays which, when the tube is absent, are squandered in space, are caused, by internal reflection, to reach the pile. You see the result : the needle, which a moment ago showed no sensible action, moves promptly to its stops.

We have now dwelt sufficiently long on the reflection of radiant heat by *plane surfaces*; let us turn for a moment to reflection from curved surfaces. I have here a concave mirror, M N (fig. 80) formed of copper, but coated with silver. I place,

FIG. 80.

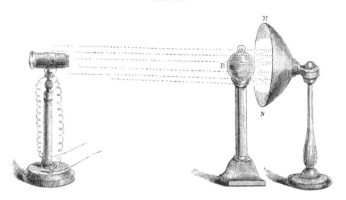

this warm copper ball, B, at a distance of eighteen inches from the pile, which has now its conical reflector removed ; you observe scarcely any motion of the needle. If I placed the reflector, M N, properly behind a candle, I should collect its rays, and send them back in a cylinder of light. I shall do the same with the calorific rays emitted by the ball B ; you cannot, of course, see the track of these obscure rays, as you can that of the luminous ones ; but you observe that while I speak, the galvanometer has revealed the action : the needle of the instrument has gone up to 90°.

I have here a pair of much larger mirrors, one of which is placed flat upon the table : now, the curvature of this

mirror is so regulated that if I place a light at this point,
which is called the focus of the mirror, the rays which fall
divergent upon the mirror are reflected upward from it
parallel. Let us make the experiment: In the focus I
place our coal-points, bring them into contact, and then
draw them a little apart; there is the electric light, and
there is a splendid vertical cylinder, cast upwards by the
reflector, and marked by the action of the light on the
dust of the room. If we reversed the experiment, and
allowed a parallel beam of light to fall upon the mirror,
the rays of that beam, after reflection, would be collected
in the focus of the mirror. We can actually make this
experiment by introducing a second mirror; here it is
suspended from the ceiling. I will now draw it up to a
height of 20 or 25 feet above the table; the vertical beam,
which before fell upon the ceiling, is now received by
the upper mirror; I have hung in the focus of the upper
mirror a bit of oiled paper, to enable you to see the collection of the rays at the focus. You observe how intensely
that piece of paper is now illuminated, not by the direct
light from below, but by the reflected light converged
upon it from above.

Many of you know the extraordinary action of light
upon a mixture of hydrogen and chlorine. I have here a
transparent collodion balloon filled with the mixed gases;
I lower my upper reflector, and suspend the balloon from
a hook attached to it, so that the little globe shall swing in
the focus; we will now draw the mirror quite up to the ceiling (fig. 81); and, as before, I place my coal-points in
the focus of the lower mirror; the moment I draw them
apart, the light gushes from them, and the gases explode.
And remember this is the action of the light; you know
collodion to be an inflammable substance, and hence
might suppose that it was the *heat* of the coal-points that
ignited it, and that it communicated its combustion to the

gases; but look here!
you see, as I speak, the
flakes of the balloon
descending on the
table; the luminous
rays went harmlessly
through it, caused the
gases to explode, and
the hydrochloric acid,
formed by their com-
bustion, 'has actually
preserved the inflam-
mable envelope from
sharing in the combus-
tion.

FIG. 81.

I lower the upper
mirror and hang in its
focus a second balloon,
containing a mixture
of oxygen and hydro-
gen, on which light
has no sensible effect;
I raise the mirror, and
in the focus of the
lower one place this
red-hot copper ball.
The calorific rays are
now reflected and con-
verged above, as the
luminous ones were re-
flected and converged
in the last experiment;
but they act upon the
*envelope*, which I have
purposely blackened a

little, so as to enable it to intercept the heat-rays; the
action is not so sudden as in the last case, but *there* is the
explosion, and you now see no trace of the balloon; the
inflammable substance is entirely dissipated.

But here, you may object, light is associated with the heat;
very well, I lower the upper mirror once more and suspend
in its focus a flask of hot water. I bring my thermo-
electric pile to the focus of the lower mirror, and first turn
the face of the pile upwards, so as to expose it to the direct
radiation of the warm flask—there is no sensible action produced by the direct rays. But I now turn my pile with its
face downwards. If light and heat behave alike, the rays
from the flask which strike the reflector will be collected at
its focus. You see that this is the case; the needle, which
was not sensibly affected by the direct rays, goes up to its
stops. I would ask you to observe the direction of that
deflection; the red end of the needle moves towards you.

I again lower the mirror, and, in the place of the flask of
hot water, suspend a second one containing a freezing
mixture. I raise the mirror and, as in the former case,
bring the pile into the focus of the lower one. Turned
directly towards the upper flask there is no action; turned
downwards, the needle moves: observe the direction of the
motion—the red end comes towards me.

Does it not appear as if this body in the upper focus
were now emitting rays of cold which are converged by the
lower mirror exactly as the rays of heat in our former·
experiment. The facts are exactly complementary, and it
would seem that we have precisely the same right to infer
from the experiments, the existence and convergence of
these cold rays, as we have to infer the existence and convergence of the heat rays. But many of you, no doubt,
have already perceived the real state of the case. The
pile is a warm body, but in the last experiment the
quantity which it lost by radiation was more than made

good by the quantity received from the hot flask above. Now the case is reversed, the quantity which the pile radiates is in excess of the quantity which it receives, and hence the pile is chilled;—the exchanges are against it, its loss of heat is only partially compensated—and the deflection due to cold is the necessary consequence.

# APPENDIX TO LECTURE VIII.

———◆◇◆———

ON THE SOUNDS PRODUCED BY THE COMBUSTION OF
GASES IN TUBES.*

IN the first volume of Nicholson's Journal, published in 1802, the sounds produced by the combustion of hydrogen in tubes are referred to as having been ' made in Italy : ' Dr. Higgins, in the same place, shows that he had discovered them in the year 1777, while observing the water formed in a glass vessel by the slow combustion of a slender stream of hydrogen. Chladni, in his ' Akustik,' published in 1802, page 74, speaks of their being mentioned, and incorrectly explained, by De Luc in his ' New Ideas on Meteorology : ' I do not know the date of the volume. Chladni himself showed that the tones produced were the same as those of an open pipe of the same length as the tube which encompassed the flame. He also succeeded in obtaining a tone and its octave from the same tube, and in one case obtained the fifth of the octave. In a paper published in the ' Journal de Physique' in 1802, G. De la Rive endeavoured to account for the sounds by referring them to the alternate contraction and expansion of aqueous vapour ; basing his opinion upon a series of experiments of great beauty and ingenuity made with the bulbs of thermometers. In 1818 Mr. Faraday took up the subject,† and showed that the tones were produced when the glass tube was enveloped by an atmosphere higher in temperature than 212° Fahr. That they were not due to aqueous vapour was further shown by the

* From the Philosophical Magazine for July, 1857. By John Tyndall, F.R.S.

† Journal of Science and the Arts, vol. v. p. 274.

fact that they could be produced by the combustion of carbonic oxide. He referred the sounds to successive explosions produced by the periodic combination of the atmospheric oxygen with the issuing jet of hydrogen gas.

I am not aware that the dependence of the pitch of the note on the size of the flame has as yet been noticed. To this point I will, in the first place, briefly direct attention.

A tube 25 inches long was placed over an ignited jet of hydrogen : the sound produced was the fundamental note of the tube.

A tube 12½ inches long was brought over the same flame, but no sound was obtained.

The flame was lowered, so as to make it as small as possible, and the tube last mentioned was again brought over it; it gave a clear melodious note, which was the octave of that obtained with the 25-inch tube.

The 25-inch tube was now brought over the same flame; it no longer gave its fundamental note, but exactly the same note as that obtained from the tube of half its length.

Thus we see, that although the speed with which the explosions succeed each other depends upon the length of the tube, the flame has also a voice in the matter : that to produce a musical sound, its size must be such as to enable it to explode in unison either with the fundamental pulses of the tube, or with the pulses of its harmonic divisions.

With a tube 6 feet 9 inches long, by varying the size of the flame, and adjusting the depth to which it reached within the tube, I have obtained a series of notes in the ratio of the numbers 1, 2, 3, 4, 5.

These experiments explain the capricious nature of the sounds sometimes obtained by lecturers upon this subject. It is, however, always possible to render the sounds clear and sweet, by suitably adjusting the size of the flame to the length of the tube.*

Since the experiments of Mr. Faraday, nothing, that I am aware of, has been added to this subject, until quite recently. In a recent number of Poggendorff's ' Annalen ' an interesting

---

* With a tube 14½ inches in length and an exceedingly minute jet of gas, I obtained, without altering the quantity of gas, a note and its octave : the flame possessed the power of changing its own dimensions to suit both notes.

experiment is described by M. von Schaffgotsch, and made the subject of some remarks by Prof. Poggendorff himself. A musical note was obtained with a jet of ordinary coal-gas, and it was found that when the voice was pitched to the same note, the flame assumed a lively motion, which could be augmented until the flame was actually extinguished. M. von Schaffgotsch does not describe the conditions necessary to the success of his experiment; and it was while endeavouring to find out these conditions that I alighted upon the facts which form the principal subject of this brief notice. I may remark that M. von Schaffgotsch's result may be produced, with certainty, if the gas be caused to issue under sufficient pressure through a very small orifice.

In the first experiments I made use of a tapering brass burner, $10\frac{1}{2}$ inches long, and having a superior orifice about $\frac{1}{20}$th of an inch in diameter. The shaking of the singing flame within the glass tube, when the voice was properly pitched, was so manifest as to be seen by several hundred people at once.

I placed a syrene within a few feet of the singing-flame, and gradually heightened the note produced by the instrument. As the sounds of the flame and syrene approached perfect unison, the flame shook, jumping up and down within the tube. The interval between the jumps became greater until the unison was perfect, when the motion ceased for an instant; the syrene still increasing in pitch, the motion of the flame again appeared, the jumping became quicker and quicker, until finally it escaped cognisance by the eye.

This experiment showed that the jumping of the flame, observed by M. von Schaffgotsch, is the optical expression of the *beats* which occur at each side of the perfect unison: the beats could be heard in exact accordance with the shortening and lengthening of the flame. Beyond the region of these beats, in both directions, the sound of the syrene produced no visible motion of the flame. What is true of the syrene is true of the voice.

While repeating and varying these experiments, I once had a silent flame within a tube, and on pitching my voice to the note of the tube, the flame, to my great surprise, instantly started into song. Placing the finger on the end of the tube, and silencing the melody, on repeating the experiment the same result was obtained.

I placed the syrene near the flame, as before. The latter was burning tranquilly within its tube. Ascending gradually from the lowest notes of the instrument, at the moment when the sound of the syrene reached the pitch of the tube which surrounded the gas flame, the latter suddenly stretched itself and commenced its song, which continued indefinitely after the syrene had ceased to sound.

With the burner which I have described, and a glass tube 12 inches long, and from $\frac{1}{2}$ to $\frac{3}{4}$ of an inch internal diameter, this result can be obtained with ease and certainty. If the voice be thrown a little higher or lower than the note due to the tube, no visible effect is produced upon the flame : the pitch of the voice must lie within the region of the audible beats.

By varying the length of the tube we vary the note produced, and the voice must be modified accordingly.

That the shaking of the flame, to which I have already referred, proceeds in exact accordance with the beats, is beautifully shown by a tuning-fork, which gives the same note as the flame. Loading the fork so as to throw it slightly out of unison with the flame, when the former is sounded and brought near the flame, the jumpings are seen at exactly the same intervals as those in which the beats are heard. When the tuning-fork is brought over a resonant jar or bottle, the beats may be heard and the jumpings seen by a thousand people at once. By changing the load upon the tuning-fork, or by slightly altering the size of the flame, the quickness with which the beats succeed each other may be changed, but in all cases the jumpings address the eye at the same moment that the beats address the ear.

With the tuning-fork I have obtained the same results as with the voice and syrene. Holding a fork over a tube which responds to it, and which contains within it a silent flame of gas, the latter immediately starts into song. I have obtained this result with a series of tubes varying from $10\frac{1}{2}$ to 29 inches in length. The following experiment could be made :—A series of tubes, capable of producing the notes of the gamut, might be placed over suitable jets of gas ; all being silent, let the gamut be run over by a musician with an instrument sufficiently powerful, placed at a distance of twenty or thirty yards. At the sound of each particular note, the gas-jet contained in the corresponding tube would instantly start into song.

I must remark, however, that with the jet which I have used, the experiment is most easily made with a tube about 11 or 12 inches long: with longer tubes it is more difficult to prevent the flame from singing spontaneously, that is, without external excitation.

The principal point to be attended to is this. With a tube, say of 12 inches in length, the flame requires to occupy a certain position in the tube in order that it shall sing with a maximum intensity. Let the tube be raised so that the flame may penetrate it to a less extent; the energy of the sound will be thereby diminished, and a point (A) will at length be attained, where it will cease altogether. Above this point, for a certain distance, the flame may be caused to burn tranquilly and silently for any length of time, but when excited by the voice it will sing.

When the flame is too near the point (A), on being excited by the voice or by a tuning-fork, it will respond for a short time, and then cease. A little above the point where this cessation occurs, the flame burns tranquilly, if unexcited, but if once caused to sing it will continue to sing. With such a flame, which is not too sensitive to external impressions, I have been able *to reverse the effect hitherto described*, and to stop the song at pleasure by the sound of my voice, or by a tuning-fork, without quenching the flame itself. Such a flame, I find, may be made to obey the word of command, and to sing or cease to sing, as the experimenter pleases.

The mere clapping of the hands, producing an explosion, shouting at an incorrect pitch, shaking of the tube surrounding the flame, are, when the arrangements are properly made, ineffectual. Each of these modes of disturbance doubtless affects the flame, but the impulses do not accumulate, as in the case where the note of the tube itself is struck. It appears as if the flame were *deaf* to a single impulse, as the tympanum would probably be, and, like the latter, needs the accumulation of impulses to give it sufficient motion. A difference of half a tone between two tuning-forks is sufficient to cause one of these to set the flame singing, while the other is powerless to produce this effect.

I have said that the voice must be pitched to the note of the tube which surrounds the flame; it would be more correct to say the note produced by the flame when singing. In all cases this note is sensibly higher than that due to the open tube which surrounds the flame; this ought to be the case, because of the high temperature of the vibrating column. An open tube, for example, which, when a tuning-fork is held over its end, gives a maximum reinforcement, produces, when surrounding a singing flame, a note higher than that of the fork. To obtain the latter note the tube must be sensibly longer.

What is the constitution of the flame of gas while it produces these musical sounds? This is the next question to which I will briefly call attention. Looked at with the naked eye, the sounding flame appears constant, but is the constancy real? Supposing each pulse to be accompanied by a physical change of the flame, such a change would not be perceptible to the naked eye, on account of the velocity with which the pulses succeed each other. The light of the flame would appear continuous, on the same principle that the troubled portion of a descending liquid jet appears continuous, although by proper means this portion of a jet can be shown to be composed of isolated drops. If we cause the image of the flame to pass speedily over different portions of the retina, the changes accompanying the periodic impulses will manifest themselves in the character of the image thus traced.

I took a glass tube 3 feet 2 inches long and about an inch and a half in internal diameter, and placing it over a very small flame of olefiant gas (common gas will also answer), obtained the fundamental note of the tube : on moving the head to and fro, the image of the sounding flame was separated into a series of distinct images ; the distance between the images depended upon the velocity with which the head was moved. This experiment is suited to a darkened lecture-room. It was still easier to obtain the separation of the images in this way, when a tube 6 feet 9 inchès in length, and a larger flame, were made use of.

The same result is obtained when an opera glass is moved to and fro before the eye.

But the most convenient mode of observing the flame is with a mirror; and it can be seen either directly in the mirror, or by projection upon a screen.

A lens of 33 centimetres focus was placed in front of a flame of common gas, upwards of an inch long, and a paper screen was hung at about 6 or 8 feet distance behind the flame. In front of the lens a small looking-glass was held, which received the light that had passed through the lens, and reflected it back upon the screen placed behind the latter. By adjusting the position of the lens, a well-defined inverted image of the flame was obtained upon the screen. On moving the mirror the image was displaced, and owing to the retention of the impression by the retina, when the movement was sufficiently speedy the image described a continuous luminous track. Holding the mirror motionless, the 6-foot 9-inch tube was placed over the flame : the latter changed its shape the moment it commenced to sound, remaining however well defined upon the screen. On now moving the mirror, a totally different effect was produced : instead of a continuous track of light, a series of distinct images of the sounding flame was observed. The distance of these images apart varied with the motion of the mirror; and, of course, could be made, by suitably turning the reflector, to form a ring of images. The experiment is beautiful, and in a dark room may be made visible to a large audience.

The experiment was also varied in the following manner :— A triangular prism of wood had its sides coated with rectangular pieces of looking glass : it was suspended by a thread with its axis vertical ; torsion was imparted to the thread, and the prism, acted upon by this torsion, caused to rotate. It was so placed that its three faces received, in succession, the beam of light sent from the flame through the lens in front of it, and threw the images upon the screen. On commencing its motion the images were but slightly separated, but became more and more so as the motion approached its maximum. This once past, the images drew closer together again, until they ended in a kind of luminous ripple. Allowing the acquired torsion to react, the same series of effects could be produced, the motion being in an opposite direction. In these experiments, that half of the tube which was turned towards the screen was coated with lamp-black, so as to cut off the direct light of the jet from the screen.*

But what is the state of the flame in the interval between two images ? The flame of common gas, or of olefiant gas, owes its

* Since these experiments were made, Mr. Wheatstone has drawn my attention to the following passage, which proves that he had already made use of the rotating mirror in examining a singing flame : ' A flame of hydrogen gas burning in the open air presents a continuous circle in the mirror; but while producing a sound within a glass tube, regular inter-missions of intensity are observed, which present a chain-like appearance,

luminousness to the solid particles of carbon discharged into it. If we blow against a luminous gas-flame, a sound is heard, a small explosion in fact, and by such a puff the light may be caused to disappear. During a windy night the exposed gas-jets in the shops are often deprived of their light, and burn blue. In like manner the common blowpipe-jet deprives burning coal-gas of its brilliant light. I hence concluded, that the explosions, the repetition of which produces the musical sound, rendered, at the moment they occurred, the combustion so perfect as to extinguish the solid carbon particles; but I imagined that the images on the screen would, on closer examination, be found united by spaces of blue, which, owing to their dimness, were not seen by the method of projection. This in many instances was found to be the case.

I was not, however, prepared for the following result :—A flame of olefiant gas, rendered almost as small as it could be, was procured. The 3-foot 2-inch tube was placed over it; the flame, on singing, became elongated, and lost some of its light, still it was bright at its top; looked at in the moving mirror, a beaded line of great beauty was observed; in front of each bead was a little luminous star, after it, and continuous with it, a spot of rich blue light, which terminated, and left, as far as I could judge, a perfectly dark space between it and the next following luminous star. I shall examine this further when time permits me, but as far as I can at present judge, the flame was actually extinguished and relighted in accordance with the sonorous pulsations.

When a silent flame, capable, however, of being excited by the voice in the manner already described, is placed within a tube, and the continuous line of light produced by it in the moving mirror is observed, I know no experiment more pretty than the resolution of this line into a string of richly luminous pearls at the instant the voice is pitched to the proper note. This may be done at a considerable distance from the jet, and with the back turned towards it.

The change produced in the line of beads when a tuning-fork, capable of giving beats with the flame, is brought over the tube,

and indicate alternate contractions and dilatations of the flame corresponding with the sonorous vibrations of the column of air.'—Phil. Trans., 1834, p. 586.

or over a resonant jar near it, is also extremely interesting to observe. I will not at present enter into a more minute description of these results. Sufficient, I trust, has been said to induce experimenters to reproduce the effects for themselves; the sight of them will give more pleasure than any description of mine could possibly do.

---

## TRANSLATION OF A PAPER ON ACOUSTIC EXPERIMENTS.*

A glass tube open at both ends, when simply blown upon by the mouth, gives its fundamental tone, i.e. the deepest tone belonging to it, as an open organ-pipe, feebly but distinctly. On placing the open hand upon one of the openings and rapidly withdrawing it, the tube yields two notes, one after the other; first the fundamental note of the closed pipe, and then the note of the open pipe, already mentioned, which is *an octave higher*. By the application of heat these fundamental tones, of which only the higher one will be taken into consideration here, are raised, as is well known; this is observed immediately on blowing upon a tube heated externally, or by a gas-flame burning in its interior. For example, a tube 242 millims. in length, and 20 millims. in diameter, heated throughout its whole length, when blown upon even before it reaches a red heat, gives a tone raised a major third, namely, the second G sharp in the treble clef, instead of the corresponding E. If a gas-flame 14 millims. in length, and 1 millim. in breadth at the bottom, is burning in the tube, the tone rises to the second treble F sharp. The same gas-flame raises the tone of a tube 273 millims. in length, and 21 millims. in width, from the second treble D to the corresponding E. These two tubes, which for brevity will hereafter be referred to as the E tube and the D tube, served for all the following experiments, the object of which was to show a well-known and by no means surprising fact, in a striking manner, namely, that the column of air in a tube is set in vibration when its fundamental tone, or one nearly allied, for example, an octave, is sounded outside the tube. The existence of the aërial vibrations

* By Count Schaffgotsch: Phil. Mag., December 1857.

was rendered perceptible by a column of smoke, a current of gas, and a gas flame.

1. A glimmering smoky taper was placed close under the E tube held perpendicularly, and the smoke passed through the tube in the form of a uniform thread. At a distance of 1·5 metre from the tube, the first treble E was sung. The smoke curled, and it appeared as if a part of it would be forced out at the upper, and the other part at the lower opening of the tube.

2. Two gas-burners, 1 millim. in the aperture, were applied near each other to the same conducting tube. Common gas flowed from both of them; one projected from below into the D tube for about one-fifth of its length; the gas flame of the other was 3 millims. in height. At a distance of 1·5 metre therefrom the first treble D was sung; the flame increased several times in breadth and height, and consequently in size generally; a larger quantity of gas therefore flowed out of the outer burner, which can only be explained by a diminution of the stream of gas in the inner burner, that is, in the one surrounded by the glass tube.

3. A burner, with an aperture of 1 millim. projecting from below into the D tube, about 80 millims., yielded a gas flame 14 millims. in length. At 5·6 metres therefrom the first treble E was sung: the flame was instantaneously extinguished. The same thing took place at 7 metres, when the flame is only 10 millims. in height, and the first treble D sharp is sung.

4. The last-mentioned flame is also extinguished by the note G sharp sounded close to it. Noises, such as the clapping of hands, pushing a chair, or shutting a book, do not produce this effect.

5. A burner with an aperture of 0·5 millim., projecting from below 60 millims. into the D tube, yielded a globular gas flame 3 to 3·5 millims. in diameter. By gradually closing a stopcock the passage of gas was more and more limited. The flame suddenly became much longer, but at the same time narrower, and nearly cylindrical, acquiring a bluish colour throughout, and from the tube a piercing second treble D was sounded; this is the phenomenon of the so-called chemical harmonica, which has been known for eighty years. When the stopcock is still further closed, the tone becomes stronger, the flame longer, narrower, and nearly spindle-shaped; at last it disappears.

An effect exactly similar to that caused by cutting off the gas is produced upon the small gas flame by a D, or the first treble D, sung or sounded from instruments; and in this case it is to be observed that the flame generally becomes the more sensitive the smaller it is, and the further the burner projects into the glass tube.

6. The flame in the D tube was 2 or 3 millims. in length; at a distance of 16·3 metres (more than 51 feet) from it, the first treble D was sounded. The flame immediately acquired the unusual form, and the second treble D sounded and continued to sound from the tube.

7. While the second treble D of the preceding experiment was sounding, the first treble D was sounded loudly close to the tube, when the flame became excessively elongated, and then disappeared.

8. The flame being only 1·5 millim. in length, the first treble D was sounded. The flame gave out the second treble D (and perhaps sometimes also a higher D) only for a moment, and disappeared. The flame is also affected by various D's of an adjustible labial pipe, by the contra D, D, D, the first treble D, and the second treble D of a harmonium, but by no single C sharp or D sharp of this powerful instrument. It is also affected by the third treble D of a clarionet, although only when quite close. The sung note also acts when it is produced by inspiration (in this case the second treble), or when the mouth is turned from the flame.

9. In immediate proximity the note G sung is effective.

Some influence is exerted by noises, but not by all, and often not by the strongest and nearest, evidently because the exciting tone is not contained in them.

10. The flame burning quietly in the interior of the D tube was about 2·5 millims. in length. In the next room, the door of which was open, the four legs of a chair were stamped simultaneously upon the wooden floor. The phenomenon of the chemical harmonica immediately occurred. A very small flame is of course extinguished, after sounding for an instant, by the noise of a chair. A tambourine, when struck, acts sometimes, but in general not.

11. The flame burning in the excited singing condition in the interior of the D tube, the latter was slowly raised as high as

possible without causing the return of the flame to the ordinary condition. The note, the first treble D, was sung strongly and *broken off suddenly* at a distance of 1·5 metre. The harmonic tone ceased, and the flame fell into a state of repose without being extinguished.

12. The same result was produced by acting upon the draught of air in the tube by a fanning motion of the open hand close above the upper aperture of the tube.

13. In the D tube there were two burners close together; one of them, 0·5 millim. in aperture, opened 5 millims below the other, the diameter of which was 1 millim. or more. Currents of gas, independent of each other, flowed out of both; that flowing from the narrower burner being very feeble, and burning when ignited, with a flame about 1·5 millim. in length, nearly invisible in the day; the first treble D was sung at a distance of three metres. The strong current of gas was immediately inflamed, because the little flame situated below it, becoming elongated, flared up into it. By a stronger action of the tone, the small flame itself is extinguished, so that an actual transfer of the flame from one burner to the other takes place. Soon afterwards the feeble current of gas is usually again inflamed by the large flame, and if the latter be again extinguished alone, everything is ready for a repetition of the experiment.

14. The same result is furnished by stamping with the chair, &c. It is evident that in this way gas-flames of any desired size and any mechanical action may be produced by musical tones and noises, if a wire stretched by a weight be passed through the glass tube in such a way that the flaring gas-flame must burn upon it.

15. If the flame of the chemical harmonica be looked at steadfastly, and at the same time the head be moved rapidly to the right and left alternately, an uninterrupted streak of light is not seen, such as is given by every other luminous body, but a series of closely approximated flames, and often dentated and undulated figures, especially when tubes of a metre and flames of a centimetre in length are employed.

This experiment also succeeds very easily without moving the eyes, when the flame is looked at through an opera-glass, the object-glass of which is moved rapidly to and fro, or in a circle; and also when the picture of the flame is observed in a hand-mirror shaken about. It is, however, only a variation of the experiment long since described and explained by Wheatstone, for which a mirror turned by watchwork was employed.

[It is perhaps but right that I should draw attention to the relation of the foregoing paper to one that I have published on the same subject. On May 6, and the days immediately following, the principal facts described in my paper were discovered; but on April 30, the foregoing results were communicated by Prof. Poggendorff to the Academy of Sciences in Berlin. Through the kindness of Mr. Schaffgotsch himself, I received his paper at Chamouni, many weeks after the publication of my own, and until then I was not aware of his having continued his experiments upon the subject.

We thus worked independently of each other, but as far as the described phenomena are common to both, all the merit of priority rests with Count Schaffgotsch.—J. T.]

# LECTURE IX.

[March 20, 1862.]

I HAVE said that the intensity of radiant heat diminishes with the distance, as light diminishes. What is the
law of diminution for light? I have here a square sheet of
paper, each side of the square measuring two feet; I fold it
thus to form a smaller square, each side of which is a foot
in length. The electric lamp now stands at a distance of sixteen feet from the screen; at a distance of eight feet, that is
exactly midway between the screen and the lamp, I hold this
square of paper; the lamp is naked, unsurrounded by its
camera, and the rays, uninfluenced by any lens, are
emitted on all sides. You see the shadow of the square
of paper on the screen. My assistant shall measure the
boundary of that shadow, and now I unfold my sheet of

paper so as to obtain the original large square; you see
by the creases, that it is exactly four times the area of the
smaller one. I place this large sheet against the screen,
and find that it exactly covers the space formerly occupied
by the shadow of the small square.

On the small square, therefore, when it stood midway
between the lamp and screen, a quantity of light fell
which, when the small square is removed, is diffused over
four times the area upon the screen. But if the same quantity of light is diffused over four times the area, it must be
diluted to one-fourth of its original intensity. Hence, by
doubling the distance from the source of light, we diminish
the intensity to one-fourth. By a precisely similar mode of
experiment we could prove, that by trebling the distance we
should diminish the intensity to one-ninth; and by quad-
rupling the distance we should reduce the intensity to one-
sixteenth: in short, we thus demonstrate the law that the
intensity of light diminishes as the square of the distance
increases. This is the celebrated law of Inverse Squares
as applied to light.

But I have said that heat diminishes according to the
same law. Observe the experiment which I am now about
to perform before you. I have here a tin vessel; narrow,
but presenting a side a square yard in area, MN (fig. 82). This
side, you observe, I have coated with lampblack. I
fill the vessel with hot water, intending to make this large
surface my source of radiant heat. I now place the conical
reflector on the thermo-electric pile, P, but instead of permitting it to remain a reflector, I push into the hollow
cone this lining of black paper, which fits exactly, and
which, instead of reflecting any heat that may fall obliquely
on it, completely cuts off the oblique radiation. The pile
is now connected with the galvanometer, and I place its
reflector close to this large radiating surface, the face of
the pile being about six inches distant from the surface.

The needle of the galvanometer moves : let it move until it takes up its final position. It now points steadily

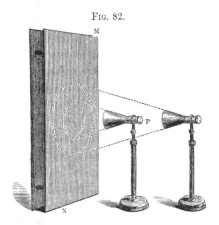

Fig. 82.

to 60°, and there it will remain as long as the temperature of the radiating surface remains sensibly constant. I will now gradually withdraw the pile from the surface, and will ask you to observe the effect upon the galvanometer. Of course you will expect that as I retreat from the source of heat, the intensity of the heat will diminish, and that the deflection of the galvanometer will diminish in a corresponding degree. I am now at double the distance, but the needle does not move; I treble the distance, the needle is still stationary; I successively quadruple, quintuple—go to ten times the distance, but the needle is rigid in its adherence to the deflection of 60° There is, to all appearance, no diminution at all of intensity with the increase of distance.

From this experiment, which might at first sight appear fatal to the law of inverse squares, as applied to heat, Melloni, in the most ingenious manner, proved the law. Mark his reasoning. I again place the pile close to the radiating surface. Imagine the hollow cone in front

of the pile prolonged ; it would cut the radiating surface
in a circle, and this circle is the only portion of the surface whose rays can reach the pile.   All the other rays are
cut off by the non-reflecting lining of the cone.   I move
the pile to double the distance ; the section of the cone
prolonged now encloses a circle of the radiating surface,
exactly four times the area of the former circle; at treble
the distance the radiating surface is augmented nine times;
at ten times the distance the radiating surface is augmented
100 times.  But the constancy of the deflection proves that
the augmentation of the radiating surface must be exactly
neutralised by the diminution of intensity ; the radiating
surface augments as the square of the distance, hence the
intensity of the heat *must diminish as the square of the
distance*; and thus the experiment, which might at first
sight appear fatal to the law, demonstrates the law in the
most simple and conclusive manner.

Let us now revert for a moment to our fundamental
conceptions regarding radiant heat.  Its origin is an
oscillatory motion of the ultimate particles of matter
—a motion taken up by the ether, and propagated
through it in waves.   The particles of ether in these waves
do not oscillate in the same manner as the particles of air
in the case of sound.   The air-particles move to and
fro, in the direction in which the sound travels, the ether
particles move to and fro, *across* the line in which the
light travels.   The undulations of the air are longitudinal,
the undulations of the ether are transversal.   The ether
waves resemble more the ripples of water than they do the
aerial pulses which produce sound ; that this is the case
has been inferred from optical phenomena.  But it is manifest that the disturbance produced in the ether must depend upon the character of the oscillating mass ; one
atom may be more unwieldy than another, and a single
atom could not be expected to produce so great a disturbance

as a group of atoms oscillating as a system. Thus, when different bodies are heated, we may fairly expect that they will not all create the same amount of disturbance in the ether. It is probable that some will communicate a greater amount of motion than others: in other words, that some will radiate more copiously than others; for radiation, strictly defined, *is the communication of motion from the particles of a heated body, to the ether in which these particles are immersed.*

Let us now test this idea by experiment. I have here a cubical vessel, c (fig. 83)—a ' Leslie's cube '—so called from its having been used by Sir John Leslie in his beautiful researches on radiant heat. The mass of the cube is pewter, but one of its sides is coated with a layer of gold, another with a layer of silver, a third with a layer of copper, while the fourth I have coated with a varnish of isinglass. I fill the cube with hot water, and keeping it at a constant distance from the thermo-electric pile, P, I

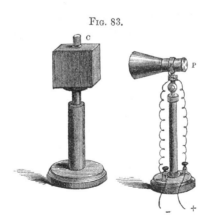

Fig. 83.

allow its four faces to radiate, in succession, against the pile. The hot gold surface, you see, produces scarcely any deflection; the hot silver is equally inoperative, the

same is the case with the copper; but when I turn this
varnished surface towards the pile, the gush of heat becomes suddenly augmented; and the needle, as you see,
moves up to its stops.  Hence we infer, that through some
physical cause or other, the molecules of the varnish, when
set in motion by the hot water within the cube, communicate more motion to the ether than the atoms of the
metals; in other words, the varnish is a better radiator
than the metals are.  I obtain a similar result when I
compare this silver teapot with this earthenware one;
filling them both with boiling water, the silver, you see,
produces but little effect, while the radiation from the earthenware is so copious as to drive the needle up to 90°.  Thus,
also, if I compare this pewter pot with this glass beaker,
when both are filled with hot water, the radiation from the
glass is much more powerful than that from the pewter.

   You have often heard of the effect of colours on radiation, and heard a good deal, no doubt, which is unwarranted
by experiment.  I have here a cube, one of whose sides
is coated with whiting, another with carmine, a third with
lampblack, while the fourth is left uncoated.  I present
the black surface first to the pile, the cube being filled
with boiling water ; the needle moves up, and now points
steadily to 65°.  The cube rests upon a little turn-table, and
by turning the support I present the white face to the pile;
the needle remains stationary, proving that the radiation
from the white surface is just as copious as that from
the black.  I turn the red surface towards the pile, there
is no change in the position of the needle.  I turn the
uncoated side, the needle instantly falls, proving the in-
feriority of the metallic surface as a radiator.  I repeat
precisely the same experiments with this cube, the sides
of which are covered with velvet; one face with black
velvet, another with white, and a third with red.    The
results are precisely the same as in the former instances;

the three velvet surfaces radiate alike, while the naked surface radiates less than any of them.  These experiments show that the radiation from the clothes which cover the human body, is independent of the colour of these clothes; the colour of an animal's fur is equally incompetent to influence the radiation.    These are the conclusions arrived at by Melloni *for obscure heat.**

But if the coated surface communicates more motion to the ether than the uncoated one, it necessarily follows that the coated vessel will cool more quickly than the uncoated one.  I have here two cubes, one of which is quite coated with lampblack, while the other is bright. At the commencement of the lecture I poured boiling water into these vessels, and placed in each a thermometer.  A short time ago both thermometers showed the same temperature, but now one of them is two degrees below the other.    The velocity of cooling in one vessel is greater than in the other, and the vessel which cools quickest is the coated one.    Here are two vessels, one of which is bright and the other closely coated with flannel.  Half an hour ago two thermometers plunged in these vessels showed the same temperature, but they show it no longer; the covered vessel has now a temperature two or three degrees lower than the naked one.  It is usual to preserve the heat of teapots by a woollen covering, but the cover must fit very loosely.    In this case, though the covering may be a good radiator, its goodness is more than counterbalanced by the difficulty encountered by the heat in reaching the outer surface of the covering.    A closely fitting cover would, as we have seen, promote the loss which it is intended to diminish, and thus do more harm than good.

One of the most interesting points connected with our

* By the application of a more powerful and delicate test than that employed by Melloni, I find that his conclusions will require modification.

subject is the reciprocity which exists between the power of a
body to communicate motion to the ether, or to radiate ; and
its capacity to accept motion from the ether, or to absorb.
As regards radiation we have already compared lampblack and chalk with metallic surfaces ; we will now compare the same substances with reference to their powers of
absorption.   I have here two sheets of tin, M N, O P (fig. 84),
one of them coated with whiting and the other left uncoated;
I place them thus parallel to each other, and at a distance
of about two feet asunder.  To the edge of each sheet I have

Fig. 84.

soldered a screw, and from one screw to the other I stretch
a copper wire, *a b*, which now connects the two sheets.  At
the back of each sheet I have soldered one end of a little
bar of bismuth, to the other end, *e*, of which a wire is
soldered, and terminated by a binding screw. To these two
binding screws I attach the two ends of the wire coming

from my galvanometer at ɢ, and you observe I have now an
unbroken circuit, in which the galvanometer is included.
You know already what the bismuth bars are intended for.
I place my warm finger on this left-hand one, a current is
immediately developed, which passes from the bismuth to
the tin, thence through the wire connecting the two sheets,
thence round the galvanometer, to the point from which
it started.   You observe the effect.   The needle of the
galvanometer moves through a large arc; the red end
going towards you.   The junction of tin and bismuth is
now cooling, the needle returns to 0°, and now I will
place my finger upon the bismuth at the back of the other
plate—you see the effect—a large deflection in the opposite
direction; the red end of the needle now comes towards
me.   I withdraw my finger, the junction cools, and once
more the needle sinks to zero.

   I set this stand exactly midway between the two sheets
of tin, and on the stand I intend to place a heated copper
ball; the ball will radiate its heat against both sheets:
on the right, however, the rays will strike upon a coated
surface, while on the left they will strike upon a naked
metallic surface.   If both surfaces drink in the radiant
heat—if both accept with equal freedom the motion of the
ethereal waves—the bismuth junctions at the backs will be
equally warmed, and one of them will neutralise the other.
But if one surface be a more powerful absorber than the
other, that which absorbs most will heat its bismuth indi-
cator most; a deflection of the galvanometer needle will be
the consequence, and the direction of the deflection will
tell us which is the best absorber.   The ball is now upon
the stand, and you see we have not long to wait for a
decision of the question.   The prompt and energetic deflection of the needle informs us that the coated surface is
the most powerful absorber.   In the same way I compare

lampblack and varnish with tin, and find the two former by far the best absorbers.*

The thinnest metallic coating furnishes a powerful defence against the absorption of radiant heat. I have here a sheet of 'gold paper,' the gold being merely copper reduced to great tenuity. Here is a red powder, the iodide of mercury, with which I coat the under surface of the gold paper. This iodide, as many of you know, has its red colour discharged by heat, the powder becoming a pale yellow. I lay the paper flat on this board with the coloured surface downwards, and on this upper metallic surface I paste pieces of paper—common letter paper will answer my purpose. A figure of any desired shape is thus formed on the surface of the copper. I now take a red-hot spatula in my hand and pass it several times over the sheet; the spatula radiates strongly against the sheet, but I apprehend that its rays are absorbed in very different degrees. The metallic surface will absorb' but little; the paper surfaces will absorb greedily; and, on turning up the sheet, you see the effect: the iodide underneath the metallic portion is perfectly unchanged, while under every bit of paper the colour is discharged, thus forming below an exact copy of the figure pasted on the opposite surface of the sheet. Here is another example of the same kind, for which I am indebted to Mr. Hill, of the establishment of Mr. Jacob Bell in Oxford Street. A hot fire sent its rays against this painted piece of wood (fig. 85), on which the number 338 was printed in gold leaf letters; the paint is blistered and charred all round the letters, but underneath the letters the wood and paint are quite unaffected. This thin film of gold has been quite sufficient to prevent the absorption, to which the destruction of the surrounding surface is due.

* Colour, according to Melloni, has no influence on the absorption of *obscure heat*: on luminous heat, such as that of the sun, it has great influence.

The luminiferous ether fills stellar space; it makes the universe a whole, and renders the intercommunication of light and energy between star and star possible.   But the subtle substance penetrates further; it surrounds the very atoms of solid and liquid substances.   Transparent bodies are such, because the ether and their atoms are so related

Fig. 85.

to each other, that the waves which excite light can pass through them, without transferring their motion to the atoms.   In coloured bodies certain waves are broken or absorbed; but those which give the body its colour pass without loss.   Through this solution of sulphate of copper, for example, the blue waves speed unimpeded, but the red waves are destroyed.   I form a spectrum upon the screen; sent through this solution you see the red end of the spectrum is cut away.   This piece of red glass, on the contrary, owes its redness to the fact that its substance can be traversed freely by the longer undulations of red, while the shorter waves are absorbed.   Interposing it in the path of this light you see it cuts the blue end of the spectrum quite away, leaving merely a vivid red band upon the screen. This blue liquid then cuts off the rays which are transmitted by the red glass; and the red glass cuts off the rays which are transmitted by the liquid; by the union of both we ought to have perfect opacity, and so we have.   When both are placed in the path of the beam, the entire spectrum

disappears; the union of these two transparent bodies produce an opacity equal to that of pitch or coal.

I have here another liquid—a solution of the perman-
ganate of potash—which I introduce into the path of the
beam.   See the effect upon the spectrum; the two ends
pass freely through, you have the red and the blue, but
between both a space of intense blackness.   The yellow of
the spectrum is pitilessly destroyed by this liquid; through
the entanglement of its atoms these yellow rays cannot pass,
while the red and the blue glide round them and get
through the inter-atomic spaces without sensible hindrance.
And hence the gorgeous colour of this liquid.   I will turn
the lamp round and project a disk of light two feet in
diameter upon the screen.   I now introduce this liquid;
can anything be more splendid than the colour of that
disk?   I again turn the lamp obliquely and introduce a
prism; here you have the components of that beautiful
colour; the violet component has slidden away from the
red.   You see two definite disks of these two colours upon
the screen, which overlap in the centre, and exhibit there
the colour of the composite light which passes through
the liquid.

Thus, as regards the waves of light, bodies exercise as
it were an elective power, singling out certain waves for
destruction, and permitting others to pass.   Transparency
to one wave does not at all imply transparency to others,
and from this we might reasonably infer, that transparency to light does not imply transparency to radiant
heat.   This conclusion is entirely verified by experiment.
I have here a tin screen, MN (fig. 86), pierced by an aperture,
behind which is soldered a small stand s.   I place this
copper ball, B, heated to dull redness, on a candlestick,
which will serve as a support for the ball.   At the other
side of the screen I place my thermo-electric pile, P; the
rays from the ball now pass through the aperture in the

screen and fall upon the pile—the needle goes up, and
finally comes to rest with a steady deflection of 80°. I
have here a glass cell, a quarter of an inch wide, which I
now fill with distilled water. I place the cell on the stand,
so that all rays reaching the pile must pass through it;

Fig. 86.

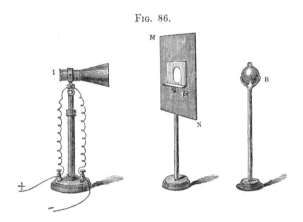

what takes place? The needle steadily sinks almost to
zero; scarcely a ray from the ball can cross this water;—
to the undulations issuing from the ball the water is practically opaque, though so extremely transparent to the
rays of light. Before removing the cell of water I place
behind it a similar cell, containing transparent bisulphide
of carbon; so that now, when I remove the water cell, the
aperture is still barred by the new liquid. What occurs?
The needle promptly moves upwards and describes a large
arc; so that the selfsame rays that found the water impenetrable, find easy access through the bisulphide of
carbon. In the same way I compare this alcohol with this
chloride of phosphorus, and find the former almost opaque
to the rays emitted by our warm ball, while the latter
permits them to pass freely.

So also as regards solid bodies; I have here a plate of

very pure glass, which I place on the stand, and, using a
cube of hot water instead of the ball B, I permit the rays
from the heated cube to pass through it, if they can. No
movement of the needle is perceptible. I now displace
the plate of glass by a plate of rocksalt of ten times the
thickness; you see how promptly the needle moves until
it is arrested by its stops. To these rays, then, the rocksalt is eminently transparent, while the glass is practically
opaque to them.

For these, and numberless similar results, we are indebted to Melloni, who may be almost regarded as the
creator of this branch of our subject. To express this power
of instantaneous transmission of radiant heat, he proposes
the word *diathermancy*. Diathermancy bears the same
relation to radiant heat that transparency does to light.
Instead of giving you determinations of my own of the
diathermancy of various bodies, I will make a selection
from the tables of the eminent Italian philosopher just
referred to. In these determinations Melloni uses four
different sources of heat, the flame of a Locatelli lamp; a
spiral of platinum wire, kept incandescent by the flame of
an alcohol lamp; a plate of copper heated to 400° Cent,
and a plate of copper heated to 100° Cent., the last
mentioned source being the surface of a copper cube
containing boiling water. The experiments were made
in the following manner:—First, the radiation of the
source, that is to say the galvanometric deflection produced by it, was determined when nothing but air
intervened between the source and the pile; then the
substance whose diathermancy was to be examined was
introduced, and the consequent deflection noted. Calling
the quantity of heat represented by the former deflection 100, the proportionate quantities transmitted by
twenty-five different substances are given in the following
table:—

| Names of substances reduced to a common thickness of $\frac{1}{10}$th of an inch (2·6 millim.) | Transmissions ; per centage of the total radiation | | | |
|---|---|---|---|---|
| | Locatelli Lamp | Incandescent Platinum | Copper at 400° C. | Copper at 100° C. |
| 1 Rocksalt . . . . . | 92·3 | 92·3 | 92·3 | 92·3 |
| 2 Sicilian sulphur . . . | 74 | 77 | 60 | 54 |
| 3 Fluor spar . . . . | 72 | 69 | 42 | 33 |
| 4 Beryl . . . . . | 54 | 23 | 13 | 0 |
| 5 Iceland spar . . . . | 39 | 28 | 6 | 0 |
| 6 Glass . . . . . | 39 | 24 | 6 | 0 |
| 7 Rock crystal (clear) . . . | 38 | 28 | 6 | 3 |
| 8 Smoky quartz . . . . | 37 | 28 | 6 | 3 |
| 9 Chromate of Potash . . | 34 | 28 | 15 | 0 |
| 10 White Topaz . . . . | 33 | 24 | 4 | 0 |
| 11 Carbonate of Lead . . . | 32 | 23 | 4 | 0 |
| 12 Sulphate of Baryta . . . | 24 | 18 | 3 | 0 |
| 13 Felspar . . . . . | 23 | 19 | 6 | 0 |
| 14 Amethyst (violet) . . . | 21 | 9 | 2 | 0 |
| 15 Artificial amber . . . | 21 | 5 | 0 | 0 |
| 16 Borate of Soda . . . | 18 | 12 | 8 | 0 |
| 17 Tourmaline (deep green) . | 18 | 16 | 3 | 0 |
| 18 Common gum . . . . | 18 | 3 | 0 | 0 |
| 19 Selenite . . . . . | 14 | 5 | 0 | 0 |
| 20 Citric acid . . . . | 11 | 2 | 0 | 0 |
| 21 Tartrate of Potash . . . | 11 | 3 | 0 | 0 |
| 22 Natural amber . . . | 11 | 5 | 0 | 0 |
| 23 Alum . . . . . | 9 | 2 | 0 | 0 |
| 24 Sugar-candy . . . . | 8 | 1 | 0 | 0 |
| 25 Ice . . . . . | 6 | 0·5 | 0 | 0 |

This table shows, in the first place, what very different transmissive powers different solid bodies possess. It shows us also that, with a single exception, the transparency of the bodies mentioned for radiant heat varies with the *quality* of the heat. Rocksalt alone is equally transparent to heat from the four sources experimented with. It must be borne in mind here that the luminous rays are also calorific rays; that the selfsame ray, falling upon the nerve of vision, produces the impression of light; while, impinging upon other nerves of the body, it produces the impression of heat. The luminous calorific rays have, however, a shorter length than the obscure rays, and knowing, as we do, how differently waves of different

lengths are absorbed by bodies, we are in a measure prepared for the results of the foregoing table.   Thus, while
glass, of the thickness specified, permits 39 per cent. of
the rays of Locatelli's lamp, and 24 per cent. of the rays
from the incandescent platinum to pass, it gives passage to
only 6 per cent. of the rays from copper, at a temperature of
400° C., while it is absolutely opaque to all rays emitted
from a source of 100° C.   We also see that limpid ice,
which is so highly transparent to light, allows to pass only
6 per cent. of the rays of the lamp, and 0·5 per cent. of the
rays emitted by the incandescent platinum, while it utterly
cuts off all rays issuing from the other two sources.   We
have here an intimation, that by far the greater portion of the rays emitted by the lamp of Locatelli must be
obscure.   Luminous rays pass through ice, of the thickness
here given, without sensible absorption, and the fact that
94 per cent. of the rays issuing from Locatelli's flame are
destroyed by the ice, proves that this proportion of these
rays *must be obscure*.   As regards the influence of transparency, clear and smoky quartz are very instructive.
Here are the two substances, one perfectly pellucid, the
other a dark brown ; still, for the luminous rays only, do
these two specimens show a difference of transmission.
The clear quartz transmits 38 per cent., and the smoky
quartz 37 per cent. of the rays from the lamp, while, for
the other three sources, the transmissions of both substances
are identical.

In the following table, which I also borrow from Melloni,
the calorific transmissions of different liquids are given.
The source of heat was an Argand lamp furnished with a
glass chimney, and the liquids were enclosed in a cell with
glass sides, the thickness of the liquid layer being 9·21
millimetres.

| Names of Liquids | Calorific transmissions; per centage of the total radiation |
|---|---|
| Bisulphide of Carbon . . . . | 63 |
| Bichloride of Sulphur . . . . | 63 |
| Protochloride of Phosphorus . . . | 62 |
| Essence of Turpentine . . . . | 31 |
| Olive Oil . . . . . . | 30 |
| Naphtha . . . . . . | 28 |
| Essence of Lavender . . . . | 26 |
| Sulphuric Ether . . . . . | 21 |
| Sulphuric Acid . . . . . | 17 |
| Hydrate of Ammonia . . . . | 15 |
| Nitric Acid . . . . . . | 15 |
| Absolute Alcohol . . . . . | 15 |
| Hydrate of Potash . . . . . | 13 |
| Acetic Acid . . . . . . | 12 |
| Pyroligneous Acid . . . . . | 12 |
| Concentrated Solution of Sugar . . | 12 |
| Solution of Rocksalt . . . . | 12 |
| White of Egg . . . . . | 11 |
| Distilled Water . . . . . | 11 |

Liquids are here shown to be as diverse in their powers of transmission as solids; and it is also worthy of remark, that water maintains its opacity, notwithstanding the change in its state of aggregation.

The reciprocity which we have already demonstrated between radiation and absorption in the case of metals, varnishes, &c., may now be extended to the bodies contained in Melloni's tables. I will content myself with one or two illustrations, borrowed from Mr. Balfour Stewart. Here is a copper vessel in which water is kept in a state of gentle ebullition. On the flat copper lid of this vessel I place plates of glass and of rocksalt, till they have assumed the temperature of the lid. I place the plate of rocksalt upon this stand, in front of the thermo-electric pile. You observe the deflection; it is so small as to be scarcely sensible. I now remove the rocksalt, and put in its place a plate of heated glass; the needle moves upwards through a large

arc, thus conclusively showing that the glass, which is the
more powerful absorber of obscure heat, is also the more
powerful radiator. Alum, unfortunately, melts at a temperature lower than that here made use of; but though
its temperature is not so high as that of the glass, you can
see that it transcends the glass as a radiator; the action
on the galvanometer is still more energetic than in the
case of the last experiment.

Absorption takes place within the absorbing body; and
it requires a certain thickness of the body to accomplish
the absorption. This is true of both light and radiant heat.
A very thin stratum of pale beer is almost as colourless as a
stratum of water, the absorption being too inconsiderable to
produce the decided colour which larger masses of the beer
exhibit. I pour distilled water into a drinking glass;
in this quantity it exhibits no trace of colour, but I have
arranged here an experiment which will show you that
this pellucid liquid, in sufficient thickness, exhibits a very
decided colour. Here is a tube fifteen feet long, A B (fig. 87),

FIG. 87.

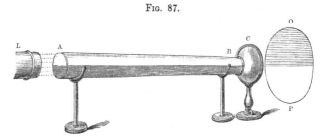

placed horizontal, the ends of which are stopped by pieces of
plate glass; at one end of the tube stands an electric lamp, L,
from which I intend to send a cylinder of light through
the tube. The tube is now half filled with water, the
upper surface of which cuts the tube in two equal parts
horizontally. Thus I send half of my beam through air
and half through water, and with this lens, c, I intend to

project a magnified image of the adjacent end of the tube, upon the screen. Here it is; you see the image, O P, composed of two semicircles, one of which is due to the light which has passed through the water, the other to the light which has passed through the air. Side by side, thus, you can compare them, and you notice that while the air semicircle is a pure white, the water semicircle is a bright and delicate blue green. Thus, by augmenting the thickness through which the light has to pass, you deepen the colour; this proves that the destruction of the light rays takes place *within* the absorbing body, and is not an effect of its surface merely.

Melloni shows the same to be true of radiant heat. In our table, at page 297, the thickness of the plates used was 2·6 millimetres, but by rendering the plate thinner we enable a greater quantity of heat to get through, and by rendering it sufficiently thin, we may, with a very opaque substance, almost reach the transmission of rocksalt. The following table shows the influence of thickness on the transmissive power of a plate of glass.

| Thickness of Plates in Milli-metres | Transmissions by Glass of different thicknesses ; per centage of the total Radiation | | | |
|---|---|---|---|---|
| | Locatelli Lamp | Incandescent Platinum | Copper at 400° C. | Copper at 100° C. |
| 2·6 | 39 | 24 | 6 | 0 |
| 0·5 | 54 | 37 | 12 | 1 |
| 0·07 | 77 | 57 | 34 | 12 |

Thus, we see, that by diminishing the thickness of the plate from 2·6 to 0·07 millimetres, the quantity of heat transmitted rises, in the case of the lamp of Locatelli, from 39 to 77 per cent.; in the case of the incandescent platinum, from 24 to 57 per cent.; in the case of copper at 400°

C. from 6 to 34 per cent.; and in the case of copper at
100° C., from absolute opacity to a transmission of 12 per
cent.

The influence of the thickness of a plate of selenite on
the quantity of heat which it transmits is exhibited in the
following table.

| Thickness of Plates in Milli- metres | Transmissions by Selenite of different thicknesses ; per centage of total radiation | | | |
|---|---|---|---|---|
| | Locatelli Lamp | Incandescent Platinum | Copper at 400° C. | Copper at 100° C |
| 2.6 | 14 | 5 | 0 | 0 |
| 0.4 | 38 | 18 | 7 | 0 |
| 0.01 | 64 | 51 | 32 | 21 |

The decomposition of the solar beam gives us the solar
spectrum ; luminous in the centre, calorific at one end,
and chemical at the other. The sun is therefore a source of
heterogenous rays, and there can scarcely be a doubt that
all other sources of heat, luminous and obscure, partake of
this heterogeniety. In general, when such mixed rays enter
a diathermic substance, some are struck down and others
permitted to pass. Supposing, then, that we take a sheaf
of calorific rays which have already passed through a diathermic plate, and permit them to fall upon a second plate
of the same material, the transparency of this second plate
to the heat incident upon it will be greater than the transparency of the first plate to the heat incident on *it*. In
fact the first plate, if sufficiently thick, has already extinguished, in great part, the rays which the substance is
capable of absorbing ; and the residual rays, as a matter of
course, penetrate a second plate of the same substance with
comparative freedom. The original beam is *sifted* by the
first plate, and the purified beam possesses, for the same substance, a higher penetrative power than the original beam.

This power of penetration has usually been taken as a test
of the *quality* of heat; the heat of the purified beam
is said to be different in quality from that of the unpurified beam.   It is not, however, that any individual ray has
changed its quality, but that from the beam, as a whole,
certain rays have been withdrawn, and that their withdrawal has altered the proportion of the incident heat transmitted by a second substance.   This, I think, is the true
meaning of the term 'quality' as applied to radiant heat.
In the path of the rays from a lamp let plates of rocksalt,
alum, bichromate of potash, and selenite be successively
placed, each plate 2·6 millimetres in thickness; let the
heat emergent from these plates fall upon a second series
of the same thickness; out of every 100 rays of this latter
heat, the following proportions are transmitted.

|  |  |
|---|---|
| Rocksalt | 92·3 |
| Alum | 90 |
| Chromate of Potash | 71 |
| Selenite | 91 |

Referring to the table, p. 297, we find that of the whole of
the rays emitted by the Locatelli lamp, only 34 per cent. are
transmitted by the chromate of potash; here we find the
percentage 71.   Of the entire radiation, selenite transmits
only 14 per cent., but of the beam which has been purified
by a plate of its own substance it transmits 91 per cent.
The same remark applies to the alum, which transmits
only 9 per cent. of the unpurified beam, and 90 per
cent. of the purified beam.   In rocksalt, on the contrary,
the transmissions of the sifted and unsifted beam are the
same, because the substance is equally transparent to rays
of all kinds.*   In these cases I have supposed the rays

* This was Melloni's conclusion; but the experiments of MM. Provostaye
and Desains, and of Mr. Balfour Stewart, prove that the conclusion is not
strictly correct.

emergent from rocksalt to pass through rocksalt; the rays
emergent from alum to pass through alum, and so of the
others; but, as might be expected, the sifting of the beam,
by any substance, will alter the proportion in which it will
be transmitted by almost any other second substance.

I will conclude these observations with an experiment
which will show you the influence of sifting in a very striking
manner.   I have here a sensitive differential air-thermo-
meter with a clean glass bulb. You see the slightest touch of
my hand causes a depression of the thermometric column.
Here is our electric lamp, and from it I will converge a
powerful beam on the bulb of that thermometer.   The
focus now falls directly on the bulb, and the air within
it is traversed by a beam of intense power; but not the
slightest depression of the thermometric column is discernible.   When I first showed this experiment to an individual here present, he almost doubted the evidence of his
senses; but the explanation is simple. The beam, before it
reaches the bulb, is already sifted by the glass lens used to
concentrate it, and having passed through 12 or 14 feet of
air, the beam contains no constituent that can be sensibly
absorbed by the air within the bulb.   Hence the hot beam
passes through both air and glass without warming either.
It is competent, however, to warm the thermo-electric pile;
exposure of the pile to it, for a single instant, suffices to drive
the needle violently aside; or let me coat with lampblack
the portion of the glass bulb struck by the beam; you see
the effect: the heat is now absorbed, the air expands, and
the thermometric column is forcibly depressed.

We use glass fire-screens, which allow the pleasant light
of the fire to pass, while they cut off the heat; the reason
is, that by far the greater part of the heat emitted by a
fire consists of obscure rays, to which the glass is opaque.
But in no case is there any loss.  The rays absorbed by
the glass go to warm the glass; the motion of the ethereal

waves is transferred to the molecules of the solid. But
you may be inclined to urge, that under these circumstances the glass screen itself ought to become a source of
heat, and that therefore we ought to derive no benefit from
its absorption. The fact is so, but the conclusion is
unwarranted. The philosophy of the screen is this:—

FIG. 88.

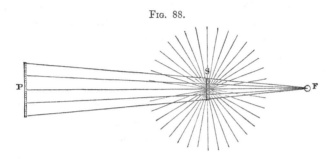

Let F (fig. 88) be a fire from which the rays proceed in
straight lines towards a person at P. Before the screen is
introduced, each ray pursues its course direct to P; but now
let a screen be placed at S. The screen intercepts the
rays of heat and becomes warmed; but instead of sending
on the rays in their original direction only, it emits them,
as a warm body, *in all directions*. Hence, it cannot restore
to the person at P all the heat intercepted. A portion
of the heat is restored, but by far the greater part is
diverted from P, and distributed in other directions.

Where the waves pursue their way unabsorbed, no motion
of heat is imparted, as we have seen in the case of the air
thermometer. A joint of meat might be roasted before a
fire, with the air around the joint as cold as ice.
The air on high mountains may be intensely cold, while
a burning sun is overhead; the solar rays which, striking on the human skin, are almost intolerable, are incompetent to heat the air sensibly, and we have only to
withdraw into perfect shade to feel the chill of the

atmosphere. I never, on any occasion, suffered so much
from solar heat as in descending from the 'Corridor' to the
Grand Plateau of Mont Blanc, on August 13, 1857; though
hip deep in snow at the time, the sun blazed against me
with unendurable power. Immersion in the shadow of
the Dome du Gouté at once changed my feelings; for
here the air was at a freezing temperature. It was not,
however, sensibly colder than the air through which the
sunbeams passed; and I suffered, not from the contact
of hot air, but from the impact of calorific rays which
had reached me through a medium icy cold.

The beams of the sun also penetrate glass without
sensibly heating it, and the reason is, that having passed
through our atmosphere, the beams have been in a great
measure deprived of those rays which can be absorbed by
glass.*  I made an experiment in a former lecture which
you will now completely understand. I sent a beam from
the electric lamp through a mass of ice without melting
the substance. I had previously sifted the beam by
sending it through a vessel of water, in which the rays
capable of being absorbed by the ice were lodged—and so
copiously lodged—that the water was raised almost to the
boiling point during the experiment. It is here worthy
of remark that the liquid water and the solid ice appear
to be pervious and impervious to the same rays; the one
may be used as a *sieve* for the other; a result which indicates that the quality of the absorption is not influenced
by the difference of aggregation between solid and liquid.
It is easy to prove that the beam which has traversed the
ice without melting it, is really a calorific beam, by allowing

* On *à priori* grounds I should conclude that the obscure solar rays which
have succeeded in getting through our atmosphere, must be able to penetrate the humours of the eye and reach the retina: the recent experiments
of M. Franz prove this. Their not producing vision is, therefore, not due to
their absorption by the humours of the eye, but to their own intrinsic incompetence to excite the retina.

it to fall upon our thermo-electric pile. Here is a beam which has passed through a layer of water; I permit it to fall upon the pile, and you instantly see its effect upon the galvanometer, causing the needle to move with energy to its stops. Here is a beam which has passed through ice, but you see that it is equally competent to affect the pile; here, finally, is a beam which has passed through both water and ice; you see it still possesses heating power.*

When the calorific rays are intercepted, they, as a general rule, raise the temperature of the body by which they are absorbed; but when the absorbing body is ice at a temperature of 32° Fahr., it is impossible to raise its temperature. How then does the heat absorbed by the ice employ itself? It produces internal liquefaction, it takes down the crystalline atoms, and thus forms those lovely liquid flowers which I showed you in a former lecture.†

We have seen that transparency to light is not at all a test of diathermancy; that a body highly transparent to the luminous undulations may be highly opaque to the nonluminous ones. I have also given you an example of the opposite kind, and showed you that a body may be absolutely opaque to light and still, in a considerable degree, transparent to heat. I set the electric lamp in action, and you see this convergent beam tracking itself through the dust of the room: you see the point of convergence of the rays here, at a distance of fifteen feet from the lamp; I will mark that point accurately by the end of this rod. Here is a plate of rocksalt coated so thickly with soot that the light, not only of every gas lamp in this room, but the electric light itself, is cut off by it. I interpose this plate of smoked salt in the path of the beam; the light is intercepted, but the rod enables me to find

* Mr. Faraday has fired gunpowder by converging the solar rays upon it by a lens of ice.

† For the bearing of these results on air and water-bubbles of ice, see Appendix to Lecture IX.

x 2

with my pile the place where the focus fell.   I place the
pile at this focus: you see no beam falling on the pile,
but the violent action of the needle instantly reveals to
the mind's eye a focus of heat at the point from which the
light has been withdrawn.

You might, perhaps, be disposed to think that the heat
falling on the pile has been first absorbed by the soot, and
then radiated from it as an independent source.   Melloni
has removed every objection of this kind; but none of his
experiments, I think, are more conclusive, as a refutation of
the objection, than that now performed before you.   For if
the smoked salt were the source, the rays could not converge
here to a focus, for the salt is *at this side* of the converging
lens, and you see when I displace my pile a little laterally,
still keeping it turned towards the smoked salt, the needle
sinks to zero.   The heat, moreover, falling on the pile is,
as shown by Melloni, practically independent of the position
of the plate of rocksalt; you may cut off the beam at a.
distance of fifteen feet from the pile, or at a distance of
one foot; the result is sensibly the same, which could not
be the case if the smoked salt itself were the source of heat.

I make a similar experiment with this black glass, and
the result, as you see, is the same.   Now the glass *reflects* a
considerable portion of the light and heat from the lamp ; if
I hold it a little oblique to the beam you can see the reflected
portion.   While the glass is in this position I will coat it
with an opaque layer of lampblack so as to cause it to absorb,
not only all the rays which are now entering it, but also the
portion which it reflects.   What is the result ?   Though the
glass plate has become the seat of augmented absorption,
it has ceased to affect the pile, the needle descends to zero,
thus furnishing additional proof that the rays which, in the
first place, acted upon the pile, came direct from the lamp,
and traversed the black glass, as light traverses a transparent
substance.

Rocksalt transmits all rays, luminous and obscure; alum, of the thickness already given, transmits only the luminous rays; hence the difference between alum and rocksalt will give the value of the obscure radiation.   Tested in this way, Melloni finds the following proportions of luminous to obscure rays for the three sources mentioned:—

| Source | Luminous | Obscure |
|---|---|---|
| Flame of Oil    .    . | 10 | 90 |
| Incandescent Platinum | 2 | 98 |
| Flame of Alcohol    .  | 1 | 99 |

Thus, of the heat radiated from the flame of oil, 90 per cent. is due to the obscure rays; of the heat radiated from incandescent platinum, 98 per cent. is due to obscure rays, while of the heat radiated from the flame of alcohol, fully 99 per cent. is due to the obscure radiations.

# APPENDIX TO LECTURE IX.

———◦◦◦———

## EXTRACT FROM A MEMOIR ON SOME PHYSICAL PROPERTIES OF ICE.*

### § I.

I AVAILED myself of the fine sunny weather with which we were favoured last September and October, to examine the effects of solar heat upon ice. The experiments were made with Wenham Lake and Norway ice. Slabs were formed of the substance, varying from one to several inches in thickness, and these· were placed in the path of a beam rendered convergent by a double convex lens, 4 inches in diameter, possessing a focal distance of $10\frac{1}{2}$ inches. The slabs were usually so placed, that the focus of parallel rays fell within the ice. Having first found the position of the focus in the air, the lens was screened; the ice was then placed in position, the screen was removed, and the effect was watched through an ordinary pocket lens.

A plate of ice an inch thick, with parallel sides, was first examined: on removing the screen the transparent mass was crossed by the sunbeams, and the path of the rays through it was instantly studded by a great number of little luminous spots, produced at the moment, and resembling shining air-bubbles. When the beam was sent through the edge of the plate, so that it traversed a considerable thickness of the ice, the path of the beam could be traced by those brilliant spots, as it is by the floating motes in a dark room.

In lake ice the planes of freezing are easily recognised by the stratified appearance which the distribution of the air-bubbles

* Phil. Trans. December 1857.

gives to the substance. A cube was cut from a perfectly transparent portion of the ice, and the solar beam was sent through the cube in three rectangular directions successively. One was perpendicular to the plane of freezing, and the other two parallel to it. The bright bubbles were formed in the ice in all three cases.

When the surfaces perpendicular to the planes of freezing were examined by a lens, after exposure to the light, they were found to be cut up by innumerable small parallel fissures, with here and there minute spurs shooting from them, which gave the fissures, in some cases, a feathery appearance. When the portions of the ice traversed by the beam were examined parallel to the surface of freezing, a very beautiful appearance revealed itself. Allowing the light from the window to fall upon the ice at a suitable incidence, the interior of the mass was found filled with little flower-shaped figures. Each flower had six petals, and at its centre was a bright spot, which shone with more than metallic brilliancy. *The petals were manifestly composed of water*, and were consequently dim, their visibility depending on the small difference of refrangibility between ice at 32° Fahr. and water at the same temperature.

For a long time I found the relation between the planes of these flowers and the planes of freezing perfectly constant. They were always parallel to each other. The developement of the flowers was independent of the direction in which the beam, traversed the ice. Hence, when an irregularly shaped mass of transparent ice was presented to me, by sending a sunbeam through it I could tell in an instant the direction in which it had been frozen.

Allowing the beam to enter the edge of a plate of ice, and causing the latter to move at right angles to the beam, so that the radiant heat traversed different portions of the ice in succession, when the track of the beam was observed through an eyeglass, the ice, which a moment before was optically continuous, was instantly starred by those lustrous little spots, and around each of them the formation and growth of its associated flower could be distinctly observed.

The maximum effect was confined to a space of about an inch from the place at which the beam first struck the ice. In this space the absorption, which resolved the ice into liquid flowers, for the most part took place, but I have traced the effect to a depth of several inches in large blocks of ice.

At a distance, however, from the point of incidence, the spaces between the flowers became greater; and it was no uncommon thing to see flowers developed in planes a quarter of an inch apart, while no change whatever was observed in the ice between these planes.

The pieces of ice experimented on appeared to be quite homogeneous, and their transparency was very perfect. Why, then, did the substance yield at particular points? Were they weak points of crystalline structure, or did the yielding depend upon the manner in which the calorific waves impinged upon the molecules of the body at these points? However these and other questions may be answered, the experiments have an important bearing upon the question of absorption. In ice the absorption which produces the flower is fitful, and not continuous; and there is no reason to suppose that in other solids the case is not the same, though their constitution may not be such as to reveal it.*

I have applied the term 'bubbles' to the little bright disks in the middle of the flowers, simply because they resembled the little air-globules entrapped in the ice; but whether they contained air or not could only be decided by experiment.

Pieces of ice were therefore prepared, through which the sunbeams were sent, so as to develope the flowers in considerable quantity and magnitude. These pieces were then dipped into warm water contained in a glass vessel, and the effect, when the melting reached the bright spots, was carefully observed through a lens. *The moment a liquid connection was established between them and the atmosphere, the bubbles suddenly collapsed, and no trace of air rose to the surface of the warm water.*

This is the result which ought to be expected. The volume of water at 32° being less than that of ice at the same temperature, the formation of each flower ought to be attended with the formation of a vacuum, which disappears in the manner described when the ice surrounding it is melted.

---

* Notwithstanding the incomparable diathermancy of the substance, M. Knoblauch finds that when plates of rocksalt are thick enough, they always exhibit an elective absorption. Effects like those above described may possibly be the cause of this,

Similar experiments were made with ice, in which true air-bubbles were enclosed. When the melting liberated the air, the bubbles rose slowly through the liquid, and floated for a time upon its surface.

Exposure for a second, or even less, to the action of the sun was sufficient to develope the flowers in the ice. The first appearance of the central star of light was often accompanied by an audible clink, as if the substance had been suddenly ruptured. The edges of the petals were at the commencement definitely curved; but when the action was permitted to continue, and sometimes even without this, when the sun was strong, the edges of the petals became serrated, the beauty of the figure being thereby augmented.

Sometimes a number of elementary flowers grouped together to form a thickly-leaved cluster resembling a rose. Here and there also amid the flowers a liquid *hexagon* might be observed, but such were of rare occurrence.

The act of crystalline dissection, if I may use the term, thus performed by the solar beams, is manifestly determined by the manner in which the crystalline forces have arranged the molecules. By the abstraction of heat the molecules are enabled to build themselves together, by the introduction of heat this architecture is taken down. The perfect symmetry of the flowers, from which there is no deviation, argues a similar symmetry in the molecular architecture; and hence, as optical phenomena depend upon the molecular arrangement, we might pronounce with perfect certainty from the foregoing experiments, that ice is, what Sir David Brewster long ago proved it to be, optically speaking, uniaxal, the axis being perpendicular to the surface of freezing.

## § II.

On September 25, while examining a perfectly transparent piece of Norway ice, which had not been traversed by the condensed sunbeams, I found the interior of the mass crowded with parallel liquid disks, varying in diameter from the tenth to the hundredth of an inch. These disks were so thin, that when looked at in section they were reduced to the finest lines. They had the exact appearance of the circular spots of oily scum which float on the surface of mutton broth, and in the pieces of ice first examined they always lay in the planes of freezing.

As time progressed, this internal disintegration of the ice appeared to become more pronounced, so that some pieces of Norway ice examined in the middle of November appeared to be reduced to a congeries of water-cells entangled in a skeleton of ice. The effect of this was rendered manifest to the hand on sawing a block of this ice, by the facility with which the saw went through it.

There seems to be no such thing as absolute homogeneity in nature. Change commences at distinct centres, instead of being uniformly and continuously distributed, and in the most apparently homogeneous substance we should discover defects, if our means of observation were fine enough. The above observations show that some portions of a mass of ice melt more readily than others. The melting temperature of the substance is set down at 32° Fahr., but the absence of perfect homogeneity, whether from difference of crystalline texture or some other cause, makes the melting temperature oscillate to a slight extent on both sides of the ordinary standard. Let this limit, expressed in parts of a degree, be $t$. Some parts of a block of ice will melt at a temperature of $32-t$, while others require a temperature of $32+t$: the consequence is, that such a block raised to the temperature of 32°, will have some of its parts liquid, and others solid.

When a mass exhibiting the water-disks was examined by a concentrated sunbeam, the six-leaved flowers before referred to *were always formed in the planes of the disks.*

\*          \*     .     \*          \*          \*          \*          \*

## § III.

What has been already said will prepare us for the consideration of an associated class of phenomena of great physical interest. The larger masses of ice which I examined exhibited layers, in which bubbles of air were collected in unusual quantity, marking, no doubt, the limits of successive acts of freezing. These bubbles were usually elongated. Between two such beds of bubbles a clear stratum of ice intervened; and a clear surface layer, which, from its appearance, seemed to have suffered more from external influences than the rest of the ice, was associated with each block. In this superficial portion I observed detached air-bubbles irregularly distributed, and associated with each vesicle of air, a bleb of water which had the appearance of a drop of clear oil within the solid. The adjacent figure will give a notion of these composite cavities : the unshaded circle represents the air-bubble, and the shaded space adjacent, the water.  When the quantity of water was sufficiently large, which was usually the case, on turning the ice round, the bubble shifted its position, rising always at the top of the bleb of water. Sometimes, however, the cell was very flat, and the air was then quite surrounded by the liquid. These composite cells often occurred in pellucid ice, which showed inwardly no other sign of disintegration.

This is manifestly the same phenomenon as that which struck M. Agassiz so forcibly during his earlier investigations on the glacier of the Aar. The same appearances have been described by the Messrs. Schlagintweit, and finally attention has been forcibly drawn to the subject in a recent paper by Mr. Huxley, published in the ' Philosophical Magazine.' *

The only explanation of this phenomenon hitherto given, and adopted apparently without hesitation, is that of M. Agassiz and the Messrs. Schlagintweit. These observers attribute the phenomenon to the diathermancy of the ice, which permits the radiant heat to pass through the substance, to heat the bubbles of air, and cause them to melt the surrounding ice.†

The apparent simplicity of this explanation contributed to ensure its general acceptance ; and yet I think a little reflection will show that the hypothesis, simple as it may appear, is attended with grave difficulties.

For the sake of distinctness I will here refer to a most interesting fact, observed first by M. Agassiz, and afterwards by

* October, 1857.

† Il est évident pour quiconque a suivi le progrès de la physique moderne, que ce phénomène est dû uniquement à la diathermanéité de la glace (Agassiz, Système, p. 157).

Das Wasser ist dadurch enstanden dass die Luft Wärmestrahlen absorbirte, welche das Eis als diathermaner Körper durchliess (Schlagintweit, Untersuchungen, S. 17).

the Messrs. Schlagintweit. In the 'Système Glaciaire' it is described in these words: 'I ought also to mention a singular property of those air-bubbles, which at first struck us forcibly, but which has since received a very satisfactory explanation. When a fragment containing air-bubbles is exposed to the action of the sun, the bubbles augment insensibly. Soon, in proportion as they enlarge, a transparent drop shows itself at some point of the bubble. This drop, in enlarging, contributes on its part to the enlargement of the cavity, and following its progress a little, it finishes by predominating over the bubble of air. The latter then swims in the midst of a zone of water, and tends incessantly to reach the most elevated point, at least if the flatness of the cavity does not hinder it.'

The satisfactory explanation here spoken of is that already mentioned: let us now endeavour to follow the hypothesis to its consequences.

Comparing equal weights of both substances, the specific heat of water being 1, that of air is 0·25. Hence to raise a pound of water one degree in temperature, a pound of air would have to lose four degrees.

Let us next compare equal volumes of the substances. The specific gravity of water being 1, that of air is $\frac{1}{770}$; hence a pound of air is 770 times the volume of a pound of water; and hence, for a quantity of air to raise *its own volume* of water one degree, it must part with $770 \times 4$, or 3,080 degrees of temperature.

Now the latent heat of water is 142·6° Fahr., hence the quantity of heat required to melt a certain weight of ice is 142·6 times the quantity required to raise the same weight of water one degree in temperature; hence, a measure of air, in order to reduce its own volume of ice to the liquid condition, must lose $3,080 \times 142\cdot6$, or 439,208 degrees of temperature.

This, then, gives us an idea of the amount of heat which, according to the above hypothesis, is absorbed by the bubble and communicated to the ice during the time occupied in melting a quantity of the latter equal in volume to the bubble, which time is stated to be brief; that is to say the quantity of heat supposed to be absorbed by the air would, if it had not been communicated to the ice, have been sufficient to raise the bubble itself to a temperature 160 times that of fused cast iron. Had air this power of absorption, it might be attended with inconvenient consequences to the denizens of the earth ; for we should dwell at the bottom of an atmospheric ocean, the upper strata of which would effectually arrest all calorific radiation.

It is established by the experiments of Delaroche and Melloni,* that a calorific beam, emerging from any medium which it has traversed for any distance, possesses, in an exalted degree, the power of passing through an additional length of the same substance. Absorption takes place, for the most part, in the portion of the medium first traversed by the rays. In the casé of a plate of glass, for example, $17\frac{1}{2}$ per cent. of the heat proceeding from a lamp, is absorbed in the first fifth of a millimetre; whereas, after the rays have passed through 6 millimetres of the substance, an additional distance of 2 millimetres absorbs less than 2 per cent. of the rays thus transmitted. Supposing the rays to have passed through a plate 25 millimetres, or an inch, in thickness, there is no doubt that the heat emerging from such a plate would pass through a second layer of glass, 1 millimetre thick, without suffering any measurable absorption. For an incomparably stronger reason, the quantity of solar heat absorbed by a bubble of air at the earth's surface, after the rays have traversed the whole thickness of our atmosphere, and been sifted in their passage through it, must be wholly inappreciable.

Such, if I mistake not, are the properties of radiant heat which modern physics have revealed ; and I think they render it evident that the hypothesis of M. Agassiz and the Messrs. Schlagintweit was accepted without due regard to its consequences.

\*        \*        \*        \*        \*        \*

## § IV.

But the question still remains, how are the water-chambers produced within the ice ? . . . One simple test will, I think, decide the question whether the liquid is, or is not, the product of melted ice. . If it be, its volume must be less than that of the ice which produced it, and the bubble associated with the water *must be composed of rarefied air*. Hence, if on establishing a

---

* La Thermochrose, p. 202,

liquid connection between this bubble and the atmosphere a diminution of volume be observed, this will indicate that the water has been produced by the melting of the ice.

From a block of Norway ice, containing such compound bubbles, I cut a prism, and immersing it in warm water, contained in a glass vessel, I carefully watched through the side of the vessel the effect of the melting upon the bubbles. *They invariably shrunk in volume at the moment the surrounding ice was melted*, and the diminished globule of air rose to the surface of the water. I then arranged matters so that the wall of the cavity might be melted away underneath, without permitting the bubble of air at the top to escape. At the moment the melting reached the cavity the air-bubble instantly collapsed to a sphere possessing, in some cases, far less than the hundredth part of its original volume. The experiments were repeated with several distinct masses of ice, and always with the same result. I think, therefore, it may be regarded as certain that the liquid cells are the product of melted ice.*

Considering the manner in which ice imported into this country is protected from the solar rays, I think we must infer that in the specimens examined by me, *the ice in contact with the bubble has been melted by heat, which has been conducted through the substance without visible prejudice to its solidity.*

Paradoxical as this may appear, I think it is no more than might reasonably be expected from *à priori* considerations. The heat of a body is referred, at the present day, to a motion of its particles. When this motion reaches an intensity sufficient to liberate the particles of a solid from their mutual attractions, the body passes into the liquid condition. Now, as regards the amount of motion necessary to produce this liberty of liquidity, the particles at the surface of a mass of ice must be very differently circumstanced from those in the interior, which are influenced and controlled on every side by other particles. But if we suppose a cavity to exist within the mass, the particles bounding that cavity will be in a state resembling that of the particles at the surface; and by the removal of all opposing action on one side, the molecules may be liberated by a force which the surrounding mass has transmitted without prejudice to

* This of course refers only to the lake ice examined as described.

its solidity. Supposing, for example, that solidity is limited by molecular vibrations of a certain amplitude, those at the surface of the internal cavity may exceed this limit, while those between the cavity and the external surface of the ice may, by their reciprocal actions, be preserved within it, just as the terminal member of a series of elastic balls is detached by a force which has been transmitted by the other members of the series without visible separation.*

Where, however, experiment is within reach we ought not to trust to speculation; and I was particularly anxious to obtain an unequivocal reply to the question whether an interior portion of a mass of ice could be melted by heat which had passed through the substance by the process of *conduction*. A piece of Norway ice, containing a great number of the liquid disks already described, and several cells of air and water, was enveloped in tinfoil and placed in a mixture of pounded ice and salt. A few minutes sufficed to freeze the disks to thin dusky circles, which appeared, in some cases, to be formed of concentric rings, and reminded me of the sections of certain agates. Looked at sideways, these disks were no thicker than a fine line. The water-cells were also frozen, and the associated air-bubbles were greatly diminished in size. I placed the mass of ice between me and a gas-light, and observed it through a lens: after some time the disks and water-cells showed signs of breaking up. The rings of the disks disappeared; the contents seemed to aggregate so as to form larger liquid spots, and finally, some of them were reduced to clear transparent disks as before.

But an objection to this experiment is, that the ice may have been liquefied by the radiation from the lamp, and I have experiments to describe which will show the justice of this objection. A rectangular slab, 1 inch thick, 3 inches long, and 2 wide, was therefore taken from a mass of Norway ice, in which the associated air and water-cells were very distinct. I enveloped it in tinfoil, and placed it in a freezing mixture. In about ten minutes the water-blebs were completely frozen within the mass. It was immediately placed in a dark room, where no radiant heat could possibly affect it, and examined every quarter of an hour. The dim frozen spots gradually broke up into little water parcels, and in two hours the water-blebs were perfectly restored in the

* Of course I intend this to help the conception merely.

centre of the slab of ice. When last examined, this plate was half an inch thick, and the drops of liquid were seen right at its centre.

A second piece, similarly frozen and wrapped up in flannel, showed the same deportment. In an hour and a half the frozen water surrounding the air-bubbles was restored to its liquid condition. Hence no doubt can remain as to the possibility of effecting liquefaction in the interior of a mass of ice, by heat which has passed by *conduction* through the substance without melting it.

I have already referred to the formation of the liquid cavities observed by M. Agassiz, when glacier ice was exposed to the sun. The same effect may be produced by exposure to a glowing coal fire. On the 21st and 22nd of November, I thus exposed plates of clear Wenham Lake ice, which contained some scattered air-bubbles. At first the bubbles were sharply rounded, and without any trace of water. Soon, however, those near the surface, on which the radiant heat fell, appeared encircled by a liquid ring, which expanded and finally became crimped at its

 border, as shown in the adjacent figure. The crimping became more pronounced as the action was permitted to continue.*

A second plate, crowded with bubbles, was held as near to the fire as the hand could bear. On withdrawing it, and examining it through a pocket lens, the appearance was perfectly beautiful. In many cases the bubbles appeared to be surrounded by a series of concentric rings, the outer ring surrounding all the others like a crimped frill.

I could not obtain these effects by placing the ice in contact with a plate of metal obscurely heated,† nor by the radiation from an obscure source. Indeed ice, as before remarked, is impervious to radiant heat from such a source.‡ The rays from a

---

* The blebs observed in glacier ice also exhibit this form: see fig. 8, plate 6, of the Atlas to the 'Système Glaciaire.' In fig. 13 we have also a close resemblance of the flower-shaped figures produced by radiant heat in lake ice.

† To develope water-cavities within ice a considerable time is necessary; more time, indeed, than was sufficient to melt the entire pieces of ice made use of in these *contact* experiments.

‡ Hence the soundness of the ice under the moraines; the sun's rays

common fire also are wholly absorbed near the surface upon which they strike, and hence the described internal liquefaction was confined to a thin layer close to this surface.

But not only does liquefaction occur in connection with the bubbles, but the ' flowers,' already described as produced by the solar beams, start by hundreds into existence, when a slab of transparent ice is placed before a glowing coal fire. They, however, are also confined to a thin stratum of the substance close to the surface of incidence. In the experiments made in this way, the central stars of the flowers were often bounded by sinuous lines of great beauty.

The foregoing considerations show that liquefaction takes place at the surface of a mass of ice at a lower temperature than that required to liquefy the interior of the solid. At the surface the temperature 32° produces a vibration, to produce which, within the ice, would necessitate a temperature of $32° + x$; the increment $x$ being the additional temperature necessary to overcome the resistance to liquefaction, arising from the action of the molecules upon each other.

Now let us suppose two pieces of ice at 32°, with moistened surfaces, to be brought into contact with each other, *we thereby virtually transfer the touching portions of these pieces from the surface to the interior*, where $32 + x$ is the melting temperature. Liquefaction will therefore be arrested at those surfaces. Before being brought together, the surfaces had the motion of liquidity, but the interior of the ice has not this motion; and as equilibrium will soon set in between the masses on each side of the liquid film and the film itself, the film will be reduced to a state of motion inconsistent with liquidity. *In other words, it will be frozen, and will cement the two surfaces of ice between which it is enclosed.*＊

If I am right here, the importance of the physical principles

are converted into obscure heat by the overlying débris; this only affects a layer of infinitesimal depth, and cannot produce the disintegration of the deeper ice, as the direct sunbeams can.

＊ It is here implied 'that the contact of the moist surfaces must be so perfect, or, in other words, the liquid film between them must be so thin, as to enable the molecules to act upon each other *across* it. The extreme tenuity of the film may be inferred from this. A thick plate of water within the ice would facilitate rather than retard liquefaction.

involved are sufficiently manifest : if I am wrong, I hope I have so expressed myself as to render the detection of my error easy. Right or wrong, my aim has been to give as explicit utterance to my meaning as the subject will admit of.

## § V.

Mr. Faraday's experiments on the freezing together of pieces of ice at 32° Fahr., and all of those recounted in the paper published by Mr. Huxley and myself, find their explanation in the principles here laid down.   The conversion of snow into névé, and of névé into glacier, is perhaps the grandest illustration of the same principle.   It has been, however, suggested to me that the sticking together of two pieces of ice may be an act of cohesion, similar to that which enables pieces of wetted glass, and other similar bodies, to stick together.   This is not the case.   There is no sliding motion possible to the ice.   When contact is broken, it breaks with the snap due to the rupture of a solid.  Glass and ice cannot be made to stick thus together, neither can glass and glass, nor alum and alum, nor nitre and nitre, at common temperatures.   I have, moreover, placed pieces of ice together over night and found them in the morning so rigidly frozen together that when I sought to separate them, the surface of fracture passed through one of them in preference to taking the surface of regelation.   Many sagacious persons have also suggested to me that the ice transported to this country from Norway and the Wenham Lake may possibly retain a residue of its cold, sufficient to freeze a thin film enclosed between two pieces of the substance.   But the facts already adverted to are a sufficient reply to this surmise.   The ice experimented on cannot be regarded as a magazine of cold, *because parcels of liquid water exist within it.*

# LECTURE X

[March 27, 1862.]

ABSORPTION OF HEAT BY GASEOUS MATTER — APPARATUS EMPLOYED — EARLY DIFFICULTIES — DIATHERMANCY OF AIR AND OF THE TRANSPARENT ELEMENTARY GASES — ATHERMANCY (OPACITY) OF OLEFIANT GAS AND OF THE COMPOUND GASES — ABSORPTION OF RADIANT HEAT BY VAPOURS — RADIATION OF HEAT BY GASES — RECIPROCITY OF RADIATION AND ABSORPTION — INFLUENCE OF MOLECULAR CONSTITUTION ON THE PASSAGE OF RADIANT HEAT.

IN our last lecture we examined the diathermancy, or transparency to heat, of solid and liquid bodies; and we then learned, that closely as the atoms of such bodies are packed together, the interstitial spaces between the atoms afford, in many cases, free play and passage to the ethereal undulations, which were transmitted without sensible hindrance among the atoms. In other cases, however, we found that the molecules stopped the waves of heat which impinged upon them; but that in so doing, they themselves became centres of oscillation. Thus we learned that while perfectly diathermic bodies allowed the waves of heat to pass through them without suffering any change of temperature, those bodies which stopped the calorific flux became heated by the absorption. Through ice, itself, we sent a powerful calorific beam; but as the beam was of such a quality as not to be intercepted by the ice, it passed through this highly sensitive substance without melting it. We have now to deal with gaseous bodies; and here the interatomic spaces are so vastly augmented, the molecules are so completely released from all mutual entanglement,

that we should be almost justified in concluding that gases
and vapours furnish a perfectly open door for the passage
of the calorific waves. This, indeed, until quite recently,
was the universal belief, and the conclusion was verified by
such experiments as had been made on atmospheric air,
which was found to give no evidence of absorption.

But each succeeding year augments our experimental
powers; our predecessors were often obliged to fight with
flints, where we may use swords, and hence the conflict
with Nature is not decided by their discomfiture. Let us,
then, test once more the diathermancy of atmospheric air.
We may make a preliminary essay in the following way:
I have here a hollow tin cylinder, A B (fig. 89), 4 feet
long, and nearly 3 inches in diameter, through which we
may send our calorific rays. We must, however, be able
to compare the passage of the rays through the air, with
their passage through a vacuum, and hence we must have
some means of stopping the ends of our cylinder, so as to
be able to exhaust it. Here we encounter our first experimental difficulty. As a general rule obscure heat is
more greedily absorbed than luminous heat, and as our
object is to make the absorption of a highly diathermic
body sensible, we are most likely to effect this object by
employing obscure heat.

Our tube, therefore, must be stopped by a substance
which permits of the free passage of such heat. Shall we
use glass for the purpose? An inspection of the table at
page 297 shows us, that for such rays plates of glass would
be perfectly opaque; we might as well stop our tube with
plates of metal. Observe here how an investigator's results
are turned to account by his successors. From one experiment buds another, and science grows by the continual
degradation of ends to means. Had not Melloni discovered
the diathermic properties of rocksalt, we should now
be utterly at a loss. For a time, however, I was ex-

tremely hampered by the difficulty of obtaining plates of
salt sufficiently large and pure to stop the ends of my
tube.   But a scientific worker does not long lack help,
and, thanks to such friendly aid, I have here plates of this
precious substance which, by means of these caps, I can

FIG. 89.

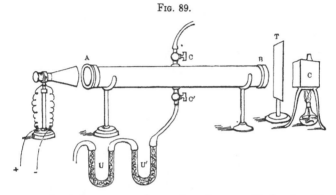

screw air-tight on to the ends of my cylinder.*   You
observe two stopcocks attached to the cylinder; this one, c,
is connected with an air-pump, by which the tube can be
exhausted ; while through this other one, c′, I can allow air
or any other gas to enter the tube.

* At a time when I was greatly in need of a supply of rocksalt, I stated
my wants in the 'Philosophical Magazine,' and met with an immediate re-
sponse from Sir John Herschel.   He sent me a block of salt, accompanied
by a note, from which, as it refers to the purpose for which the salt
was originally designed, I will make an extract.   I have not yet been
able to examine the extremely remarkable point to which the eminent
writer directs my attention.   I am also greatly indebted to Dr. Szabo,
the Hungarian Commissioner to the International Exhibition, by whom
I have been lately raised to comparative opulence, as regards the possession of rocksalt.   To the Messrs. Fletcher, of Northwich, and to Mr.
Corbett, of Bromsgrove, my best thanks are also due for their obliging
kindness.

Here follows the extract from Sir J. Herschel's note :—'After the publication of my paper in the Phil. Trans., 1840, I was very desirous to disengage
myself from the influence of glass prisms and lenses, and ascertain, if possible, whether in reality my insulated heat spots $\beta \gamma \delta \epsilon$ in the spectrum
were of solar or terrestrial origin.   Rocksalt was the obvious resource,
and after many and fruitless endeavours to obtain sufficiently large and

At one end of the cylinder I place this Leslie's cube c, containing boiling water; and which is coated with lamp-black, to augment its power of radiation. At the other end of the cylinder stands our thermo-electric pile, from which wires lead to the galvanometer. Between the end of the cylinder and the source of heat I have introduced a tin screen, T, which, when withdrawn, will allow the calorific rays to pass through the tube to the pile. We first exhaust the cylinder, then draw the screen a little aside, and now the rays are traversing a vacuum and falling upon the pile. The tin screen, you observe, is only partially withdrawn, and the steady deflection produced by the heat at present transmitted is 30 degrees.

Let us now admit dry air: I can do so by means of the cock c', from which a piece of flexible tubing leads to the bent tubes υ, υ', the use of which I will now explain; υ is filled with fragments of pumice stone moistened with a solution of caustic potash; it is destined to withdraw whatever carbonic acid may be contained in the air; υ' is a similar tube, filled with fragments of pumice stone moistened with sulphuric acid; it is intended to absorb the aqueous vapour of the air. Thus the air reaches the cylinder deprived both of its aqueous vapour and

pure specimens, the late Dr. Somerville was so good as to send me (as I understood from a friend in Cheshire) the very fine block which I now forward. It is, however, much cracked, but I have no doubt pieces large enough for lenses and prisms (especially if cemented together) might be got from it.

'But I was not prepared for the working of it—evidently a very delicate and difficult process, (I proposed to *dissolve* off the corners, &c., and, as it were, *lick* it into shape) and though I have never quite lost sight of the matter, I have not yet been able to do anything with it; meanwhile, I put it by. On looking at it a year or two after, I was dismayed to find it had lost much by deliquescence. Accordingly, I potted it up in salt in an earthen dish, with iron rim, and placed it on an upper shelf in a room with an Arnott stove, where it has remained ever since.

'If you should find it of any use I would ask you, if possible, to repeat my experiment as described, and settle that point, which has always struck me as a very important one.'

its carbonic acid. It is now entering,—the mercury-gauge of the pump is descending, and as it enters I would beg of you to observe the needle. If the entrance of the air diminish the radiation through the cylinder — if air be a substance which is competent to destroy the waves of ether in any sensible degree — this will be declared by the diminished deflection of the galvanometer. The tube is now full, but you see no change in the position of the needle, nor could you see any change even if you were close to the instrument. The air thus examined seems as transparent to radiant heat as the vacuum itself.

By changing the screen I can alter the amount of heat falling upon the pile; thus, by withdrawing it, I can cause the needle to stand at 40°, 50°, 60°, 70°, and 80° in succession; and while it occupies each position I can repeat the experiment which I have just performed before you. In no instance could you recognise the slightest movement of the needle. The same is the case if I push the screen forward, so as to reduce the deflection to 20 and 10 degrees.

The experiment just made is a question addressed to Nature, and her silence might be construed into a negative reply. But a natural philosopher must not lightly accept a negative, and I am not sure that we have put our question in the best possible language. Let us analyse what we have done, and first consider the case of our smallest deflection of 10 degrees. Supposing that the air is *not* perfectly diathermic; that it really intercepts a small portion—say the thousandth part of the heat passing through the tube—that out of every thousand rays it struck down one; should we be able to detect this execution? This absorption, if it took place, would lower the deflection the thousandth part of ten degrees, or the hundredth part of one degree, a diminution which it would be impossible for you to see, even if you were close to the

galvanometer.*  In the case here supposed, *the total quantity of heat falling upon the pile is so inconsiderable,
that a small fraction of it, even if absorbed, might well
escape detection.*

But we have not confined ourselves to a small quantity
of heat; the result was the same when the deflection was
80° as when it was 10°. Here I must ask you to sharpen
your attention and accompany me, for a time, over rather
difficult ground. I want now to make clearly intelligible
to you an important peculiarity of the galvanometer.

The needle being at zero, let us suppose a quantity of heat
to fall upon the pile, sufficient to produce a deflection of one
degree. Suppose that I afterwards augment the quantity
of heat, so as to produce deflections of two degrees, three
degrees, four degrees, five degrees; I then know that the
quantities of heat which produce these deflections stand to
each other in the ratios of 1 : 2 : 3 : 4 : 5 ; the quantity
of heat which produces a deflection of 5° being exactly
five times that which produces a deflection of 1°. But
this proportionality exists only so long as the deflections
do not exceed a certain magnitude. For, as the needle
is drawn more and more aside from zero, the current acts
upon it at an ever augmenting disadvantage. The case is
illustrated by a sailor working a capstan ; he always
applies his strength at right angles to the lever, for, if he
applied it obliquely, only a portion of that strength would
be effective in turning the capstan round. And in the
case of our electric current, when the needle is very
oblique to the current's direction, only a portion of its
force is effective in moving the needle round. Thus it
happens, that though the quantity of heat may be, and, in
our case, *is*, accurately expressed by the strength of the
current which it excites, still the larger deflections, inas-

* It will be borne in mind that I am here speaking of *galvanometric*, not
of *thermometric* degrees.

much as they do not give us the action of the whole current, but only of a part of it, cannot be a true measure of the amount of heat falling upon the pile.

The galvanometer now before you is so constructed that the angles of deflection, up to 30° or thereabouts, are pro-portional to the quantities of heat; the quantity necessary to move the needle from 30° to 31° is nearly the same as that required to move it from 0° to 1°. But beyond 30° the proportionality ceases. The quantity of heat required to move the needle from 40° to 41° is three times that necessary to move it from 0° to 1°; to deflect it from 50° to 51° requires five times the heat necessary to move it from 0° to 1°; to deflect it from 60° to 61° requires about ten times the heat necessary to move it from 0° to 1°; to deflect it from 70° to 71° requires nearly twenty times, while to move it from 80° to 81° requires more than fifty times the heat necessary to move it from 0° to 1°. Thus, the higher we go, the greater is the quantity of heat represent-ed by a degree of deflection; the reason being, that the force which then moves the needle is only a fraction of the force really circulating in the wire, and hence represents only a fraction of the heat falling upon the pile.

By a certain process, which I will not stop here to des-cribe,* I can express the higher degrees in terms of the lower ones; I thus learn, that while deflections of 10°, 20°, 30°, respectively, express quantities of heat represented by the numbers 10, 20, 30, a deflection of 40° represents a quantity of heat expressed by the number 47; a deflec-tion of 50° expresses a quantity of heat expressed by the number 80; while the deflections 60°, 70°, 80°, express quantities of heat which increase in a much more rapid ratio than the deflections themselves.

What is the upshot of this analysis? It will drive us, I think, to a better method of questioning Nature. It

* See Appendix to Lecture X.

leads to the reflection that, when we make our angles
*small*, the quantity of heat falling on the pile is so inconsiderable, that even if a fraction of it were absorbed, it
might escape detection; while, if we make our deflections
large, by employing a powerful flux of heat, the needle is
in a position from which it would require a considerable
addition or abstraction of heat, to move it. The 1,000th
part of the whole radiation in the one case would be too
small, absolutely, to be measured; the 1,000th part in the
other case might be something considerable, without, however, being considerable enough to affect the needle in any
sensible degree. When, for example, the deflection is over
80°, an augmentation or diminution of heat, equivalent to
15 or 20 of the lower degrees of the galvanometer, would
be scarcely measurable.

We are now face to face with our problem; it is this,
to work with a flux of heat so large that a small fractional
part of it will not be infinitesimal, and still to keep our
needle in its most sensitive position. If we can accomplish this we shall augment indefinitely our experimental
power. If a fraction of the heat, however small, be
intercepted by the gas, *we can augment the absolute
value of that fraction by augmenting the total of which
it is a fraction.*

The problem, happily, admits of an effective practical
solution. You know that when we allow heat to fall upon
the opposite faces of the thermo-electric pile, the currents
generated neutralise each other more or less; and, if the
quantities of heat falling upon the two faces be perfectly
equal, the neutralisation is complete. Our galvanometer
needle is now deflected to 80° by the flux of heat passing
through the tube; I uncover the second face of the pile,
furnish it with its conical reflector, and place a second
cube of boiling water in front of it; the needle, as you
see, descends instantly.

By means of a proper adjusting screen I can so regulate the quantity of heat falling upon the posterior face of the pile, that it shall exactly neutralise the heat incident upon its other face: this is now effected; and the needle points to zero.

Here, then, we have two powerful and perfectly equal fluxes of heat, falling upon the opposite faces of the pile, one of which passes through our exhausted cylinder. If I allow air to enter the cylinder, and if this air exert any appreciable action upon the rays of heat, the equality now existing will be destroyed; a portion of the rays passing through the tube being struck down by the air, the second source of heat will triumph; the needle, now in its most sensitive position, will be deflected; and from the magnitude of the deflection we can accurately calculate the absorption.

I have thus sketched, in rough outline, the apparatus by which our researches on the relation of radiant heat to gaseous matter must be conducted. The necessary tests are, however, at the same time so powerful and so delicate, that a rough apparatus like that just described would not answer our purpose. But you will now experience no difficulty in comprehending the construction and application of the more perfect apparatus, with which the experiments on gaseous absorption and radiation have been actually made. See Plate I., at the end of the volume.

Between s and s′ stretches the *experimental cylinder*, a hollow tube of brass, polished within; at s, and s′, are the plates of rock salt which close the cylinder air-tight; the length from s to s′, in the experiments to be first recorded, is 4 feet. c, the source of heat, is a cube of cast copper, filled with water, which is kept continually boiling by the lamp L. Attached to the cube c by brazing is the short cylinder F, of the same diameter as the experimental cylinder, and capable of being connected air-tight with the latter at s. Thus between the source c and the end s′ of the experimental tube, we have *the front chamber* F, from which the air can be removed, so that the rays from the source will enter the cylinder s s′ unsifted.    To prevent the heat from the source c passing by conduction to the plate at s, the chamber F is caused to pass through the vessel v, in which a stream of cold water continually circulates, entering through the pipe *i i*, which dips to the bottom of the vessel, and escaping through the waste-pipe *e e*. The experimental tube and the front chamber are connected, independently, with the air-pump A A, so that either of them may be exhausted or filled without interfering with the other. I may remark that in later arrangements the experimental cylinder was supported apart from the pump, being connected with the latter by a flexible tube.    The tremulous motion of the pump, which occurred when the connection was rigid, was thus completely avoided.

P is the thermo-electric pile, placed on its stand at the end of the experimental tube, and furnished with its two conical reflectors.    c′ is the *compensating cube,* used to neutralise the radiation from c; H is the *adjusting screen,* which is capable of an exceedingly fine motion to and fro.    N N is a delicate galvanometer connected with the pile P, by the wires *w w′.*    The graduated tube o o (to the right of the plate), and the appendage M K (attached to the centre of the experimental tube) shall be referred to more particularly by and by.

I should hardly sustain your interest in stating the difficulties which at first beset the investigation conducted with this apparatus, or the numberless precautions which the exact balancing of the two powerful sources of heat, here resorted to, rendered necessary.    I believe the experiments made with atmospheric air alone might be numbered by tens of thousands.    Sometimes for a week, or even for a fortnight, coincident and satisfactory results would be obtained;

the strict conditions of accurate experimenting would appear to be found, when an additional day's experience would destroy the superstructure of hope, and necessitate a recommencement, under changed conditions, of the whole enquiry. It is this which daunts the experimenter; it is this preliminary fight with the entanglements of a subject, so dark, so doubtful, so uncheering; without any knowledge whether the conflict is to lead to anything worth possessing, that renders discovery difficult and rare. But the experimenter, and particularly the *young* experimenter, ought to know, that as regards his own moral manhood, he cannot but win if he only contend aright. Even with a negative result, the consciousness that he has gone fairly to the bottom of his subject, as far as his means allowed—the feeling that he has not shunned labour, though that labour may have resulted in laying bare the nakedness of his case—reacts upon his own mind, and gives it firmness for future work.

But to return;—I first neglected atmospheric vapour and carbonic acid altogether, concluding, as others did afterwards, that the quantities of these substances being so small, their effect upon radiant heat must be quite inappreciable; after a time, however, I found this assumption leading me quite astray. I first used chloride of calcium as a drying agent, but had to abandon it. I next used pumice stone moistened with sulphuric acid, and had to give it up also. I finally resorted to pure glass broken to small fragments, wetted with sulphuric acid, and inserted by means of a funnel into a U tube. I found this arrangement best, but even here the greatest care was needed. It was necessary to cover each column with a layer of dry glass fragments, for I found that the smallest particle of dust from the cork, or a quantity of sealing wax not more than the twentieth-part of a pin's head in size, was quite sufficient, if it reached the acid, to vitiate the results. The drying-

tubes moreover had to be frequently changed, as the
organic matter of the atmosphere, infinitesimal though it
was, soon introduced disturbance.

To remove the carbonic acid, pure Carrara marble was
broken into fragments, wetted with caustic potash, and
introduced into a U tube. These, then, are the agents for
drying the gas and removing the carbonic acid which are
used at present; but previous to their final adoption, I
employed, to dry the air, the arrangement shown in Plate I.,
where the glass tubes marked Y Y, each three feet long, were
filled with chloride of calcium, after which were placed two
U tubes R Z, filled with pumice stone and sulphuric acid.
Hence, the air, in the first place, had to pass over 18
feet of chloride of calcium, and afterwards through the
sulphuric acid tubes, before it entered the experimental
tube s s'. A gas-holder, G G, was employed for other gases
than atmospheric air. In the investigation on which I am
at present engaged, this arrangement, as I have said, is
abandoned, a simpler one being found more effectual.

My assistant has now exhausted both the front chamber F and the experimental tube s s'. The rays are
passing from the source c through the front chamber;
across the plate of rocksalt at s, through the experimental
tube, across the plate at s', afterwards impinging upon the
anterior surface of the pile P. This radiation is neutralised
by that from the compensating cube c'. The needle, you
will observe, is at zero. We will commence our experiments by applying this powerful test to dry air. It is
now entering the experimental cylinder; but, at your distance, you see no motion of the needle, and thus our more
powerful mode of experiment fails to detect any absorption
on the part of the air. Its atoms, apparently, are incompetent to shatter a single calorific wave; *it is a practical
vacuum as regards the rays of heat.* Were you quite
near, however, you would see a deflection of the needle

amounting to about one degree. Oxygen, hydrogen, and nitrogen, when carefully purified, exhibit the action of atmospheric air; they are almost neutral.

But the neutral quality of atmospheric air was thought to extend to transparent gases generally. Let us see whether this is correct. I have here a gasholder of olefiant gas,—common coal gas would also answer my purpose. I discharge a little of the olefiant gas in the air, but you see nothing; the gas is perfectly transparent. The experimental tube is exhausted, and the needle points to zero; and now we will allow the olefiant gas to enter. Observe the effect. The needle moves in a moment; the transparent gas strikes down the rays wholesale—the final and permanent deflection, when the tube is full, amounting to 70 degrees.

I will now interpose a metal screen between the pile ꜰ and the end s′ of the experimental tube, thus entirely cutting off the radiation through the tube. The face of the pile turned towards the metal screen wastes its heat speedily by radiation; it is now at the temperature of this room, and the radiation from the compensating cube alone acts on the pile, producing a deflection of 75 degrees. But at the commencement of the experiment the radiations from both cubes were equal, hence the deflection 75° corresponds to the *total radiation* through the experimental tube, when the latter is exhausted.

Taking as unit the quantity of heat necessary to move the needle from 0° to 1°, the number of units expressed by a deflection of 75° is

360.

The number of units expressed by a deflection of 70° is

290.

Out of a total, therefore, of 360, olefiant gas has struck down 290; that is about seven-ninths of the whole, or about 81 per cent.

Does it not seem to you as if an opaque layer had been

suddenly precipitated on our plates of salt, when the gas
entered?  The substance, however, deposits no such layer.
I discharge a current of the dried gas against a polished
plate of salt, but you do not perceive the slightest dimness.
The rocksalt plates, moreover, though necessary for exact
measurements, are not necessary to show the destructive
powers of this gas.  Here is an open tin cylinder which I
interpose between the pile and our radiating source; I force
olefiant gas gently into the cylinder from this gas-holder
and you see the needle fly up to its stops.  Observe the
smallness of the quantity of gas which I shall next use.
I cleanse the open tube by forcing a current of air
through it; the needle is now at zero; and I will simply
turn this cock on and off, as speedily as I can.  A mere
bubble of the gas enters the tube in this brief interval;
still you see that its presence causes the needle to swing
to 70°.  I next abolish the open tube, and leave nothing
but the free air between the pile and source; from the
gasometer I discharge olefiant gas into this space.  You
see nothing in the air; but the swing of the needle through
an arc of 60° declares the presence of this invisible barrier
to the calorific rays.

    Thus, it is shown that the ethereal undulations which
glide among the atoms of oxygen, nitrogen, and hydrogen,
without hindrance, are powerfully absorbed by the molecules of olefiant gas.  We shall find other transparent
gases also almost immeasurably superior to air.  We can
limit at pleasure the number of the gaseous atoms,
and thus vary the amount of destruction of the ethereal
waves.  In this respect gaseous bodies possess a great advantage over liquids and solids, in experiments on radiation.  Attached to the air-pump is a barometric tube, by
means of which I can admit measured portions of the gas.
The experimental cylinder is now exhausted, and turning
this cock slowly on, and observing the mercury gauge, I

allow the olefiant gas to enter, till the mercurial column has been depressed an inch. I observe the galvanometer and read the deflection. Determining thus the absorption produced by one inch, another inch is added, and the absorption effected by two inches of the gas is determined. Proceeding thus we obtain for tensions from 1 to 10 inches the following absorptions :—

### Olefiant Gas.

| Tensions in inches | | | | | | | Absorption |
|---|---|---|---|---|---|---|---|
| 1 | . | . | . | . | . | . | . 90 |
| 2 | . | . | . | . | . | . | . 123 |
| 3 | . | . | . | . | . | . | . 142 |
| 4 | . | . | . | . | . | . | . 157 |
| 5 | . | . | . | . | . | . | . 168 |
| 6 | . | . | . | . | . | . | . 177 |
| 7 | . | . | . | . | . | . | . 182 |
| 8 | . | . | . | . | . | . | . 186 |
| 9 | . | . | . | . | . | . | . 190 |
| 10 | . | . | . | . | . | . | . 193 |

The unit here used is the amount of heat absorbed when a *whole atmosphere* of dried air is allowed to enter the tube. The table, for example, shows that one-thirtieth of an atmosphere of olefiant gas exercises ninety times the absorption of a whole atmosphere of air.

The table also informs us that each additional inch of olefiant gas produces less destruction than the preceding one. A single inch, at the commencement, strikes down 90 rays, but a second inch strikes down only 33, while the addition of an inch, when nine inches are already in the tube, effects the destruction of only 3 rays. This is what might reasonably be expected. The number of rays emitted is finite, and the discharge of the first inch of olefiant gas amongst them has so thinned their ranks that the execution produced by the second inch is naturally less than that of the first. This execution must diminish, as the number of rays capable of being

z

destroyed by the gas, becomes less; until, finally, all absorb-
able rays being removed, the residual heat would pass
through the gas unimpeded.

But supposing the quantity of gas first introduced to be
so inconsiderable, that the number of rays extinguished by
it is a vanishing quantity, compared with the total number
capable of being destroyed, we might then reasonably
expect that, for some time at least, the quantity of
execution done would be proportional to the quantity of
gas present. That a double quantity of gas would produce
a double effect, a treble quantity a treble effect; or, in
general terms, that the absorption would, for a time, be
found proportional to the density.

To test this idea we will make use of a portion of the
apparatus omitted in the general description. o o (Plate I.)
is a graduated glass tube, the end of which dips into the
basin of water B. The tube is closed above by means of the
stopcock $r$; $d\ d$ is a tube containing fragments of chloride
of calcium. The tube o o is first filled with water up to
the cock $r$, and the water is afterwards carefully displaced
by olefiant gas admitted in bubbles from below. The gas is
admitted into the experimental cylinder by the cock $r$, and
as it enters, the water rises in o o, each of whose divisions
represents a volume of $\frac{1}{50}$th of a cubic inch. Successive
measures of this capacity are permitted to enter the
tube, and the absorption in each particular case is determined.

In the following table the first column contains the
quantity of gas admitted into the tube; the second contains
the corresponding absorption; the third column contains
the absorption, calculated on the supposition that it is proportional to the density.

## Olefiant Gas.

Unit-measure $\frac{1}{50}$th of a cubic inch.

| Measures of Gas | Observed Absorption | Calculated Absorption |
|---|---|---|
| 1 | 2·2 | 2·2 |
| 2 | 4·5 | 4·4 |
| 3 | 6·6 | 6·6 |
| 4 | 8·8 | 8·8 |
| 5 | 11·0 | 11·0 |
| 6 | 12·0 | 13·2 |
| 7 | 14·8 | 15·4 |
| 8 | 16·8 | 17·6 |
| 9 | 19·8 | 19·8 |
| 10 | 22·0 | 22·0 |
| 11 | 24·0 | 24·2 |
| 12 | 25·4 | 26·4 |
| 13 | 29·0 | 28·6 |
| 14 | 30·2 | 29·8 |
| 15 | 33·5 | 33·0 |

This table proves the correctness of the surmise, that when very small quantites of the gas are employed, the absorption is sensibly proportional to the density. But consider for a moment the tenuity of the gas with which we have here operated. The volume of our experimental tube is 220 cubic inches; imagine $\frac{1}{50}$th of a cubic inch of gas diffused in this space, and you have the atmosphere through which the calorific rays passed in our first experiment. This atmosphere possesses a tension not exceeding $\frac{1}{11000}$th of that of ordinary air. It would depress the mercurial column connected with the air-pump not more than $\frac{1}{367}$th of an English inch. Its action, however, upon the calorific rays is perfectly measurable.

But the absorptive energy of olefiant gas, extraordinary as it is shown to be by the foregoing experiments, is exceeded by that of various vapours, the action of which I will now endeavour to illustrate. Here is a glass flask, G (fig. 90), provided with a brass cap, into which a stopcock can be screwed air-tight. I pour a small quantity of sulphuric ether into the flask, and completely remove, in the first

place, the air which fills the flask above the liquid. I
attach the flask to the experimental tube, which is now

FIG. 90. exhausted—the needle pointing to zero—and
permit the vapour from the flask to enter the
experimental tube. The mercury of the gauge
sinks, and now that it is depressed one inch I
will stop the further supply of vapour. The
moment the vapour entered, the needle moved,
and it now points to 65° I can add another inch,
and again determine the absorption, a third inch
and do the same. The absorptions effected by
four inches, introduced in this way, are given in
the following table. For the sake of comparison
I place the corresponding absorptions of olefiant gas in
the third column.

### Sulphuric Ether.

| Tensions in inches | | | | | Absorption | | | | Corresponding absorption of Olefiant Gas |
|---|---|---|---|---|---|---|---|---|---|
| 1 | . | . | . | . | . 214 | . | . | . | . 90 |
| 2 | . | . | . | . | . 282 | . | . | . | . 123 |
| 3 | . | . | . | . | . 315 | . | . | . | . 142 |
| 4 | . | . | . | . | . 330 | . | . | . | . 154 |

For these tensions the absorption of radiant heat by the
vapour of sulphuric ether is about two and two-third times
the absorption of olefiant gas. There is, moreover, no
proportionality between the quantity of vapour and the
absorption.

But reflections similar to those which we have already
applied to olefiant gas are also applicable to the ether.
Supposing we make our unit-measure small enough, the
number of rays first destroyed will vanish in comparison
with the total number, and, for a time, the fact will probably manifest itself, that the absorption is directly proportional to the density. To examine whether this is the case,
the other portion of the apparatus, omitted in the general
description, was made use of. K is one of the small flasks

already described, with a brass cap, which is closely screwed on to the stopcock $c'$.   Between the cocks $c'$ and $c$, which latter is connected with the experimental tube, is the chamber M, the capacity of which was accurately determined.   The flask $k$ was partially filled with ether, and the air above the liquid removed.   The stopcock $c'$ being shut off and $c$ turned on, the tube s s' and the chamber M are exhausted.   The cock $c$ is now shut off, and $c'$ being turned on, the chamber M becomes filled with pure ether vapour.   By turning $c'$ off and $c$ on, this quantity of vapour is allowed to diffuse itself through the experimental tube, where its absorption is determined; successive measures are thus sent into the tube, and the effect produced by each is noted.

In the following table the unit measure made use of had a volume of $\frac{1}{100}$th of a cubic inch.

### Sulphuric Ether.

| | Absorption | |
|---|---|---|
| Measures | Observed | Calculated |
| 1 | 5·0 | 4·6 |
| 2 | 10·3 | 9·2 |
| 4 | 19·2 | 18·4 |
| 5 | 24·5 | 23·0 |
| 6 | 29·5 | 27·0 |
| 7 | 34·5 | 32·2 |
| 8 | 38·0 | 36·8 |
| 9 | 44·0 | 41·4 |
| 10 | 46·2 | 46·2 |
| 11 | 50·0 | 50·6 |
| 12 | 52·8 | 55·2 |
| 13 | 55·0 | 59·8 |
| 14 | 57·2 | 64·4 |
| 15 | 59·4 | 69·0 |

We here find that the proportion between density and absorption holds sensibly good for the first eleven measures, after which the deviation from proportionality gradually augments.

No doubt, for smaller measures than $\frac{1}{100}$th of a cubic inch the above law holds still more rigidly true; and in a

suitable locality it would be easy to determine, with perfect
accuracy, $\frac{1}{10}$th of the absorption produced by the first
measure; this would correspond to $\frac{1}{1000}$th of a cubic inch
of vapour. But, before entering the tube, the vapour had
only the tension due to the temperature of the laboratory,
namely 12 inches. This would require to be multiplied
by 2·5 to bring it up to that of the atmosphere. Hence
the $\frac{1}{1000}$th of a cubic inch would, on being diffused
through a tube possessing a capacity of 220 cubic inches,
have a tension of $\frac{1}{220} \times \frac{1}{2.5} \times \frac{1}{1000} = \frac{1}{500\,000}$th of an atmosphere!

These experiments with ether and olefiant gas show
that not only do gaseous bodies, at the ordinary tension of
the atmosphere, offer an impediment to the transmission
of radiant heat; not only are the interstitial spaces of
such gases incompetent to allow the ethereal undulations
free passage; but, also, that their density may be reduced
vastly below that which corresponds to the atmospheric
pressure, and still the door thus opened is not wide enough
to let the undulations through. There is something in
the constitution of the individual molecules, thus sparsely
scattered, which enables them to destroy the calorific waves.
The destruction, however, is merely one of form; there is
no absolute loss. Through dry air the heat rays pass without sensibly warming it; through olefiant gas and ether
vapour they cannot pass thus freely; but every wave withdrawn from the radiant sheaf produces its equivalent
motion in the body of the absorbing gas, and raises its
temperature. It is a case of transference, not of annihilation.
I might extend the experiments to all available volatile
liquids, and show you that the same rule holds good for
the vapours of all.

Before changing the source of heat here made use of, I
wish to direct your attention for a moment to the action
of a few of the permanent gases on radiant heat. To

measure the quantities introduced into the experimental tube, the mercury gauge of the air-pump was made use of.   In the case of carbonic oxide, the following absorptions correspond to the tensions annexed to them, the action of a full atmosphere of air, which, as you remember, produces a deflection of 1°, being taken as unit:—

Carbonic Oxide.

| Tension in inches | Absorption | |
|---|---|---|
| | Observed | Calculated |
| 0·5 | 2·5 | 2·5 |
| 1·0 | 5·6 | 5·0 |
| 1·5 | 8·0 | 7·5 |
| 2·0 | 10·0 | 10·0 |
| 2·5 | 12·0 | 12·5 |
| 3·0 | 15·0 | 15·0 |
| 3·5 | 17·5 | 17·5 |

As in former cases, the third column is calculated on the assumption that the absorption is directly proportional to the density of the gas; and we see that for seven measures, or up to a tension of 3·5 inches, the proportionality holds strictly good.   But for large quantities this is not the case; when, for instance, the unit measure is 5 inches instead of half-an-inch, we obtain the following result:—

| Tension in inches | Absorption | |
|---|---|---|
| | Observed | Calculated |
| 5 | 18 | 18 |
| 10 | 32·5 | 36 |
| 15 | 45 | 54 |

The case of carbonic oxide is therefore similar to that of olefiant gas.   Carbonic acid, sulphide of hydrogen, nitrous oxide, and other gases, though differing in the energy of their absorption, and all of them exceeding carbonic oxide, exhibit, when small and large quantities are used, a similar deportment towards radiant heat.

Thus, then, in the case of some gases, we find an almost absolute incompetence on the part of their atoms to be shaken by the ethereal waves.   They remain practically at rest when the undulations speed amongst them, while the atoms of other gases, struck by these same undulations, absorb their motion, and become themselves centres of heat. We have now to examine what gaseous bodies are competent to do in this latter capacity; we have to enquire whether these atoms and molecules, which can accept motion from the ether in such very different degrees, are not also characterised by their competency to *impart* motion to the ether in different degrees; or, to use the common language, having learned something of the power of different gases, as *absorbers* of radiant heat, we have now to enquire into their capacities as *radiators*.

I have here an arrangement, by means of which we can put the necessary question, which has hitherto received only a negative reply. P (fig. 91) is the thermo-electric pile with

FIG. 91.

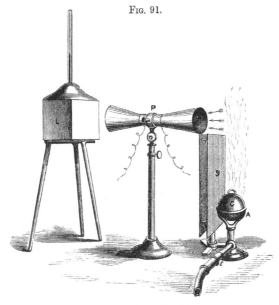

its two conical reflectors; s is a double screen of polished tin; A is an argand burner, consisting of two concentric perforated rings; c is a copper ball, which, during the experiments, is heated under redness; while the tube $t\,t$ leads to a gas holder. When the hot ball c is placed on the burner it warms the air in contact with it; an ascending current is thus established, which, to some extent, acts upon the pile. To neutralise this action a large Leslie's cube, L, filled with water, a few degrees above the air in temperature, is placed before the opposite face of the pile. The needle being thus brought to zero, the gas is forced, by a gentle water pressure, through the orifices of the burner; it meets the ball c, glides along its surface, and ascends, in a warm current, in front of the pile. The rays from the heated gas gush forth in the direction of the arrows against the pile, and the consequent deflection of the galvanometer needle indicates the magnitude of the radiation.

The results of the experiments are given in the second column of the following table, the numbers there recorded marking the extreme limit to which the needle swung, when the rays from the gas fell upon the pile :—

|  | Radiation | Absorption |
|---|---|---|
| Air | 0 | 0·2 |
| Oxygen | 0 | 0·2 |
| Nitrogen | 0 | 0·2 |
| Hydrogen | 0 | 0·2 |
| Carbonic oxide | 12 | 18·0 |
| Carbonic acid | 18 | 25·0 |
| Nitrous oxide | 29 | 44·0 |
| Olefiant gas | 53 | 61·0 |

In order to compare the radiation with the absorption, I have placed in the third column the deflections due to the absorption of the same gases, at a common tension of 5 inches. We see that radiation and absorption go hand in hand; that the molecule which shows itself competent to *intercept* a calorific flux, shows itself competent, in a proportionate degree, to *generate* a calorific flux. That, in short, a capacity to accept motion from the ether, and to impart motion to the ether, by gaseous bodies, are correlative properties.

And here, be it remarked, we are relieved from all considerations regarding the influence of cohesion, on the results. In solids and liquids the particles are more or less in thrall, and cannot be considered as individually free. The difference in point of radiative and absorptive power, between alum and rocksalt, for example, might be fairly regarded as due to their character as aggregates, held together by crystallising force. But the difference between olefiant gas and atmospheric air cannot be explained in this way; it is a difference dependent on the individual molecules of these substances, and thus our experiments with gases and vapours probe the question of atomic constitution to a depth, quite unattainable with solids and liquids.

I have refrained thus far from giving you as full a tabular statement of the absorptive powers of gases and vapours as the experiments made with the apparatus already described would enable me to do, knowing that I had in reserve results, obtained with another apparatus,

FIG. 92.

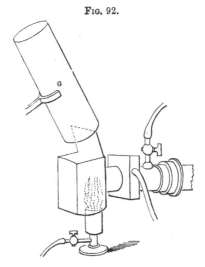

which would better illustrate the subject. This second arrangement is the same in principle as the first; only two changes of importance have been made in it. The first is, that instead of making a cube of boiling water my source of heat, I employ a plate of copper, against which a thin steady gas-flame from a Bunsen's burner is caused to play; the heated plate forms the back of my new front chamber, which latter can be exhausted independently, as before. This portion of the apparatus is sketched in fig. 92, the chimney G being added. The second alteration is the substitution of a tube of glass of the same diameter, and 2 feet 8 inches long, for the tube of brass s s', Plate I. All the other parts of the apparatus remain as before. The gases were introduced in the manner already described into the experimental tube, and from the galvanometric deflection, consequent on the entrance of each gas, its absorption was calculated.

The following table gives the relative absorptions of several gases, at a common tension of one atmosphere :—

| Name | Absorption at 30 inches tension |
|---|---|
| Air . . . . . . | 1 |
| Oxygen . . . . . | 1 |
| Nitrogen . . . . . | 1 |
| Hydrogen . . . . . | 1 |
| Chlorine . . . . . | 39 |
| Hydrochloric acid . . . . . | 62 |
| Carbonic oxide . . . . . | 90 |
| Carbonic acid . . . . . | 90 |
| Nitrous oxide . . . . . | 355 |
| Sulphide of hydrogen . . . . | 390 |
| Marsh gas . . . . . | 403 |
| Sulphurous acid . . . . | 710 |
| Olefiant gas . . . . . | 970 |
| Ammonia . . . . . | 1195 |

The most powerful and delicate tests that I have been able to apply have not yet enabled me to establish a difference between oxygen, nitrogen, hydrogen, and air. The absorption of these substances is exceedingly small —

probably even smaller than I have made it. The more perfectly the above-named gases are purified, the more closely does their action approach to that of a vacuum. And who can say that the best drying apparatus is perfect? I cannot even say that sulphuric acid, however pure, may not yield a modicum of vapour to the gases passing through it, and thus make the absorption by those gases appear greater than it ought. Stopcocks also must be greased, and hence may contribute an infinitesimal impurity to the air passing through them. But however this may be, it is certain that if any further advance should be made in the purification of the more feebly acting gases, it will only serve to augment the enormous differences of absorption exhibited by the foregoing table.

Ammonia, at the tension of an atmosphere, exerts an absorption at least 1,195 times that of the air. If I interpose this metal screen between the pile and the experimental tube, the needle will move a little, but so little that you entirely fail to see it. What does this experiment mean? Why, that this ammonia which, within our glass tube, is as transparent to light as the air we breathe, is so opaque to the heat radiating from our source, that the addition of a plate of metal hardly augments the opacity. I have reason to believe that it does not augment it at all, and that this light transparent gas is really as black, at the present moment, to the calorific rays, as if the experimental tube were filled with ink, pitch, or any other impervious substance.

In the case of oxygen, nitrogen, hydrogen, and air, the action of a whole atmosphere is so small that it would be quite useless to attempt to determine the action of a fractional part of an atmosphere. Could we, however, make such a determination, the difference between them and the other gases would come out still more forcibly than in the last table. In the case of the energetic gases, we know that the calorific rays are most copiously absorbed by the portion of gas which first enters the experimental tube, the quantities which enter last producing, in many cases, a merely infinitesimal effect.   If, therefore, instead of comparing the gases at a common tension of one atmosphere, we were to compare them at a common tension of an inch, we should doubtless find the difference between the least absorbent and the most absorbent gases greatly augmented.   We have already learned that for small quantities, the heat absorbed is proportional to the amount of gas present.   Assuming this to be true for air and the other feeble gases referred to; taking, that is, their absorption at 1 inch of tension to be $\frac{1}{30}$th of that at 30 inches, we have the following comparative effects.   It will be understood that in every case, except the first four, the absorption of 1 inch of the gas was determined by direct experiment.

| Name | Absorption at 1 inch tension |
|---|---|
| Air . . . . . . | 1 |
| Oxygen . ` . . . . | 1 |
| Nitrogen . . . . . | 1 |
| Hydrogen . . . . . | 1 |
| Chlorine . . . . . | 60 |
| Bromine . . . . . | 160 |
| Carbonic oxide . . . . | 750 |
| Hydrobromic acid . . . | 1005 |
| Nitric oxide . . . . | 1590 |
| Nitrous oxide . . . . | 1860 |
| Sulphide of hydrogen . . . | 2100 |
| Ammonia . . . . . | 7260 |
| Olefiant gas . . . . | 7950 |
| Sulphurous acid . . . . | 8800 |

What extraordinary differences in the constitution and character of the ultimate particles of various gases do the above results reveal!  For every individual ray struck down by the air, oxygen, hydrogen, or nitrogen—the ammonia strikes down a brigade of 7,260 rays; the olefiant gas a brigade of 7,950;  while the sulphurous acid destroys 8,800.

With these results before us, we can hardly help attempting to visualise the atoms themselves, trying to discern,
with the eye of intellect, the actual physical qualities on
which these vast differences depend.  These atoms are
particles of matter, plunged in an elastic medium, accepting
its motions and imparting their motions to it.   Is the
hope unwarranted, that we may be able finally to make
radiant heat such a *feeler* of atomic constitution, that we
shall be able to infer from their action upon it, the mechanism of the ultimate particles of matter themselves?

Have we even now no glimpse of the atomic qualities
necessary to form a good absorber?   You remember our
experiments with gold, silver, and copper; you recollect how
feebly they radiate, and how feebly they absorb.  We heated
them by boiling water; that is to say, we imparted, by the
contact of the water, motion to their atoms; still this
motion was imparted with extreme slowness to the ether
in which those atoms swung.   That their particles glide
through the ether with scarcely any resistance may also be
inferred from the length of time which they require to
cool in vacuo.    But we have seen that when the motion
which the atoms of the above bodies possess, and which
they are incompetent to transfer to the ether, is imparted,
by contact, to a coat of varnish, or to a coat of chalk or
lampblack, or even to flannel or velvet, these bodies soon
waste the motion on the ether.   The same we found true
for glass and earthenware.

In what respect do those good radiators differ from the
metals referred to?   In one profound particular — the
metals *are elements*; the others *are compounds*.   In the
metals the atoms swung singly; in the varnish, velvet,
earthenware and glass, they swung in groups.   And now,
in bodies as diverse from the metals as can possibly
be conceived, we find the same significant fact making
its appearance.   Oxygen, hydrogen, nitrogen, and air,

are elements, or mixtures of elements, and, both as regards radiation and absorption, their feebleness is declared. They swing in the ether with scarcely any loss of moving force. They bear the same relation to the compound gases as a smooth cylinder turning in water does to a paddle-wheel. They create a small comparative disturbance.

We may push these considerations still further. It is impossible not to be struck by the position of chlorine and bromine in the last table. Chlorine is an extremely dense and coloured gas; bromine is a far more densely-coloured vapour; still we find them, as regards perviousness to the heat of our source, standing above every transparent compound gas in the table. The act of combination with hydrogen produces, in the case of each of these substances, a transparent compound; but the chemical act, which augments the transparency to light, augments the opacity to heat; hydrochloric acid absorbs more than chlorine; and hydrobromic acid absorbs more than bromine.

Further, I have here the element bromine in the liquid condition; I enclose a portion of it in this glass cell; the layer is of a thickness sufficient to extinguish utterly the flame of a lamp or candle. But I place a candle in front of the cell, and a thermo-electric pile behind it; the prompt movement of the needle declares the passage of radiant heat through the bromine. This consists entirely of the obscure rays of the candle, for the light, as I have stated, is utterly cut off. I remove the candle, and put in its place our copper ball, heated not quite to redness. The needle at once flies to its stops, showing the transparency of the bromine to the heat emitted by the ball. I cannot use iodine in a solid state, but, happily, it dissolves in bisulphide of carbon. I have here the densely coloured liquid in this glass cell. I throw the parallel electric beam upon the screen; this solution of iodine completely cuts the light off; but

if I bring my pile into the path of the beam, the violence
of the needle's motion shows how copious is the transmission of the obscure rays.  It is impossible, I think, to
close our eyes upon this convergent evidence that the free
atoms swing with ease in the ether, while when grouped
to oscillating systems, they cause its waves to swell,
imparting to it, as compound molecules, an amount of
motion which was quite beyond their power to communicate, as long as they remained uncombined.

But it will occur to you, no doubt, that lampblack,
which is an elementary substance, is one of the best
absorbers and radiators in nature.  Let us examine this
substance a little :  ordinary lampblack contains many
impurities; it has various hydro-carbons condensed within
it, and these hydro-carbons are all powerful absorbers and
radiators.   Lampblack, therefore, as hitherto applied,
can hardly be considered an element at all.   I have, however, had these hydro-carbons in great part removed,
by carrying through red hot lampblack a current of
chlorine gas; but the substance has continued to be both
a powerful radiator and a powerful absorber.   Well, what
*is* lampblack ?   Chemists will tell you that it is an allotropic form of the diamond: here, in fact, is a diamond
reduced to charcoal by intense heat.   Now, the allotropic
condition has long been defined as due to a difference in
the arrangement of a body's particles ; hence, it is conceivable that this arrangement, which causes such a marked
physical difference between lampblack and diamond, may
consist of an atomic grouping, which causes the body to
act on radiant heat as if it were a compound.   I say
such an arrangement of an element, though exceptional,
is quite conceivable ; and I shall show you this to be eminently the case as regards an allotropic form of our highly
ineffectual oxygen.

But, in reality, lampblack is not so impervious as you

might suppose it to be.  Melloni has shown it to be transparent, in an unexpected degree, to radiant heat emanating from a low source, and I have prepared an experiment which will corroborate his.  Here is a plate of
rock-salt, which, by holding it over a smoky lamp, has
been so thickly coated with soot that it does not allow a
trace of light from the most brilliant gas jet to pass
through it.  I place the plate upon its stand, and between
it and this vessel of boiling water, which is to serve as our
source of heat, I place a screen.  The thermo-electric pile
is at the other side of the smoked plate.  The needle is
now at zero, and I withdraw my screen; instantly the
needle moves, and its final and permanent deflection is 52°.
I now cleanse the salt perfectly, and determine the radiation through the unsmoked plate,—it is 71°.  Now, the
value of the deflection 52°, expressed with reference to
our usual unit, is 90, and the value of 71°, or the total
radiation, is about 300.  Hence, the radiation through the
soot is to the whole radiation as

$$90 : 300$$
$$\text{or as } 30 : 100$$

that is to say, 30 per cent. of the incident heat has been
transmitted by the layer of lampblack.

I have shown you the action of gases upon radiant heat,
with our glass experimental tube and our new source of heat;
let us complete this lecture by reference to the action of
vapours.  Here, you see, I have several glass flasks, each furnished with a brass cap, to which a stopcock can be screwed.
Into each I pour a quantity of a volatile liquid, reserving
a flask for each liquid, so as to render admixture of the
vapours impossible.  From each flask I remove the air,—
not only the air above the liquid, but the air dissolved
in it.  This latter bubbles freely away when the flask
is exhausted; now I attach my flask to the exhausted

A A

experimental tube, and allow the vapour to enter, without
permitting any ebullition to occur. The mercury column
of the pump sinks, and when the required depression
has been obtained, I cut off the supply of vapour. In
this way the vapours of the substances mentioned in the
next table have been examined, at tensions of 0·1, 0·5, and
1 inch, respectively.

| Name | Absorption of Vapours at Tensions | | |
|---|---|---|---|
| | 0·1 | 0·5 | 1·0 |
| Bisulphide of carbon . . . | 15 | 47 | 62 |
| Iodide of methyl . . . | 35 | 147 | 242 |
| Benzol . . . . . | 66 | 182 | 267 |
| Chloroform . . . . . | 85 | 182 | 236 |
| Methylic alcohol . . . . | 109 | 390 | 590 |
| Amylene . . . . . | 182 | 535 | 823 |
| Sulphuric ether . . . . | 300 | 710 | 870 |
| Alcohol . . . . . | 325 | 622 | |
| Formic ether . . . . | 480 | 870 | 1075 |
| Acetic ether . . . . | 590 | 980 | 1195 |
| Propionate of ethyl . . . | 596 | 970 | |
| Boracic ether . . . . | 620 | | |

These numbers refer to the absorption of a whole atmosphere of dry air as their unit; that is to say, $\frac{1}{10}$th of an
inch of bisulphide of carbon vapour does 15 times the
execution of 30 inches of atmospheric air; while $\frac{1}{10}$th of
an inch of boracic ether vapour does 620 times the execution of a whole atmosphere of atmospheric air. Comparing
air at a tension of 0·01 with boracic ether at the same
tension, the absorption of the latter is probably 180,000
times that of the former.

355

# APPENDIX TO LECTURE X.

I GIVE here the method of calibrating the galvanometer which Melloni recommends, as leaving nothing to be desired as regards facility, promptness, and precision. His own statement of the method, translated from La Thermochrose, page 59, is as follows :—

Two small vessels, v v, are half-filled with mercury, and connected, separately, by two short wires, with the extremities G G of the galvanometer. The vessels and wires thus disposed make no change in the action of the instrument; the thermo-electric current being freely transmitted, as before, from the pile to the galvanometer. But if, by means of a wire F, a communication be established between the two vessels, part of the current will pass through this wire and return to the pile. The quantity of electricity circulating in the galvanometer will be thus diminished, and with it the deflection of the needle.

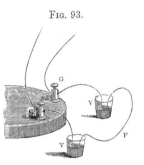

FIG. 93.

Suppose, then, that by this artifice we have reduced the galvanometric deviation to its fourth or fifth part; in other words, supposing that the needle being at 10 or 12 degrees, under the action of a constant source of heat, placed at a fixed distance from the pile, that it descends to 2 or 3 degrees when a portion of the current is diverted by the external wire; I say that by causing the source to act from various distances, and observing in each case the *total* deflection, and the *reduced* deflection, we have all the data necessary to determine the ratio of the deflections of the needle, to the forces which produce these deflections.

To render the exposition clearer, and to furnish, at the same time, an example of the mode of operation, I will take the numbers

A A 2

relating to the application of the method to one of my thermo-
multipliers.

The external circuit being interrupted, and the source of heat
being sufficiently distant from the pile to give a deflection not
exceeding 5 degrees of the galvanometer, let the wire be placed
from v to v; the needle falls to 1°·5.   The connection between
the two vessels being again interrupted, let the source be brought
near enough to obtain successively the deflections :—

$$5°, 10°, 15°, 20°, 25°, 30°, 35°, 40°, 45°.$$

Interposing after each the same wire between v and v we obtain
the following numbers :—

$$1°·5, 3°, 4°·5, 6°·3, 8°·4, 11°·2, 15°·3, 22°·4, 29°·7.$$

Assuming the force necessary to cause the needle to describe
each of the first degrees of the galvanometer to be equal to
unity, we have the number 5 as the expression of the force corresponding to the first observation.   The other forces are easily
obtained by the proportions :—

$$1·5 : 5 = a : x = \tfrac{5}{1·5} \, a = 3·333 \, a.*$$

where $a$ represents the deflection when the exterior circuit is
closed.   We thus obtain

$$5, 10, 15·2, 21, 28, 37·3.$$

for the forces, corresponding to the deflections,

$$5°, 10°, 15°, 20°, 25°, 30°.$$

In this instrument, therefore, the forces are sensibly proportional to the arcs, up to nearly 15 degrees.   Beyond this, the
proportionality ceases, and the divergence augments as the arcs
increase in size.

The forces belonging to the intermediate degrees are obtained
with great ease either by calculation or by graphical construction,
which latter is sufficiently accurate for these determinations.

By these means we find,

| | |
|---|---|
| Degrees     .     . | 13°, 14°, 15°, 16°, 17°, 18°, 19°, 20°, 21°. |
| Forces      .     . | 13, 14·1, 15·2, 16·3, 17·4, 18·6, 19·8, 21, 22·3. |
| Differences .     . | 1·1, 1·1, 1·1, 1·1, 1·2, 1·2, 1·2, 1·3. |

* That is to say, one reduced current is to the total current to which
it corresponds, as any other reduced current is to its corresponding total
current.

Degrees . . 22°, 23°, 24°, 25°, 26°, 27°, 28°, 29°, 30°.
Forces . . 23·5, 24·9, 26·4, 28, 29·7, 31·5, 33·4, 35·3, 37·3.
Differences . . 1·4, 1·5, 1·6, 1·7, 1·8, 1·9, 2.

In this table we do not take into account any of the degrees preceding the 13th, because the force corresponding to each of them possesses the same value as the deflection.

The forces corresponding to the first 30 degrees being known, nothing is easier than to determine the values of the forces corresponding to 35, 40, 45 degrees, and upwards.

The reduced deflections of these three arcs are,

$$15°·3, 22°·4, 29°·7.$$

Let us consider them separately; commencing with the first. In the first place, then, 15 degrees, according to our calculation, are equal to 15·2; we obtain the value of the decimal 0·3 by multiplying this fraction by the difference 1·1 which exists between the 15th and 16th degrees; for we have evidently the proportion

$$1 : 1·1 = 0·3 : x = 0·3.$$

The value of the reduced deflection corresponding to the 35th degree, will not, therefore, be 15°·3, but 15°·2 + 0°·3 = 15°·5. By similar considerations we find 23°·5 + 0°·6 = 24°·1, instead of 22°·4, and 36°·7 instead of 29°·7 for the reduced deflections of 40 and 45 degrees.

It now only remains to calculate the forces belonging to these three deflections, 15°·5, 24°·1, and 36°·7, by means of the expression $3·333\ a$; this gives us,

the forces, 51·7, 80·3, 122·3.
for the degrees, 35, 40, 45.

Comparing these numbers with those of the preceding table, we see that the sensitiveness of our galvanometer diminishes considerably when we use deflections greater than 30 degrees.

# LECTURE XI.

[April 3, 1862.]

ACTION OF ODOROUS SUBSTANCES UPON RADIANT HEAT — ACTION OF
OZONE UPON RADIANT HEAT — DETERMINATION OF THE RADIATION
AND ABSORPTION OF GASES AND VAPOURS WITHOUT ANY SOURCE
OF HEAT EXTERNAL TO THE GASEOUS BODY — DYNAMIC RADIATION
AND ABSORPTION — RADIATION THROUGH THE EARTH'S ATMOSPHERE
— INFLUENCE OF THE AQUEOUS VAPOUR OF THE ATMOSPHERE ON
RADIANT HEAT — CONNECTION OF THE RADIANT AND ABSORBENT
POWER OF AQUEOUS VAPOUR WITH METEOROLOGICAL PHENOMENA.

APPENDIX :—FURTHER DETAILS OF THE ACTION OF HUMID AIR.

SCENTS and effluvia generally have long occupied the
attention of observant men, and they have formed
favourite illustrations of the 'divisibility of matter.' No
chemist ever weighed the perfume of a rose; but in radiant heat we have a test more refined than the chemist's
balance. The results brought before you in our last lecture would enable you to hear me without surprise, were I
to assert that the quantity of volatile matter removed from
a hartshorn bottle by any person in this room, by a single
act of inhalation, would exercise a more potent action on
radiant heat, than the whole body of oxygen and nitrogen
which the room contains. Let us apply this test to other
odours, and see whether they also, notwithstanding their
almost infinite attenuation, do not exercise a measurable
influence on radiant heat.

I will operate in this simple way: here is a number of
small and equal squares of bibulous paper, which I roll up
thus, to form little cylinders, each about two inches in
length. I moisten the paper cylinder by dipping one end

of it into an aromatic oil; the oil creeps by capillary attraction through the paper, and the whole of the cylinder is now moist. I introduce the rolled papers thus into a glass tube, of such a diameter that the cylinder fills it without being squeezed, and between my drying apparatus and the experimental cylinder I place the tube containing the scented paper. The experimental cylinder is now exhausted, and the needle at zero; turning this cock on, I allow dry air to pass gently through the folds of the saturated paper. Here the air takes up the perfume of the aromatic oil, and carries it into the experimental tube. The absorption of an atmosphere of dry air we know to be unity; it produces a deflection of one degree; hence, any additional absorption which these experiments reveal, must be due to the scent which accompanies the air.

The following table will give a condensed view of the absorption of the substances mentioned in it; air at the tension of one atmosphere being regarded as unity:—

## *Perfumes.*

| Name of Perfume | Absorption |
|---|---|
| Pachouli | 30 |
| Sandal Wood | 32 |
| Geranium | 33 |
| Oil of Cloves | 33·5 |
| Otto of Roses | 36·5 |
| Bergamot | 44 |
| Neroli | 47 |
| Lavender | 60 |
| Lemon | 65 |
| Portugal | 67 |
| Thyme | 68 |
| Rosemary | 74 |
| Oil of Laurel | 80 |
| Camomile Flowers | 87 |
| Cassia | 109 |
| Spikenard | 355 |
| Aniseed | 372 |

The number of atoms of air here in the tube must be regarded as almost infinite in comparison with those of the odours; still the latter, thinly scattered as they are, do, in the case of pachouli, 30 times the execution of the air; otto of roses does upwards of 36 times the execution of the air; thyme, 74 times; spikenard, 355 times; and aniseed 372 times the execution of the air. It would be idle to speculate on the quantities of matter implicated in these results. Probably they would have to be multiplied by millions to bring them up to the tension of ordinary air. Thus,—

> The sweet south
> That breathes upon a bank of violets,
> Stealing and giving odour,

owes its sweetness to an agent, which, though almost infinitely attenuated, may be more potent, as an intercepter of terrestrial radiation, than the entire atmosphere from 'bank' to sky.

In addition to these experiments on the essential oils, others were made on aromatic herbs. A number of such were obtained from Covent Garden Market; they were dry, in the common acceptation of the term; that is to say, they were not green, but withered. Still I fear the results obtained with them cannot be regarded as pure, on account of the probable admixture of aqueous vapour. The aromatic parts of the plants were stuffed into a glass tube eighteen inches long and a quarter of an inch in diameter. Previous to connecting them with the experimental tube, they were attached to a second air-pump, and dry air was carried over them for some minutes. They were then connected with the experimental cylinder, and treated as the essential oils; the only difference being that a length of eighteen inches, instead of two, was occupied by the herbs.

Thyme, thus examined, gave an action thirty-three times that of the air which passed over it.

Peppermint exercised thirty-four times the action of the air.

Spearmint exercised thirty-eight times the action of the air.

Lavender exercised thirty-two times the action of the air.

Wormwood exercised forty-one times the action of the air.

Cinnamon exercised fifty-three times the action of the air.

As already hinted, I fear that these results may be complicated with the action of aqueous vapour : its quantity, however, must have been infinitesimal.

There is another substance of great interest to the chemist, but the attainable quantities of which are so minute as almost to elude measurement, to which we may apply the test of radiant heat. I mean that extraordinary substance, ozone. This body is known to be liberated at the oxygen electrode, when water is decomposed by an electric current. To investigate its action I had constructed three different decomposing cells. In the first, which I shall call No. 1, the platinum plates used as electrodes had about four square inches of surface; the plates of the second (No. 2) had two square inches of surface; while the plates of the third (No. 3) had only one square inch of surface, each.

My reason for using electrodes of different sizes was this:— On first applying radiant heat to the examination of ozone, I constructed a decomposing cell, in which, to diminish the resistance of the current, very large platinum plates were used. The oxygen thus obtained, and which ought to have embraced the ozone, showed scarcely any of the reactions of this substance. It hardly discoloured iodide of potassium, and was almost without action on radiant heat. A second decomposing apparatus, with smaller plates, was tried, and here I found both the action on iodide of potassium, and on radiant heat, very decided. Being unable to refer these differences to any other cause than the different magnitudes of the plates, I formally attacked the subject by operating with the three cells above described. Calling the action of the main body of the electrolytic oxygen unity; that of the ozone which accompanied it, in the respective cases, is given in the following table :—

| Number of Cell | | | | | | | Absorption |
|---|---|---|---|---|---|---|---|
| No. 1 | . | . | . | . | . | . | . 20 |
| No. 2 | . | . | . | . | . | . | . 34 |
| No. 3 | . | . | . | . | . | . | . 47 |

Thus the modicum of ozone which accompanied the oxygen, and in comparison to which it is a vanishing quantity, exerted, in the case of the first pair of plates, an action twenty times that of the oxygen itself, while with the third pair of plates the ozone was forty-seven times more energetic than the oxygen. The influence of the size of the plates ; or, in other words, of the *density* of the current where it enters the liquid, on the production of the ozone, is rendered strikingly manifest by these experiments.

I then cut away portions of the plates of cell No. 2, so as to make them smaller than those of No. 3. The reduction of the plates was accompanied by an augmentation of the action upon radiant heat ; the absorption rose at once from 34 to

65.

The reduced plates of No. 2 here transcend those of No. 3, which, in the first experiments, gave the largest action.

The plates of No. 3 were next reduced, so as to make them smallest of all. The ozone now generated by No 3, effected an absorption of

85.

Thus we see that the action upon radiant heat advances as the size of the electrodes is diminished.

Heat is known to be very destructive of ozone, and suspecting the developement of heat at the small electrodes of the cell last made use of, I surrounded the cell with a mixture of pounded ice and salt. Kept thus cool, the absorption of the ozone generated rose to

136.

These experiments on the action of ozone upon radiant heat were made, before I was acquainted with the researches of MM. De la Rive, Soret, and Meidinger, on this substance. There is a perfect correspondence in our results, though there is no resemblance between our modes of experiment. Such a correspondence is calculated to augment our confidence in radiant heat, as an investigator of molecular condition.*

* M. Meidinger commences his paper by showing the absence of agreement between theory and experiment in the decomposition of water, the difference showing itself very decidedly in a deficiency of oxygen *when the current was strong.* On heating his electrolyte, he found that this difference disappeared, the proper quantity of oxygen being then liberated. He at once surmised that the defect of oxygen might be due to the formation of ozone; but how did the substance act to produce the diminution of the oxygen?. If the defect were due to the great density of the ozone, the destruction of this substance, by heat, would restore the oxygen to its true volume. Strong heating, however, which destroyed the ozone, produced no alteration of volume, hence M. Meidinger concluded that the effect which he observed was not due to the ozone which remained mixed with the oxygen itself. He finally concluded, and justified his conclusion by satisfactory experiments, that the loss of oxygen was due to the formation, in the water, of peroxide of hydrogen by the ozone; the oxygen being thus withdrawn from the tube to which it belonged. He also, as M. De la Rive had previously done, experimented with electrodes of different sizes, and found the loss of oxygen much more considerable when a small electrode was used than with a large one; whence he inferred that the formation of ozone was facilitated by *augmenting the density of the current at the place where electrode and electrolyte meet.* The same conclusion is deduced from the above experiments on radiant heat. No two things could be more diverse than the two modes of proceeding. M. Meidinger sought for the oxygen which had disappeared, and found it in the liquid; I examined the oxygen actually liberated, and found that the ozone mixed with it augments in quantity as the electrodes diminish in size. It may be added that since the perusal of M. Meidinger's paper I have repeated his experiments with my own decomposition cells, and

The quantities of ozone with which the foregoing experiments were made, must be perfectly unmeasurable by ordinary means. Still its action upon radiant heat is so energetic, as to place it beside olefiant gas, or boracic ether, as an absorbent—bulk for bulk it might transcend either. No *elementary gas* that I have examined behaves at all like ozone. In its swing through the ether it must powerfully disturb the medium. If it be oxygen, it must, I think, be oxygen atoms packed into groups. I sought to decide the question whether it is oxygen, or a compound of hydrogen, in the following way. Heat destroys ozone. If it were oxygen only, heat would convert it into the common gas; if it were the hydrogen compound, which some chemists consider it to be, heat would convert it into oxygen, plus aqueous vapour. The gas alone, admitted into my tube, would give the neutral action of oxygen, but the gas, plus the aqueous vapour, I hoped might give a sensibly greater action. The dried electrolytic gas was caused to pass through a glass tube heated to redness, and thence direct into the experimental tube. It was next, after heating, made to pass through a drying tube into the experimental tube. Hitherto I have not been able to establish, with certainty, a difference between the dried and undried gas. If, therefore, the act of heating develope aqueous vapour, the experimental means which I have employed have not yet enabled me to detect it. For the present, therefore, I hold the belief, that ozone is produced by the packing of the atoms of elementary oxygen into oscillating groups; and that heating dissolves the bond of union, and allows the atoms to swing singly, thus disqualifying them for either intercepting or generating the motion, which, as systems, they are competent to intercept and generate.

found that those which gave me the greatest absorption, also showed the greatest deficiency in the amount of oxygen liberated.

I have now to direct your attention to a series of facts which surprised and perplexed me when I first observed them. While experimenting last November (1861), on one occasion I permitted a quantity of alcohol vapour, sufficient to depress the mercury gauge 0·5 of an inch, to enter the experimental tube; it produced a deflection of 72°. While the needle pointed to this high figure, and previously to pumping out the vapour, I allowed dry air to stream into the tube, and happened, as it entered, to keep my eye upon the galvanometer.

The needle, to my astonishment, sank speedily to zero, and went to 25° on the opposite side. The entry of the almost neutral air, not only neutralised the absorption previously observed, but left a considerable balance in favour of the face of the pile turned towards the source. A repetition of the experiment brought the needle down from 70° to zero, and sent it to 38° on the opposite side. In like manner, a very small quantity of the vapour of sulphuric ether produced a deflection of 30°; on allowing dry air to fill the tube, the needle descended speedily to zero, and swung to 60° at the opposite side.

My first thought, on observing these extraordinary effects, was, that the vapours had deposited themselves in opaque films on the plates of rock-salt, and that the dry air on entering had cleared these films away, and allowed the heat from the source free transmission.

But a moment's reflection dissipated this supposition. The clearing away of such a film could, at best, but restore the state of things existing prior to the entrance of the vapour. It might be conceived to bring the needle again to 0°, but it could not possibly produce the negative deflection. Nevertheless, I dismounted the tube, and subjected the plates of salt to a searching examination. No such deposit as that above surmised was observed. The salt remained perfectly transparent while in contact with

the vapour.    How, then, are the effects to be accounted
for?

We have already made ourselves acquainted with the
thermal effects produced when air is permitted to stream
into a vacuum (page 30). We know that the air is warmed
by its collision against the sides of the receiver. Can it be
the heat thus generated, imparted by the air to the alcohol
and ether vapours, and radiated by them against the pile,
that was more than sufficient to make amends for the absorption? The *experimentum crucis* at once suggests itself
here. If the effects observed be due to the heating of the
air on entering the partial vacuum in which the vapour
was diffused, we ought to obtain the same effects when the
sources of heat made use of hitherto are entirely abolished.
We are thus led to the consideration of the novel and at
first sight utterly paradoxical problem—namely, to determine the radiation and absorption of a gas or vapour *without any source of heat external to the gaseous body itself.*

Let us, then, erect our apparatus, and omit our two
sources of heat. Here is our glass tube, stopped at one
end by a plate of glass, for we do not now need the passage
of the heat through this end; and at the other end by a
plate of rock-salt. In front of the salt is placed the pile,
connected with its galvanometer. Though there is now no
special source of heat acting upon the pile, you see the
needle does not come quite to zero; indeed, the walls of
this room, and the people who sit before me, are so
many sources of heat, to neutralise which, and thus to
bring the needle accurately to zero, I must slightly warm
the defective face of the pile. This is done without any
difficulty by a cube of lukewarm water, placed at a distance; the needle is now at zero.

The experimental tube being exhausted, I allow air to
enter, till the tube is filled; the horizontal column of air
at present in the tube is warmed; every atom of the air is

oscillating; and if the atoms possessed any sensible power of communicating their motion to the luminiferous ether, we should have from each atom a train of waves impinging on the face of the pile. But you observe scarcely any motion of the galvanometer, and hence may infer that the quantity of heat radiated by the air is exceedingly small. The deflection produced is 7 degrees.

But these 7° are not really due to the radiation of the air. To what, then? I open one of the ends of the experimental tube, and place a bit of black paper as a lining within it; the paper merely constitutes a ring which covers the interior surface of the tube for a length of 12 inches. I close the tube and repeat the last experiment. The tube has been exhausted and the air is now entering, but mark the needle—it has already flown through an arc of 70°. You see here exemplified the influence of this bit of paper lining; it is warmed by the air, and it radiates towards the pile in this copious way. *The interior surface of the tube itself must do the same*, though in a less degree, and to the radiation from this surface, and not from the air itself, the deflection of 7° which we have just obtained is, I believe, to be ascribed.

Removing the bit of lining from the tube, instead of air I allow nitrous oxide to stream into it; the needle swings to 28°, thus showing the superior radiative power of this gas. I now work the pump, the gas within the experimental tube becomes chilled, and into it the pile pours its heat; a swing of 20° in the opposite direction is the consequence.

Instead of nitrous oxide, I now allow olefiant gas to stream into the exhausted tube. We have already learned that this gas is highly gifted with the power of radiation. Its atoms are here warmed, and everyone of them asserts its power; the needle swings through an arc of 67°. Let it waste its heat, and let the needle come to zero. I now

pump out, and the consequent chilling of the gas, within
the tube, produces a deflection of 40° on the side of cold.
We have certainly here a key to the solution of the
enigmatical effects observed with the alcohol and ether
vapour.

For the sake of convenience we may call the heating of
the gas on entering the vacuum *dynamic heating*; its
radiation I have called *dynamic radiation*, and its absorption, when chilled by pumping out, *dynamic absorption*. These terms being understood, the following table
explains itself. In each case the extreme limit to which
the needle swung, on the entry of the gas into the experimental tube, is recorded.

### Dynamic Radiation of Gases.

| Name | Limit of 1st impulsion |
|---|---|
| | ° |
| Air . . . . . . . | 7 |
| Oxygen . . . . . . | 7 |
| Hydrogen . . . . . | 7 |
| Nitrogen . . . . . | 7 |
| Carbonic oxide . . . . | 19 |
| Carbonic acid . . . . . | 21 |
| Nitrous oxide . . . . . | 31 |
| Olefiant gas . . . . . | 63 |

We observe that the order of the radiative powers,
determined in this novel way, is the same as that already
obtained from a totally different mode of experiment. It
must be borne in mind that the discovery of dynamic
radiation is quite recent, and that the conditions of perfect
accuracy have not yet been developed; it is, however,
certain, that the mode of experiment is susceptible of the
last degree of precision.

Let us now turn to our vapours, and while dealing with
them I shall endeavour to unite two effects which, at first
sight, might appear utterly incongruous. We have already
learned that a polished metal surface emits an extremely
feeble radiation; but that when the same surface is coated

with varnish the radiation is copious.  In the communication of motion to the ether the atoms of the metal need
a mediator, and this they find in the varnish.  They communicate their motion to the molecules of the varnish, and
the latter are so related to the luminiferous ether * that
they can communicate their motion to it.  *You may
varnish a metallic surface by a film of a powerful gas.*
I have here an arrangement which enables me to cause a

FIG. 94.

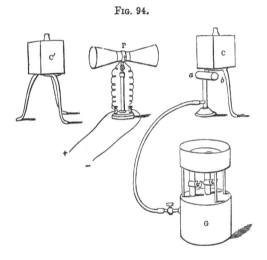

thin stream of olefiant gas from the gasholder G (fig. 94)
to pass through the slit tube *a b*, and over the heated surface of the cube c.  The radiation from c is now neutralised
by that from c′; but I allow the gas to flow over the cube
c; and though the surface is actually cooled by the passage
of the gas, for the gas has to be warmed by the metal, you
see the effect is to augment considerably the radiation :

* If we could change either the name given to the interstellar medium,
or that given to certain volatile liquids by chemists, it would be an
advantage.  It is difficult to avoid confusion in the use of the same name
for objects so utterly diverse.

as soon as the gas begins to flow the needle begins to move, and reaches an amplitude of 45°.

We have here varnished a metal by a gas, but a more interesting and subtle effect is *the varnishing of one gaseous body by another.* I have here a flask containing some acetic ether; a volatile, and, as you know, a highly absorbent substance. I attach the flask to the experimental tube, and permit the vapour to enter the tube, until the mercury column has been depressed half an inch. There is now vapour possessing half an inch of tension in the tube. I intend to use that vapour as my varnish; and I intend to use the element oxygen instead of the element gold, silver, or copper, as the substance to which my vapour varnish is to be applied. At the present moment the needle is at zero, and I now permit dry oxygen to enter the tube : the gas is dynamically heated, and we have seen its incompetence to radiate its heat; but now it comes into contact with the acetic ether vapour, and, communicating its motion to the vapour by direct collision, the latter is able to send on the motion to the pile. Observe the needle — it is caused to swing through an arc of 70° by the radiation from the vapour particles. I need not insist upon the fact that in this experiment the vapour bears precisely the same relation to the oxygen, that the varnish does to the metal in our former experiments.

Let us wait a little, and allow the vapour to pour away the heat: it is the discharger of the calorific force generated by the oxygen — the needle is again at zero. I work the pump, the vapour within the tube becomes chilled, and now you observe the needle swing nearly 45° on the other side of zero. In this way the dynamic radiation and absorption of the vapours mentioned in the following table have been determined; air, however, instead of oxygen, being the substance employed to heat the vapour. The limit of the first swing of the needle is noted as before.

*Dynamic Radiation and Absorption of Vapours.*

| | Deflections | |
|---|---|---|
| | Radiation | Absorption |
| | ° | ° |
| 1. Bisulphide of carbon . . | 14 . . . | 6 |
| 2. Iodide of methyl . . | 19·5 . . | 8 |
| 3. Benzol . . . . | 30 . . . | 14 |
| 4. Iodide of ethyl . . | 34 . . . | 15·5 |
| 5. Methylic alcohol . . | 36 . . . | |
| 6. Chloride of amyl . . | 41 . . . | 23 |
| 7. Amylene . . . . | 48 . . . | |
| 8. Alcohol . . . . | 50 . . . | 27·5 |
| 9. Sulphuric ether . . | 64 . . . | 34 |
| 10. Formic ether . . . | 68·5 . . . | 38 |
| 11. Acetic ether . . . | 70 . . . | 43 |

We have here used eleven different kinds of vapour as varnish for our air, and we find that the dynamic radiation and absorption augment exactly in the order established by experiments with external sources of heat. We also see how beautifully dynamic radiation and absorption go hand in hand, the one augmenting and diminishing with the other.

The smallness of the quantity of matter concerned in some of these actions on radiant heat has been often referred to; and I wish now to describe an experiment which shall furnish a more striking example of this kind than any hitherto brought before you. The absorption of boracic ether vapour, as given at page 354, exceeds that of any other substance there referred to; and its dynamic radiation may be presumed to be commensurate. I exhaust the experimental tube as perfectly as possible, and introduce into it a quantity of boracic ether vapour sufficient to depress the mercury column $\frac{1}{10}$th of an inch. The barometer stands to-day at 30 inches; hence the tension of the ether vapour now in our tube is $\frac{1}{300}$th of an atmosphere.

I send dry air into the tube; the vapour is warmed, and the dynamic radiation produces the deflection 56°.

I work the pump until I reduce the residue of air within it to a tension of 0·2 of an inch, or $\frac{1}{150}$th of an atmosphere. A residue of the boracic ether vapour remains of course in the tube, the tension of this residue being the $\frac{1}{150}$th part of that of the vapour when it first entered the tube. I let in dry air, and find the dynamic radiation of the residual vapour expressed by the deflection 42°.

I again work the pump till the tension of the air within it is 0·2 of an inch; the quantity of ether vapour now in the tube being $\frac{1}{150}$th of that present in the last experiment. The dynamic radiation of this residue gives a deflection of 20°.

Two additional experiments, conducted in the same way, gave deflections of 14° and 10° respectively. The question now is, what was the tension of the boracic ether vapour when this last deflection was obtained? The following table contains the answer to this question:—

### Dynamic Radiation of Boracic Ether.

| Tension in parts of an atmosphere | Deflection ° |
|---|---|
| $\frac{1}{300}$ | 56 |
| $\frac{1}{150} + \frac{1}{300} = \frac{1}{45000}$ | 42 |
| $\frac{1}{150} \times \frac{1}{150} \times \frac{1}{300} = \frac{1}{6750000}$ | 20 |
| $\frac{1}{150} \times \frac{1}{150} \times \frac{1}{150} \times \frac{1}{300} = \frac{1}{1012500000}$ | 14 |
| $\frac{1}{150} \times \frac{1}{150} \times \frac{1}{150} \times \frac{1}{150} \times \frac{1}{300} = \frac{1}{151875000000}$ | 10 |

The air itself, warming the interior of the tube, produces, as we have seen, a deflection of 7°; hence the entire deflection of 10° was not due to the radiation of the vapour. Deducting 7°, it would leave a residue of 3°. But supposing we entirely omit the last experiment, we can then have no doubt that at least half the deflection 14° is due to the residue of boracic ether vapour; this residue we find, by strict measurement, would have to be multiplied by one thousand millions to bring it up to the tension of ordinary atmospheric air.

Another reflection here presents itself, which is worthy of our consideration. We have measured the dynamic radiation of olefiant gas, by allowing the gas to enter our tube, until the latter was quite filled. What was the state of the warm radiating column of olefiant gas in this experiment? It is manifest that the portions of the column most distant from the pile must radiate *through the gas in front of them*, and, in this forward portion of the column of gas, a large quantity of the rays emitted by its hinder portion will be absorbed. In fact, it is quite certain that if we made our column sufficiently long, the frontal portions would act as a perfectly impenetrable screen to the radiation of the hinder ones. Thus, by cutting off the part of the gaseous column most distant from the pile, we might diminish only in a very small degree the amount of radiation which reaches the pile.

Let us now compare the dynamic radiation of a vapour with that of olefiant gas. In the case of vapour we use only 0·5 of an inch of tension, hence the radiating molecules of the ether are much wider apart than those of the olefiant gas, which have 60 times the tension; and consequently the radiation of the hinder portions of the column of vapour will have a comparatively open door through which to reach the pile. These considerations render it manifest that in the case of the vapour *a greater length of tube* is available for radiation than in the case of olefiant gas. This leads to the conclusion, that if we shorten the tube, we shall diminish the radiation in the case of the vapour more considerably than in the case of the gas. Let us now bring our reasoning to the test of experiment.

We have found the dynamic radiation of the following four substances, when the radiating column was 2 feet

9 inches long, to be represented by the annexed deflections :—

| | |
|---|---:|
| Olefiant gas | $\overset{\text{o}}{63}$ |
| Sulphuric ether | 64 |
| Formic ether | 68·5 |
| Acetic ether | 70 |

olefiant gas giving here the least dynamic radiation.

Experiments made in precisely the same manner with a
tube 3 inches long, or $\frac{1}{11}$th of the former length, gave the
following deflections :—

| | |
|---|---:|
| Olefiant gas | $\overset{\text{o}}{39}$ |
| Sulphuric ether | 11 |
| Formic ether | 12 |
| Acetic ether | 15 |

The verification of our reasoning is therefore complete.
It is proved, that in the long tube the dynamic radiation
of the vapour exceeds that of the gas, while in a short one
the dynamic radiation of the gas exceeds that of the
vapour. The result proves, if proof were needed, that
though diffused in air, the vapour molecules are really the
centres of the radiation.

Up to the present point, I have purposely omitted all
reference to the most important vapour of all, as far as our
world is concerned—I mean, of course, the vapour of
water. This vapour, as you know, is always diffused
through the atmosphere. The clearest day is not exempt
from it : indeed, in the Alps, the purest skies are often the
most treacherous, the blue deepening with the amount of
aqueous vapour in the air. It is needless, therefore, to
remind you, that when I speak of aqueous vapour, I mean
nothing visible ; it is not fog ; it is not cloud ; it is not
mist of any kind. These are formed of vapour which has
been condensed to water ; but the true vapour with which
we have to deal is an impalpable transparent gas. It is

diffused everywhere throughout the atmosphere, though in very different proportions.

To prove the existence of aqueous vapour in the air of this room, I have placed in front of the table a copper vessel, which was filled an hour ago with a mixture of pounded ice and salt. The surface of the vessel was then black, but it is now white—furred all over with hoar-frost— produced by the condensation, and subsequent congelation upon its surface of the aqueous vapour. I can scrape off this white substance, and collect it in my hand. As I remove the frozen vapour, the black surface of the vessel reappears; and now I have collected a sufficient quantity to form a respectable snow-ball. Let us go one step further. I place this snow in a mould, and squeeze it before you into a cup of ice—there is the cup; and thus, without quitting this room, we have experimentally illustrated the manufacture of glaciers from beginning to end. On the plate of glass which I have used to cover the vessel the vapour is not congealed, but it is condensed so copiously, that when I hold the plate edgeways the water runs off it in a stream.

The quantity of this vapour is small. Oxygen and nitrogen constitute about $99\frac{1}{2}$ per cent. of our atmosphere; of the remaining 0·5, about 0·45 is aqueous vapour; the residue is carbonic acid. Had we not been already acquainted with the action of almost infinitesimal quantities of matter on radiant heat, we might well despair of being able to establish a measurable action on the part of the aqueous vapour of our atmosphere. Indeed, I quite neglected the action of this substance for a time, and could hardly credit my first result, which made the action of the aqueous vapour of our laboratory fifteen times that of the air in which it was diffused. This, however, by no means expresses the true relation between aqueous vapour and dry air.

I will make an experiment before you which shall
illustrate this. Here, you see, I have resumed our first
arrangement, as shown in Plate I., with a brass tube, and
with two sources of heat acting on the opposite faces of
the pile. I exhaust the experimental tube, and repeat
to-day the experiment with dry air, which I made at the
commencement of the last lecture. The needle does not
move sensibly. If close to it you would, as I have already
stated, observe a motion through about one degree. Probably, could we get our air quite pure, its action would be
even less than this. I now pump out, and allow the air
of this room to enter the experimental cylinder direct,
without permitting it to pass through the drying apparatus.
The needle, you observe, moves as the air enters, and the
final deflection is 48°. The needle will steadily point
to this figure as long as the sources of heat remain constant, and as long as the air continues in the tube. These
48° correspond to an absorption of 72; that is to say, the
aqueous vapour contained in the atmosphere of this room
to-day exerts an action on the radiant heat, 72 times
more powerful than that of the air itself.

This result is obtained with perfect ease, still not
without due care. In comparing dry with humid air, it
is perfectly essential that the substances be pure. You
may work for months with an imperfect drying apparatus and fail to obtain air, which shows this almost total
absence of action on radiant heat. An amount of organic
impurity, too small to be seen by the eye, is sufficient
to augment fiftyfold the action of the air. Knowing the
effect which an almost infinitesimal amount of matter,
in certain cases, can produce, you are better prepared for
such facts than I was when they first forced themselves
upon my attention. But let us be careful in our enquiries.
The experimental result which we have just obtained will,
if true, have so important an influence on the science of

meteorology, that, before it is admitted, it ought to be subjected to the closest scrutiny. First of all, look at this piece of rocksalt brought in from the next room, where it has stood for some time near a tank, but not in contact with visible moisture. The salt is wet; it is a hygroscopic substance, and freely condenses moisture upon its surface. Here, also, is a polished plate of the substance, which is now quite dry; I breathe upon it, and instantly its affinity for moisture causes the vapour of my breath to overspread the surface in a film which exhibits beautifully the colours of thin plates. Now we know from the table, at page 299, how opaque a solution of rocksalt is to the calorific rays, and hence arises the question whether, in the above experiment with undried air, we may not in reality be measuring the action of a thin stratum of such a solution, deposited on our plates of salt, instead of the pure action of the aqueous vapour of the air.

If you operate incautiously, and, more particularly, if it be your actual intention to wet your plates of salt, you may readily obtain the deposition of moisture. This is a point on which any competent experimenter will soon instruct himself; but the essence of good experimenting consists in the exclusion of circumstances which would render the pure and simple questions which we intend to put to Nature, impure and composite ones. The first way of replying to the doubt here raised is to examine our plates of salt; if the experiments have been properly conducted, no trace of moisture is found upon the surface. To render the success of this experiment more certain, I will slightly alter the arrangement of our apparatus. Hitherto we have had the thermo-electric pile and its two reflectors entirely *outside* the experimental cylinder. I now take this reflector from the pile, and removing this terminal plate of rocksalt, I push the reflector into the cylinder. The hollow reflecting cone is 'sprung' at its

base $a\,b$ (fig. 95), (our former arrangement, with the single
exception that one of the reflectors of the pile P is now

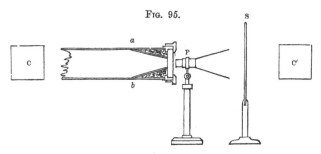

FIG. 95.

within the tube) so that it is held tightly by its own pressure against the inner surface of the cylinder.   The space
between the outer surface of the reflector and the inner
surface of the tube I fill with fragments of fused chloride
of calcium, which are prevented from falling out by a little
screen of wire gauze.   I now reattach my plate of salt,
against the inner surface of which abuts the narrow end of
the reflector; bring the face of the pile close up to the
plate, though not into actual contact with it, and now our
arrangement is complete.

In the first place it is to be remarked, that the plate of
salt nearest to the source of heat c is never moistened, unless
the experiments are of the grossest character.   Its proximity to the source makes it the track of a flux of heat,
powerful enough to chase away every trace of humidity
from its surface.   The distant plate is the one in danger,
and now we have the circumferential portions of this
plate kept perfectly dry by the chloride of calcium; no
moist air can at all reach the rim of the plate; while upon
its central portion, measuring about a square inch in
area, *we have converged our entire radiation*. On *à priori*
grounds we should conclude that it is quite impossible
that a film of moisture could collect there; and this conclusion is justified by fact.   I test, as before, the dried air

and the undried air of this room, and find, as in the former instance, that the latter produces seventy times the effect of the former. The needle is now deflected by the absorption of the undried air; allowing this air to remain in the tube, I unscrew my plate of salt, and examine its surface. I even use a lens for this purpose, taking care, however, that my breath does not strike the plate. It was carefully polished when attached to the tube; it is perfectly polished now. Glass, or rockcrystal, could not show a surface more exempt from any appearance of moisture. I place a dry handkerchief over my finger, and draw it along the surface: it leaves no trace behind. There is not the slightest deposition of moisture; still we see that absorption has taken place. This experiment is conclusive against the hypothesis that the effects observed are due to a film of brine instead of to aqueous vapour.

The doubt may, however, linger, that although we are unable to detect the film of moisture, it may still be there. This doubt is answered in the following way :—I detach the experimental tube from the front chamber, and remove the two plates of rocksalt; the tube is now *open at both ends,* and my aim will be to introduce dry and moist air into this open tube, and to compare their effects upon the radiation from our source. And here, as in all other cases, the practical tact of the experimenter must come into play. The source, on the one hand, and the pile on the other, are now freely exposed to the air; and a very slight agitation acting upon either would disturb, and might, indeed, altogether mask the effect we seek. The air, then, must be introduced into the open tube, without producing any commotion either near the source or near the pile. The length of the experimental tube is now 4 feet 3 inches; at c (fig. 96) is a cock connected with an India-rubber bag containing common air, and subjected by a weight to gentle pressure; at D is a second cock

connected by a flexible tube, $t$, with an air-pump ; between
the cock c and the India-rubber bag our drying tubes are

FIG. 96.

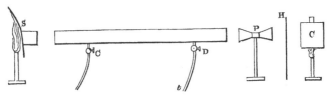

introduced ; when a cock near the bag is opened, the
air is forced gently through the drying tubes into the
experimental cylinder. The air-pump is slowly worked
at the same time, and the dry air thereby drawn towards
D. The distance of c from the source s is 18 inches, and
the distance of D from the pile P is 12 inches, the compensating cube c, and the screen H, serve the same purpose
as before. By thus isolating the central portion of the
tube, we can displace dry air by moist, or moist air by
dry, without permitting any agitation to reach either the
source or the pile.

At present the tube is filled with the common air of the
laboratory, and the needle of the galvanometer points
steadily to zero. I now allow air to pass through the
drying apparatus and to enter the open tube at c, the
pump being worked at the same time. Mark the effect.
When the dry air enters the needle commences to move,
and the direction of its motion shows that more heat is
now passing than before. The substitution of dry air for the
air of the laboratory has rendered the tube more transparent to the rays of heat. The final deflection thus
obtained is 45 degrees. Here the needle steadily remains,
and beyond this point it cannot be moved by any further
pumping in of the dry air.

I now shut off the supply of dry air and cease working
the pump ; the needle sinks, but with great slowness, in-

dicating a correspondingly slow diffusion of the aqueous vapour of the adjacent air into the dry air of the tube. If I work the pump I hasten the removal of the dry air, and the needle sinks more speedily,—it now points to zero. The experiment may be made a hundred times in succession without any deviation from this result; on the entrance of the dry air the needle invariably goes up to 45°, showing augmented transparency; on the entrance of the undried air the needle sinks to 0°, showing augmented absorption.

But the atmosphere to-day is not saturated with moisture; hence, if I saturate the air, I may expect to get a greater action. I remove the drying apparatus and put in its place a U tube, which is filled with fragments of glass moistened by distilled water. Through this tube I force the air from the India-rubber bag, and work the pump as before. We are now displacing the humid air of the laboratory by still more humid air, and see the consequence. The needle moves in a direction which indicates augmented opacity, the final deflection being 15°.

Here then we have substantially the same result as that obtained when we stopped our tube with plates of rock-salt; hence the action cannot be referred to a hypothetical film of moisture deposited upon the surface of the plates. And be it remarked that there is not the slightest caprice or uncertainty in these experiments when properly conducted. They have been executed at different times and seasons; the tube has been dismounted and remounted; the suggestions of eminent men who have seen the experiments, and whose object it was to test the results, have been complied with; but no deviation from the effects just recorded has been observed. The entrance of each kind of air is invariably accompanied by its characteristic action; the needle is under the most complete control: in short, no experiments hitherto made with solid and liquid bodies, are more certain in their execution, than the fore-going experiments on dry and humid air.

We can easily estimate the per centage of the entire radiation absorbed by the common air between the points c and d.

Introducing this tin screen between the experimental cylinder and the pile, I shut off one of the sources of heat. The deflection produced by the other source indicates the *total radiation*.

This deflection corresponds to about 1,200 of the units which have been adopted throughout these lectures ; that unit being the quantity of heat necessary to move the needle from 0° to 1°. The deflection of 45° corresponds to 50 units; out of 1,200, therefore, 50 in this instance have been destroyed by the moist air. The following statement gives us the absorption per hundred :—

$$1200 : 100 = 50 : 4{\cdot}2$$

An absorption of at least 4·2 per cent. was, therefore, effected by the atmospheric vapour which occupied the tube between c and d. Air *perfectly saturated* gives an absorption of more than 5 per cent.

This absorption took place notwithstanding the partial *sifting* of the heat in its passage from the source to c, and from d to the pile. The moist air, moreover, was, probably, only in part displaced by the dry. In other experiments I found, with a tube 4 feet long, and polished within, that the atmospheric vapour, on a day of average dryness, absorbed over 6 per cent. of the radiation from our source. Regarding the earth as a source of heat, no doubt, *at least 10 per cent. of its heat is intercepted within ten feet of the surface.** This single fact suggests the enormous influence which this newly developed pro-

---

* Under some circumstances the absorption, I have reason to believe, considerably exceeds this amount.

perty of aqueous vapour must have in the phenomena of
meteorology.

But we have not yet disposed of all objections. It has
been intimated to me that the air of our laboratory might
be impure; and the suspended carbon particles of the
London air have also been referred to, as a possible cause
of the absorption, ascribed to aqueous vapour.

I reply: 1st. The results were obtained when the
apparatus was removed from the laboratory—they are
obtainable in this room. 2ndly. Air was brought from the
following localities in impervious bags:—Hyde Park,
Primrose Hill, Hampstead Heath, Epsom Downs (near the
Grand Stand); a field near Newport, Isle of Wight; St.
Catharine's Down, Isle of Wight; the sea beach near
Black-gang Chine. *The aqueous vapour of the air from
all these localities, examined in the usual way, exerted
an absorption seventy times that of the air in which the
vapour was diffused.*

Again, I experimented thus. The air of the laboratory
was dried and purified until its absorption fell below
unity; this purified air was then led through a U tube,
filled with fragments of perfectly clean glass moistened
with distilled water. Its neutrality, when dry, showed that
all prejudicial substances had been removed from it, and
in passing through the U tube, it could take up nothing
but the pure vapour of water. The vapour thus carried
into the experimental tube produced an action ninety
times greater than that of the air which carried it.

But fair and philosophic criticism does not end even
here. The tube with which these experiments were made
is polished within, and it was surmised that the vapour
of the humid air might, on entering, have deposited itself
upon the interior surface of the tube, thus diminishing
its reflective power, and producing an effect apparently
the same as absorption. But why, I would ask, should

such a deposition of moisture take place?  On many
of the days when these experiments were made the air
was at least 25 per cent. under its point of saturation.  It
can hardly be assumed that such air would deposit its
moisture on a metallic surface, *against which, moreover,*
*the rays from our source of heat were at the time*
*impinging.*  The mere consideration of the objection
must deprive it of weight.  Further, the absorption is
exerted when only a small fraction of an atmosphere
is introduced into the tube, and it is proportional to the
quantity of air present.  This is shown by the following
table, which gives the absorption, by humid air, at tensions
varying from 5 to 30 inches of mercury.

*Humid Air.*

| Tension in inches | | | | | | Absorption | | |
|---|---|---|---|---|---|---|---|---|
| | | | | | | Observed | | Calculated |
| 5 | . | . | . | . | . | 16 | . . | 16 |
| 10 | . | . | . | . | . | 32 | . . | 32 |
| 15 | . | . | . | . | . | 49 | . . | 48 |
| 20 | . | . | . | . | . | 64 | . . | 64 |
| 25 | . | . | . | . | . | 82 | . . | 80 |
| 30 | . | . | . | . | . | 98 | . . | 96 |

The third column of this table is calculated on the assumption that the absorption is proportional to the quantity of
vapour in the tube, and the agreement of the calculated
and observed results show this to be the case, within the
limits of the experiment.  It cannot be supposed that
effects so regular as these, and agreeing so completely with
those obtained with small quantities of other vapours, and
even with small quantities of the permanent gases, can be
due to the condensation of the vapour on the interior surface.
When, moreover, five inches of air were in the tube, less
than $\frac{1}{6}$th of the vapour necessary to saturate the space was
present.  The driest day would make no approach to this dryness.  Condensation under these circumstances is impossible,
and more especially a condensation which should destroy, by

its action upon the inner reflector, quantities of heat so accurately proportional to the quantities of matter present.

My desire, however, was to take this important question quite out of the domain of mere reasoning, however strong this might appear.   I therefore resolved to abandon, not only the plates of rocksalt but also the experimental tube itself, and to displace one portion of the free

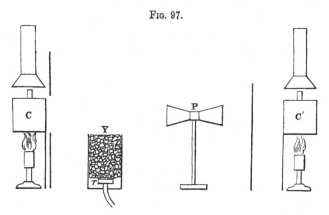

atmosphere by another.   With this view the following arrangement was made:— c (fig. 97), a cube of boiling water, is our source of heat.   Y is a hollow brass cylinder set upright, 3·5 inches wide, and 7·5 inches high.   P is the thermo-electric pile, and c′ a compensating cube, between which and P is an adjusting screen, to regulate the amount of radiation falling on the posterior surface of the pile.   The whole arrangement was surrounded by a hoarding, the space within which was divided into compartments by sheets of tin, and these spaces were stuffed loosely with paper or horsehair. These precautions, which required time to be learned, were necessary to prevent the formation of local air-currents, and also to intercept the irregular action of the external air.   The effect to be measured here

is very small, and hence the necessity of removing all
causes of disturbance which could possibly interfere with
its clearness and purity.

A rose-burner $r$ was placed at the bottom of the cylinder Y, and from it a tube passed to an India-rubber bag
containing air. The cylinder Y was first filled with fragments of rock-crystal, moistened with distilled water. On
subjecting the India-rubber bag to pressure, the air
from it was gently forced up among the fragments of
quartz, and having there charged itself with vapour it was
discharged in the space between the cube C and the pile.
Previous to this the needle stood at zero; but on the
emergence of the saturated air from the cylinder, the needle
moved and took up a final deflection of 5 degrees. The
direction of the deflection showed that the opacity of
the space between the source C and the pile was augmented
by the presence of the saturated air.

The quartz fragments were now removed, and the cylinder was filled with fragments of fresh chloride of calcium,
through which the air was gently forced, exactly as in the
last experiment. Now, however, in passing through the
chloride of calcium, it was in great part robbed of its
aqueous vapour, and the air, thus dried, displaced the
common air between the source and pile. The needle
moved, declaring a permanent deflection of 10 degrees;
the direction of the deflection showed that the transparency
of the space was augmented by the presence of the dry
air. By properly timing the discharges of the air, the
swing of the needle could be augmented to 15 or 20
degrees. Repetition showed no deviation from this result;
the saturated air always augmented the opacity, the dry
air always augmented the transparency of the space between
the source and the pile. Not only, therefore, have the
plates of rocksalt been abandoned, but also the experimental tube itself, and the results are all perfectly con-

current as regards the action of aqueous vapour upon radiant heat.

Were this subject less important I should not have dwelt upon it so long. I thought it right to remove every objection, so that meteorologists might apply, without the faintest misgiving, the results of experiment. The applications of these results to their science must be innumerable ; and here I cannot but regret that the incompleteness of my knowledge prevents me from making the proper applications myself. I would, however, ask your permission to refer to such points as I can now call to mind, with which the facts just established appear to be more or less intimately connected.

And, first, it is to be remarked, that the vapour which absorbs heat thus greedily, radiates it copiously. This fact must, I imagine, come powerfully into play in the tropics. We know that the sun raises from the equatorial ocean enormous quantities of vapour, and that immediately under him, in the region of calms, the rain, due to the condensation of the vapour, descends in deluges. Hitherto, this has been ascribed to the chilling which accompanies the expansion of the ascending air, and no doubt this, as a true cause, must produce its proportional effect. But I cannot help thinking that the radiation from the vapour itself is also influential. Imagine a column of saturated air ascending from the equatorial ocean ; for a time the vapour entangled in this air, is surrounded by air almost fully saturated. Its vapour radiates, but it radiates into vapour, and the vapour into it. To the radiation from any vapour, a screen of the same vapour is particularly opaque. Hence, for a time, the radiation from our ascending column is intercepted, and in great part returned by the surrounding vapour ; condensation under such circumstances cannot occur. But the quantity of aqueous vapour in the air diminishes speedily as we ascend ; the decrement of tension, as proved by the observations of Hooker, Strachy, and Welsh, is much more speedy than that of the air; and, finally, our vaporous column finds itself elevated beyond the protecting screen which, during the first portion of its ascent, was spread out above it. It is now in the presence of pure space, and into space it pours its heat without stoppage or requital. To the loss of heat thus endured, the condensation of the vapour, and its torrential descent to the earth, must certainly be in part ascribed.

Similar remarks apply to the formation of cumuli in our own latitudes; they are the heads of columnar bodies of vapour which rise from the earth's surface, and are precipitated as soon as they reach a certain elevation. Thus the visible cloud forms the capital. of an invisible pillar of saturated air. Certainly the top of such a column, raised above the lower vapour screen which clasps the earth, and offering itself to space, must be chilled by radiation; in this action alone we have a physical cause for the generation of clouds.

Mountains act as condensers, but how? Partly, no doubt, by the coldness of their own masses; which coldness they owe to their elevation. Above them spreads no vapour screen of sufficient density to intercept their heat, which consequently gushes unrequited into space. When the sun is withdrawn, this loss is shown by the quick and large descent of the thermometer. This descent is not due to radiation from the air, but to radiation from the earth, or from the thermometer itself. Thus the difference between a thermometer which, properly confined, gives the true temperature of the night air, and one which is permitted to radiate freely towards space, must be greater at high elevations than at low ones. This conclusion is entirely confirmed by observation. On the Grand Plateau of Mont Blanc, for example, MM. Martins and Bravais found

the difference between two such thermometers to be 24°
Fahr.; when a difference of only 10° was observed at
Chamouni.

But mountains also act as condensers by the deflection
upwards of moist winds, and their consequent expansion;
the chilling thus produced is the same as that which accompanies the direct ascent of a column of warm air into the
atmosphere; the elevated air performs work, and its heat
is correspondingly consumed. But in addition to these
causes, I think we must take into account the radiant
power of the moist air when thus tilted upwards. It is
thereby lifted beyond the protection of the aqueous layer
which lies close to the earth, and therefore pours its heat
freely into space, thus effecting its own condensation. No
doubt, I think, can be entertained, that the extraordinary
energy of water as a radiant, *in all its states of aggregation*, must play a powerful part in the condensation
of a mountain region. As vapour it pours its heat into
space and promotes condensation; as liquid it pours its
heat into space and promotes congelation; as snow it
pours its heat into space and thus converts the surfaces on
which it falls into more powerful condensers than they otherwise would be. Of the numerous wonderful properties
of water, not the least important is this extraordinary power
which it possesses, of discharging the motion of heat upon
the interstellar ether.

A freedom of escape similar to that from bodies of
vapour at great elevations would occur at the earth's surface generally, were the aqueous vapour removed from
the air above it, for the body of the atmosphere is a
practical vacuum as regards the transmission of radiant
heat. The withdrawal of the sun from any region over
which the atmosphere is dry must be followed by quick
refrigeration. The moon would be rendered entirely un-
inhabitable by beings like ourselves through the operation

of this single cause; with an outward radiation uninterrupted by aqueous vapour, the difference between her
monthly maxima and minima must be enormous.   The
winters of Thibet are almost unendurable from the
same cause.  Witness how the isothermal lines dip from
the north into Asia, in winter, as a proof of the low
temperature of this region.  Humboldt has dwelt upon
the 'frigorific power' of the central portions of this
continent, and controverted the idea that it was to be explained by reference to its elevation, for there were vast
expanses of country, not much above the sea level, with an
exceedingly low temperature.  But not knowing the influence which we are now studying, Humboldt, I imagine,
omitted one of the most important of the causes which
contributed to the observed result.  Even the absence of
the sun at night causes powerful refrigeration when the
air is dry.  The removal, for a single summer night, of the
aqueous vapour from the atmosphere which covers England,
would be attended by the destruction of every plant which
a freezing temperature could kill.  In Sahara, where 'the
soil is fire and the wind is flame,' the refrigeration at night
is often painful to bear.  Ice has been formed in this region
at night.  In Australia, also, the *diurnal range* of temperature is very great, amounting, commonly, to between
40 and 50 degrees.  In short, it may be safely predicted,
that wherever the air is *dry*, the daily thermometric range
will be great.  This, however, is quite different from saying
that when the air is *clear* the thermometric range will be
great.   Great clearness to light is perfectly compatible
with great opacity to heat; the atmosphere may be charged
with aqueous vapour while a deep blue sky is overhead,
and on such occasions the terrestrial radiation would,
notwithstanding the ' clearness,' be intercepted.

    And here we are led to an easy explanation of a fact
which evidently perplexed Sir John Leslie.  This cele-

brated experimenter constructed an instrument which he named an *æthrioscope,* the function of which was to determine the radiation against the sky. It consisted of two glass bulbs united by a vertical glass tube, so narrow that a little column of liquid was supported in the tube by its own adhesion. The lower bulb D (fig. 98) was protected by a metallic envelope, and gave the temperature of the air; the upper bulb B, was blackened, and was surrounded by a metallic cup C, which protected the bulb from terrestrial radiation.

'This instrument,' says its inventor, 'exposed to the open air in clear weather will at all times, both during the day and the night, indicate an impression of cold shot downwards from the higher regions. . . . The sensibility of the instrument is very striking, for the liquor incessantly falls and rises in the stem, with every passing cloud. But the cause of its variations does not always appear so obvious. Under a fine blue sky the *æthrioscope* will sometimes indicate a cold of 50 millesimal degrees; yet on other days, *when the air seems equally bright,* the effect is hardly 30°.' This anomaly is simply due to the difference in the quantity of aqueous vapour present in the atmosphere. Indeed, Leslie himself connects the effect with aqueous vapour in these words, 'The pressure of hygrometric moisture in the air probably affects the instrument.' It is not, however, the 'pressure' that is effective; the presence of invisible vapour intercepted the radiation from the æthrioscope, while its absence opened a door for the escape of this radiation into space.

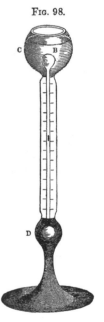

FIG. 98.

As regards experiments on terrestrial radiation, a new definition will have to be given for 'a clear day;' it is manifest, for example, that in experiments with the pyrheliometer,* two days of equal visual clearness may give totally different results. We are also enabled to account for the fact that the radiation from this instrument is often intercepted when no cloud is seen. Could we, however, make the constituents of the atmosphere, its vapour included, objects of vision, we should see sufficient to account for this result.

Another interesting point on which this subject has a bearing is Melloni's theory of *sérein*. 'Most authors,' writes this eminent philosopher, 'attribute to the cold, resulting from the radiation of the air, the excessively fine rain which sometimes falls in a clear sky, during the fine season, a few moments after sunset.' 'But,' he continues, 'as no fact is yet known which directly proves the emissive power of pure and transparent elastic fluids, it appears to me more conformable,' &c., &c. If the difficulty here urged against the theory of *sérein* be its only one, the theory will stand, for transparent elastic fluids are now proved to possess the power of radiation which the theory assumes. It is not, however, to radiation from the *air* that the chilling can be ascribed, but to radiation from the body itself, whose condensation produces the *sérein*.

Let me add the remark, that as far as I can at present judge, aqueous vapour and liquid water absorb the same class of rays; this is another way of stating that the colour of pure water is shared by its vapour. In virtue of aqueous vapour the atmosphere is therefore a blue medium. I believe it has been remarked that the colour of the firmamental blue, and of distant hills, *deepens* with the amount of aqueous vapour in the air; but the

---

* The instrument is described in Lecture XII.

substance which produces a *variation* of depth must be effective as an *origin* of colour. Whether the azure of the sky—the most difficult question of meteorology,—is to be thus accounted for, I will not at present venture to enquire.

NOTE.

The fear of being led too far from my subject causes me to withhold all speculation as to the cause of atmospheric polarisation. I may, however, remark that the polarisation of heat was illustrated in the lectures by means of the mica piles with which Professor (now Principal) J. D. Forbes first succeeded in establishing the fact of polarisation.

In connection with the investigation of the radiation and absorption of heat by gases and vapours, it gives me pleasure to refer to the prompt and intelligent aid rendered me by Mr. Becker, of the firm of Elliotts', 30 West Strand.

From the more energetic gases and vapours a series of very striking class experiments may be derived, interesting alike to the chemist and the natural philosopher. Mr. Becker has constructed a cheap form of apparatus suitable for the experiments. Where quantitative results are not required, two cubes of hot water, an open tin tube, a thermo-electric pile, and a galvanometer, magnetised, as described in the Appendix to Lecture I., will suffice to illustrate the action of the stronger gases and vapours. A current of air from a common bellows will carry the vapour into the tube.

# APPENDIX TO LECTURE XI.

EXTRACT FROM A PAPER IN THE 'PHILOSOPHICAL TRANSACTIONS FOR 1862,' ON THE ABSORPTION AND RADIATION OF HEAT BY GASEOUS MATTER.

' I WAS engaged in experiments on aqueous vapours when my other duties compelled me to close this enquiry for a time. I believe, however, I may safely say that not only is the action of aqueous vapour on radiant heat measurable, but *this action may be made use of as a measure of atmospheric moisture, the tube used in my experiments being thus converted into a hygrometer of surpassing delicacy.* Unhappily, as in other cases touched upon in this memoir, I have been unable to give this subject the developement I could wish; but the results which I am in a position to record are nevertheless interesting.

' On a great number of occasions I compared the air sent directly from the laboratory into the experimental tube with the same air after it had been passed through the drying apparatus. Calling the action of the dry air unity, or supposing it rather to oscillate about unity (for the temperature of my source varied a little from day to day), on the following days the annexed absorptions were observed with the undried air of the laboratory:—

## Absorptions by Undried Air.

| | | |
|---|---|---|
| October 23rd . . . 63 | November 1st . . 50 |
| „ 24th . . . 62 | „ 4th. . . 58 |
| „ 29th . . . 65 | „ 8th. . . 49 |
| „ 31st . . . 56 | „ 12th . . 62 |

' Nearly $\frac{9}{10}$ths of the above effects are due to aqueous vapour; which, therefore, in some instances *exerted nearly sixty times the action of the air in which it was diffused.*

' The experiments which I have made on aqueous vapour have been very numerous and varied. Differing as I did from so cautious and able an experimenter, I spared no pains to secure myself against error. I have experimented with air moistened in various ways, sometimes by allowing small bubbles of it to ascend through water, sometimes dividing it, by sending it through the pores of common cane immersed in water. Between the drying apparatus and the experimental tube I have introduced tubes containing fragments of glass moistened with water, and allowed the air to pass over them; large effects were in all such cases obtained, the absorption being usually *more than eighty times that of dried air*. Fragments of unwetted glass, which had been merely exposed to the air of the laboratory, had dried air led over them into the experimental tube; the absorption was fifteen times that of dried air. A roll of bibulous paper, taken from one of the drawers of the laboratory, and to all appearance perfectly dry, was enclosed in a glass tube, and dry air carried between its leaves. The experiment was made five times in succession with the same paper, and the following absorptions were observed : —

|  |  | Absorption |
|---|---|---|
| No. 1 | . . . . . . | . 72 |
| No. 2 | . . . . . . | . 62 |
| No. 3 | . . . . . . | . 62 |
| No. 4 | . . . . . . | . 47 |
| No. 5 | . . . . . . | . 47 |

' In fact, the action of aqueous vapour is exactly such as might be expected from the vapour of a liquid which Melloni found to be the most powerful absorber of radiant heat of all he had examined.

' Every morning, on commencing my experiments, I had an interesting example of the power of glass to gather a film of moisture on its surface. Suppose the tube mounted, and the air of the laboratory removed, as far as the air-pump was capable of removing it. On allowing dry air to enter for the first time, the needle would move from 0° to 50°. On pumping out it would return to 0°, and on letting in dry air a second time it would swing almost to 40°. Repeated exhaustions would cause this action to sink almost to nothing. These results were entirely due to the moisture collected during the night in an invisible film on the inner surface of the tube, and which was removed by the air on entering, and diffused through the tube. If the dry air entered at the end of the tube nearest to the source of heat, on the first and second admissions, and sometimes even on

a third, the vapour carried from the warm end to the cold end
of the tube was precipitated as a mist upon the latter, for a
distance sometimes of nearly a foot. The mist always disappeared on pumping out. It is needless to remark that facts of
this character, of which I could cite many, were not calculated
to promote incautiousness or rashness on my part. I saw very
clearly how easy it was to fall into the gravest errors, and I took
due precautions to prevent myself from doing so. . . . . .

'But, to place the matter beyond all doubt, I abolished the
plates of rocksalt altogether, and operated thus :—An India-rubber bag, B (fig. 99), was filled with air, and to its nozzle a T-piece,
with the cocks QQ', was attached. The cock Q' was connected
with two tubes, U'U', each of which was filled with fragments
of glass moistened with distilled water. The cock Q was connected with the tubes UU, each of which was filled with fragments of glass moistened by sulphuric acid. The other ends of
these two series of tubes were connected with the cocks O O'; and
from the T-piece between these cocks a tube led to the end E' of
the open experimental tube T. The cock A at the other end
of the experimental tube was placed in connection with an air-
pump. The pile P, the screen S, and the compensating cube C'
were used as in the other experiments. E is the end of the front
chamber, and C the source of heat. In some experiments I had
the end E closed by a plate of rocksalt, in others it was allowed
to remain open, a distance of about 12 inches intervening
between the radiating surface and the open end E' of the experimental tube.

'Closing the cocks Q and O, and opening Q' and O', gentle
pressure being applied to the bag B, a current of moist air was
slowly discharged at the end of E' of the experimental tube. The
pump in connection with A was then worked, and thus by degrees
the air was sucked into the tube T. The deflection of the galvanometer was 30°, when the moist air filled the tube as completely
as the arrangement permitted — this deflection being due to
the predominance of the compensating cube over the radiating
source C.

'The cocks Q' and O' were now closed, and Q and O opened;
proceeding as before, a current of *dry* air was discharged at E',
and this air was drawn into the tube T in the manner just described. The moist air was thus displaced by dry; and, while

Fig. 99.

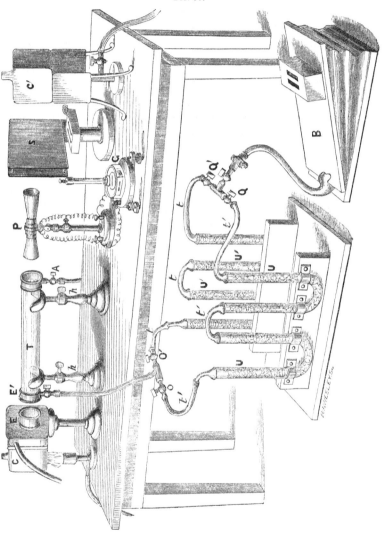

the displacement was going on, the galvanometer was observed through the distant telescope. The needle soon commenced to sink, and slowly went down to zero, proving that a greater quantity of heat passed through the dry than through the moist air. The wet air was substituted for the dry, and the dry for the wet, twenty times in succession, with the same constant result: the entrance of the humid air always caused the needle to move from 0° to 30°, while the entrance of dry air caused it to fall from 30° to 0°. The air-pump was resorted to, because I found that when I attempted to displace the air by the direct force of the current from B, the temperature of the pile, or of the source, was so affected by the fresh air as to confuse the result. I may remark that not only have I operated thus for days with aqueous vapour, but every result which I have obtained with vapours generally has been thus confirmed, so that all doubt as to the applicability of the rock-salt plates to researches of this nature may, I think, be abandoned.'

# LECTURE XII.

[April 10, 1862.]

WE have learned that our atmosphere is always more or less charged with aqueous vapour, the condensation of which forms our clouds, fogs, hail, rain, and snow. I have now to direct your attention to one particular case of condensation, of great interest and beauty — one, moreover, regarding which erroneous notions were for a long time entertained. I refer to the phenomenon of Dew. The aqueous vapour of our atmosphere is a powerful radiant, but it is diffused through air which usually exceeds its own mass more than one hundred times. Not only, then, its own heat, but the heat of the large quantity of air which surrounds it, must be discharged by the vapour, before it can sink to its point of condensation. The retardation of chilling due to this cause enables good solid radiators, at the earth's surface, to outstrip the vapour in their speed of refrigeration; and hence upon these bodies aqueous vapour may be condensed to liquid, or even congealed to hoar-frost, while at a few feet above the surface it still maintains its gaseous state. This is actually the case in the beautiful phenomenon which we have now to examine.

We are indebted to a London physician for a true theory of dew. In 1818 Dr. Wells published his admirable Essay upon this subject. He made his experiments in a garden in Surrey, at a distance of three miles from Blackfriars Bridge. To collect the dew, he used little bundles of wool, which, when dry, weighed 10 grains each; and having exposed them during a clear night, the amount of dew deposited on them was determined by the augmentation of their weight. He soon found that whatever interfered with the view of the sky from his piece of wool, interfered also with the deposition of dew. He supported a board on four props; *on* the board he laid one of his wool parcels, and *under* it a second similar one; during a clear calm night, the former gained 14 grains in weight, while the latter gained only 4. He bent a sheet of pasteboard like the roof of a house, and placed underneath it a bundle of wool on the grass: by a single night's exposure the wool gained 2 grains in weight, while a similar piece of wool exposed on the grass, but quite unshaded by the roof, collected 16 grains of moisture.

Is it steam from the earth, or is it fine rain from the heavens, that produces this deposition of dew? Both of these notions have been advocated. That it does not arise from the earth is, however, proved by the observation, that more moisture was collected on the propped board than on the earth's surface under it. That it is not a fine rain is proved by the fact, that the most copious deposition occurred on the clearest nights.

Dr. Wells next exposed-thermometers, as he had done his wool-bundles, and found that *at those places where the dew fell most copiously the temperature sank lowest*. On the propped board already referred to, he found the temperature 9° lower than under it; beneath the pasteboard roof the thermometer was 10° warmer than on the open grass. He also found that when he laid his thermometer upon a grass plot, on a clear night, it sank sometimes 14° lower than a similar thermometer suspended in free air at a height of 4 feet above the grass. A bit of cotton, placed beside the former, gained 20 grains; a similar bit, beside the latter, only 11 grains in weight. *The lowering of the temperature and the deposition of the dew went hand in hand.* Not only did the shade of artificial screens interfere with the lowering of the temperature and the formation of the dew, but a cloud-screen acted in the same manner. He once observed his thermometer, which, as it lay upon the grass, showed a temperature 12° lower than the air a few feet above the grass, rise, on the passage of some clouds, until it was only 2° colder than the air. In fact, as the clouds crossed his zenith, or disappeared from it, the temperature of his thermometer rose and fell.

A series of such experiments, conceived and executed with singular clearness and skill, enabled Dr. Wells to propound a Theory of Dew, which has stood the test of all subsequent criticism, and is now universally accepted.

It is an effect of chilling by radiation. 'The upper parts of the grass radiate their heat into regions of empty space, which, consequently, send no heat back in return; its lower parts, from the smallness of their conducting power, transmit little of the earth's heat to the upper parts, which, at the same time, receiving only a small quantity from the atmosphere, and none from any other lateral body, must remain colder than the air, and condense into dew its watery vapour, if this be sufficiently abundant in respect to the decreased temperature of the grass.' Why the vapour itself, being a powerful radiant, is not as quickly chilled as the grass, I have already explained, on the ground that the vapour has not only its own heat to discharge, but also that of the large mass of air by which it is surrounded.

Dew being the result of the condensation of atmospheric vapour, on substances which have been sufficiently cooled by radiation, and as bodies differ widely in their radiative powers, we may expect corresponding differences in the deposition of dew. This Wells proved to be the case. He often saw dew copiously deposited on grass and painted wood, when none could be observed on gravel walks adjacent. He found plates of metal, which he had exposed, quite dry, while adjacent bodies were covered with dew: *in all such cases the temperature of the metal was found to be higher than that of the dewed substances.* This is quite in accordance with our knowledge that metals are the worst radiators. On one occasion he placed a plate of metal upon grass, and upon the plate he laid a glass thermometer; the thermometer, after some time, exhibited dew, while the plate remained dry. This led him to suppose that the instrument, though lying on the plate, did not share its temperature. He placed a second thermometer, with a *gilt bulb*, beside the first; the naked glass thermometer — a good radiator — remained 9° colder than its companion. To determine the true temperature of a body is, I may remark, a difficult task: a glass thermometer, suspended in the air, will not give the temperature of the air; its own power as a radiant or an absorbent comes into play. On a clear day, when the sun shines, the thermometer will be warmer than the air; on a clear night, on the contrary, the thermometer will be colder than the air. We have seen that the passage of a cloud can raise the temperature of a thermometer 10 degrees in a few minutes. This augmentation, it is manifest, does not indicate a corresponding augmentation of the temperature of the air, but merely the interception and reflection, by the cloud, of the rays of heat emitted by the thermometer.

Dr. Wells applied his principles to the explanation of many curious effects, and to the correction of many popular errors. Moon blindness he refers to the chill produced by radiation into clear space, the shining of the moon being merely an accompaniment to the clearness of the atmosphere. The putrefying influence ascribed to the moonbeams is really due to the deposition of moisture, as a kind of dew, on the exposed animal substances. The nipping of tender plants by frost, even when the air of the garden is some degrees above the freezing temperature, is also to be referred to chilling by radiation. A cobweb screen would be sufficient to preserve them from injury.*

Wells was the first to explain the formation, artificially, of ice in Bengal, where the substance is never formed naturally. Shallow pits are dug, which are partially filled with straw, and on the straw flat pans, containing water which had been boiled, is exposed to the clear firmament. The water is a powerful radiant, and sends off its ·heat copiously into space. The heat thus lost cannot be supplied from the earth—this source being cut off by the non-conducting straw. Before sunrise a cake of ice is

---

* With reference to this point we have the following beautiful passage in the Essay of Wells :—' I had often, in the pride of half knowledge, smiled at the means frequently employed by gardeners to protect tender plants from cold, as it appeared to me impossible that a thin mat, or any such flimsy substance could prevent them from attaining the temperature of the atmosphere, by which alone I thought them liable to be injured. But when I had learned that bodies on the surface of the earth become, during a still and serene night, colder than the atmosphere, by radiating their heat to the heavens, I perceived immediately a just reason for the practice which I had before deemed useless.'

formed in each vessel. This is the explanation of Wells, and it is, no doubt, the true one. I think, however, it needs supplementing. It appears, from the description, that the condition most suitable for the formation of ice, is not only a clear air, but a *dry* air. The nights, says Sir Robert Barker, most favourable for the production of ice, are those which are clearest and most serene, and *in which very little dew appears after midnight.* I have italicised a very significant phrase. To produce the ice in abundance, the atmosphere must not only be clear, but it must be comparatively free from aqueous vapour. When the straw in which the pans were laid became wet, it was always changed for dry straw, and the reason Wells assigned for this was, that the straw, by being wetted, was rendered more compact, and efficient as a conductor. This may have been the case, but it is also certain that the vapour rising from the wet straw, and overspreading the pans like a screen, would check the chill, and retard the congelation.

With broken health Wells pursued and completed this beautiful investigation; and, on the brink of the grave, he composed his Essay. It is a model of wise enquiry and of lucid exposition. He made no haste, but he took no rest till he had mastered his subject, looking steadfastly into it until it became transparent to his gaze. Thus he solved his problem, and stated its solution in a fashion which renders his work imperishable.*

Since his time various experimenters have occupied themselves with the question of nocturnal radiation; but, though valuable facts have been accumulated, if we except a supplement contributed by Melloni, nothing of importance has been added to the theory of Wells. Mr. Glaisher, M. Martins, and others, have occupied themselves with the

---

* The tract of Wells is preceded by a personal memoir written by himself. It has the solidity of an essay of Montaigne.

subject. The following table contains some results obtained by Mr. Glaisher, by exposing thermometers at different heights above the surface of a grass field. The chilling observed, when the thermometer was exposed on long grass, is represented by the number 1,000; while the succeeding numbers represent the relative chilling of the thermometers placed in the positions indicated :—

### Radiation.

| | | | |
|---|---|---|---|
| Long grass . . . . | | | . 1000 |
| One inch above the points of the grass | | . | 671 |
| Two inches | ,, | ,, | . 570 |
| Three inches | ,, | ,, | . 477 |
| Six inches | ,, | ,, | . 282 |
| One foot | ,, | ,, | . 129 |
| Two feet | ,, | ,, | . 86 |
| Four feet | ,, | ,, | . 69 |
| Six feet | ,, | ,, | . 52 |

It may be asked why the thermometer, which is a good radiator, is not, when suspended in free air, just as much chilled as at the earth's surface. Wells has answered the question. It is because the thermometer, when chilled, cools the air in immediate contact with it; this air contracts, becomes heavy, and trickles downwards, thus allowing its place to be taken by warmer air. In this way the free thermometer is prevented from falling very low beneath the temperature of the air. Hence, also, the necessity of a still night for the copious formation of dew; for, when the wind blows, fresh air continually circulates amid the blades of grass, and prevents any considerable chilling by radiation.

When a radiator is exposed to a clear sky it tends to keep a certain thermometric distance, if I may use the term, between its temperature and that of the surrounding air. This distance will depend upon the energy of the body as a radiator, but it is to a great extent independent of the temperature of the air.

Thus, M. Pouillet has proved that in the month of April,
when the temperature of the air was 3°·6 C., swansdown
fell by radiation to—3°·5 : the whole chilling, therefore,
was 7°·1.  In the month of June, when the temperature
of the air was 17°·75 C., the temperature of the radiating
swansdown was 10°·54, the chilling of the swansdown by
radiation is here 7°·21 ; almost precisely the same as that
which occurred in April.  Thus, while the general temperature varies within wide limits, the *difference* of temperature between the radiating body and the surrounding
air, remains sensibly constant.

These facts enabled Melloni to make an important
addition to the theory of dew.  He found that a glass
thermometer, placed on the ground, is never chilled more
than 2° C. below an adjacent thermometer, *with silvered
bulb*, which hardly radiates at all.  These 2° C., or thereabouts, mark the thermometric distance above referred to,
which the glass tends to preserve between it and the
surrounding air.  But Six, Wilson, Wells, Parry, Scoresby,
Glaisher, and others, have found differences of more than
10° C., between a thermometer on grass, and a second
thermometer hung a few feet above the grass.  How is
this to be accounted for?  Very simply, according to
Melloni, thus:—The grass blades first chill themselves
by radiation, 2° C. below the surrounding air : the air is
then chilled by contact with the grass, and forms around
it a cold aërial bath.  But the tendency of the grass is
to keep the above constant difference between its own
temperature   and   that   of   the   surrounding   medium.
It therefore sinks lower.  The air sinks in its turn,
being still further chilled by contact with the grass; the
grass, however, again seeks to re-establish the former difference ; it is again followed by the air, and thus, by a
series of actions and reactions, the entire stratum of air
in contact with the grass becomes lowered far below the

temperature which corresponds to the actual radiative energy of the grass.

So much for terrestrial radiation ; that of the moon will not occupy us so long. Many futile attempts have been made to detect the warmth of the moon's beams. No doubt is entertained that every luminous ray is also a heat ray ; but the light-giving power is not even an approximate measure of the calorific energy of a beam. With a large polyzonal lens, Melloni converged an image of the moon upon his pile ; but he found the cold of his lens far more than sufficient to mask the heat, if such there were, of the moon. He screened off his lens from the heavens, placed his pile in the focus of the lens, waited until the needle came to zero, and then suddenly removing his screen allowed the concentrated light to strike his pile. The slight air-drafts in the place of experiment were sufficient to disguise the effect. He then stopped the tube in front of his pile with glass screens, through which the light went freely to the blackened face of the pile, where it was converted into heat. *This heat could not get back through the glass screen,* and thus Melloni, following the example of Saussure, accumulated his effects, and obtained a deflection of 3° or 4°. The deflection indicated warmth, and this is the only experiment which gives us any positive evidence as to the calorific action of the moon's rays. Incomparably less powerful than the solar rays in the first instance, their action is first enfeebled by distance, and, secondly, by the fact that *the obscure heat of the moon* is almost wholly absorbed by our atmospheric vapour. Even such obscure rays as might happen to reach the earth would be utterly cut off by such a lens as Melloni made use of. It might be worth while to make the experiment with a metallic reflector, instead of with a lens. I have myself tried a conical reflector of very large dimensions, but have hitherto been defeated by the unsteadiness of the London air.

We have now to turn our thoughts to the source from
which all terrestrial and lunar heat is derived. This
source is the sun; for if the earth has ever been a molten
sphere, which is now cooling, the quantity of heat which
reaches its surface from within, has long ceased to be
sensible. First, then, let us enquire what is the constitution of this wondrous body, to which we owe both light
and life.

Let us approach the subject gradually and prepare our
minds, by previous discipline, for the treatment of this
noble problem. You already know how the spectrum of
the electric light is formed. Here you have one upon the
screen, two feet wide and eight long, with all its magnificent
gradations of colour, one fading into the other, without
solution of continuity. The light from which this spectrum is derived, is emitted from the incandescent carbon
points within our electric lamp. All other solids give a
similar spectrum. When I raise this platinum wire to
whiteness by an electric current, and examine its light by a
prism, I find the same gradation of colours, and no gap
whatever between one colour and the other. But by intense
heat,—by the heat of the electric lamp, for example,—I
can volatilise that platinum, and throw upon the screen,
not the spectrum of the incandescent solid, but of
its *incandescent vapour*. The spectrum is now changed;
instead of being a continuous gradation of colours, it
consists of a series of brilliant lines, separated from each
other by spaces of darkness.

I have arranged my pieces of carbon thus: — the
lower one is now a cylinder, about half an inch in
diameter, in the top of which I have scooped a small
hollow; in this hollow I place the metal which I wish to
examine — say this piece of zinc, — and bring down upon
it the upper point. The current passes; I draw the points
apart, and you see the magnificent arc that now unites

them; here is its magnified image upon the screen, a
fine stream of purple light 18 inches long.  That coloured
space contains the particles of the zinc discharged across
from carbon to carbon; these particles are now oscillating
in certain definite periods, and the colour which we perceive
is the mixture of impressions due to these oscillations.  Let
us separate, by a prism, the coloured stream into its components; here they are, splendid bands of red and blue.  Pray
remember the character and position of these bands, as I
shall have to refer to them again immediately.

I interrupt the current; eject the zinc, and put in its
place a bit of copper.  Here you see a stream of green
light between the carbons, which we will analyse as we
did the light of the zinc.  You can see that the spectrum
of the copper is different from that of the zinc: here you
have bands of brilliant green, which were absent from the
zinc.  We may therefore infer, with certainty, that the
atoms of copper, in the voltaic arc, swing in periods different from those of zinc.  Let us now see whether these
different periods create any confusion, when we operate
upon a substance composed of zinc and copper,—the familiar substance brass.  Its spectrum is now before you,
and if you have retained the impression made by our two
last experiments, you will recognise here a spectrum formed
by the superposition of the two separate spectra of zinc
and copper.  The alloy emits, without confusion, the rays
peculiar to the metals of which it is composed.

Every metal emits its own system of bands, which are as
characteristic of it as those physical and chemical qualities
which give it its individuality.  By a method of experiment
sufficiently refined we can measure, accurately, the position
of the bright lines of every known metal.  Acquainted
with such lines we should, by the mere inspection
of the spectrum of any single metal, be able at once to
declare its name.  And not only so, but in the case of a

mixed spectrum, we should be able to declare the constituents of the mixture from which it emanated.

This is true, not only of the metals themselves, but
also of their compounds, if they be volatile. I place a
bit of sodium on my lower cylinder and cause the voltaic
discharge to pass from it to the upper coal-point;
here is the spectrum of the sodium : a single band of
brilliant yellow. If I operated with sufficient delicacy I
should divide that band into two, with a narrow dark
interval between them. I eject the sodium from the lamp
and put in its place a little common salt, or chloride of
sodium. At this high temperature the salt is volatile,
and you see the exact yellow band produced by the salt
that was given by the metal. Thus, also, by means of
the chloride of strontium I produce the bands of the
metal strontium; by the chlorides of calcium, magnesium,
and lithium, I produce the spectra of these respective
metals.

Here, finally, I have a carbon cylinder perforated with
holes, into which I have stuffed a mixture of all the
compounds just mentioned; and *there* is the spectrum
of the mixture upon the screen. Surely nothing more
magnificent can be imagined. Each substance gives out its
own peculiar rays, and thus they cut transversely, the whole
eight feet of the spectrum into splendid parallel bars of
coloured light. Having previously made yourselves acquainted with the lines emitted by all the metals, you
would be able to unravel this spectrum, and to tell me
what substances I have employed in its production.

I make use of the voltaic arc simply because its light
is so intense as to be visible to a large audience like
the present, but I might make the same experiments
with a common blow-pipe flame, which is nearly deprived
of light by the admixture of air or oxygen. The introduction of sodium, or chloride of sodium, turns the flame

yellow; strontium turns it red; copper green, &c. The flames thus coloured, when examined by a prism, show the exact bands which I have displayed before you on the screen.

We have already learned that gases and vapours absorb the rays of heat, the heat that we employed being obscure. I have no doubt that if those rays could make an impression upon the eye—if I could spread them out before you like the colours of the spectrum—you would find certain classes of rays selected, in each case, for destruction, the others being allowed free passage through the vapours. A famous experiment of Sir David Brewster's, which I will throw into a form suited to the lecture room, will enable me to illustrate this power of selection in the case of light. Into this cylinder, the ends of which are stopped by plates of glass, I introduce a quantity of nitrous acid gas, the presence of which is now indicated by its rich brown colour. I project a spectrum on the screen, eight feet long and nearly two in width, and I place this cylinder, containing the brown gas, in the path of the beam as it issues from the lamp. You see the effect; the continuous spectrum is now furrowed by numerous dark bands, the rays answering to which are struck down by the nitric gas, while it permits the intervening bands of light to pass without hindrance.

We must now take a step in advance of the principle of reciprocity, which I have already enunciated. Hitherto we have found in gases, liquids, and solids, that the good absorber is the good radiator; we must now go further and state, that *a gas, or vapour, absorbs those precise rays which it can itself emit*; the atoms which swing at a certain rate intercept the waves excited by atoms swinging at the same rate. The atoms which vibrate red light will stop red light; the atoms that oscillate yellow will stop

yellow; those that oscillate green will stop green, and so
of the rest. Absorption, you know, is a transference of
motion from thê ether to the particles immersed in it, and
the absorption of any atom is exerted chiefly upon those
waves which arrive in periods that correspond with the
atom's own rate of oscillation.

Let us endeavour to prove this experimentally. We
already know that a sodium flame, when analysed, gives us
a brilliant double band of yellow. Here is a flat vessel containing a mixture of alcohol and water; I warm the mixture
and ignite it: it gives a flame which is so feebly luminous
as to be scarcely visible. I now mix salt with the liquid,
and again ignite it; the flame, which a moment ago was
scarcely to be seen, is now a brilliant yellow. I project a
continuous spectrum upon the screen, and in the track of the
beam, as it issues from the electric lamp, I place the yellow
sodium flame. Observe the spectrum narrowly: you see
a flickering gray band in the yellow of the spectrum;
sometimes it is shaded deeply enough to show you all that
the flame has, at least in part, intercepted the yellow band
of the spectrum: it has partially absorbed the precise
light which it can itself emit.

But I wish to make the effect plainer, and therefore
abandon the alcohol light, and proceed thus: here is
a Bunsen's burner, the flame of which is intensely hot,
though it hardly emits any light. I place the burner in
front of the lamp, so that the beam, whose decomposition
is to form our spectrum, shall pass through the flame. I
have here a little net of platinum wire, in which I place a
bit of the metal sodium, about the size of a pea. I also set up
a pasteboard shade, which shall cut off the light emitted by
the sodium, from the screen on which the spectrum falls.
And now I am ready to make the experiment. Here, then,
in the first place, is the spectrum. I now introduce the
platinum net in front of the lamp; the sodium instantly

colours the flame intensely yellow, and you see a shadow
coming over the yellow of the spectrum. But the effect
is not yet at its maximum. The sodium now suddenly
bursts into intensified combustion, and there you see the
yellow dug utterly out of the spectrum, and a bar of intense darkness in its place. This violent combustion will
endure for a little time. I withdraw the flame, the yellow
reappears upon the screen; I reintroduce it, the yellow
band is cut out. This I can do ten times in succession, and
in the whole range of optics I do not think there is a more
striking experiment. Here, then, we have conclusively
proved, that the light which the sodium flame absorbs is
the precise light which it can emit.

Let me be still more precise in my experiment. The
yellow of the spectrum spreads over a widish interval;
and I wish now to show you that it is the particular portion of the yellow which the sodium emits, that is absorbed
by its flame. I place a little salt solution on the ends of
my coal points; you now see the continuous spectrum
with the yellow band of the sodium brighter than the rest
of the yellow. It is thus clearly defined before your
eyes. I again place the sodium flame in front, and that
particular band which now stands out from the spectrum
is cut away—a space of intense gloom occupying its place.

You have already seen a spectrum, derived from a mixture of various substances, and which was composed of a
succession of sharply defined and brilliant bars, separated
from each other by intervals of darkness. Could I take the
mixture which produced that striped spectrum, and raise it,
by means of Bunsen's burner, to a temperature sufficiently
intense to render its vapours incandescent; on placing
its flame in the path of a beam producing a continuous
spectrum, I should cut out of the latter the precise rays
emitted by the components of my mixture. I should thus,
instead of furrowing my spectrum by a single dark band,

as in the case of sodium, furrow it by a series of dark bands, equal in number to the bright bands produced, when the mixture itself was the source of light.

I think we now possess knowledge sufficient to raise us to the level of one of the most remarkable generalisations of our age. When the light of the sun is properly decomposed, the spectrum is seen furrowed by innumerable *dark* lines. A few of these were observed, for the first time, by Dr. Wollaston; but they were investigated with profound skill by Fraunhofer, and called, after him, Fraunhofer's lines. It has long been supposed that these dark spaces were caused by the absorption of the rays which correspond to them, in the atmosphere of the sun; but nobody knew how. Having once proved that an incandescent vapour absorbs the precise rays which it can itself emit, and knowing that the body of the sun is surrounded by an incandescent photosphere, the supposition at once flashes on the mind, that this photosphere may cut off those rays of the central incandescent orb, which the photosphere itself can emit. We are thus led to a theory of the constitution of the sun, which renders a complete account of the lines of Fraunhofer.

The sun consists of a central orb, liquid or solid, of exceeding brightness, which, of itself, would give a continuous spectrum, or in other words, which emits all kinds of rays. These, however, have to pass through the photosphere, which wraps the sun like a flame, and this vaporous envelope cuts off those particular rays of the nucleus which it can itself emit—the lines of Fraunhofer marking the position of these failing rays. Could we abolish the central orb, and obtain the spectrum of the gaseous envelope, we should obtain a striped spectrum, each bright band of which would coincide with one of Fraunhofer's dark lines. These lines, therefore, are spaces of relative, not of absolute darkness; upon them the rays of the absorbent photosphere fall ; but these, not being sufficiently intense to make good the light intercepted, the spaces which they illuminate are dark, in comparison to the general brilliancy of the spectrum.

It has long been supposed that sun and planets have had a common origin, and that hence the same substances are more or less common to them all. Can we detect the presence of any of our terrestrial substances in the sun ? I have said that the bright bands of a metal are characteristic of the metal; that we can, without seeing the metal, declare its name from the inspection of the bands. The bands are, so to speak, the *voice* of the metal declaring its presence. Hence, if any of our terrestrial metals be contained in the sun's atmosphere, the dark lines which they produce ought to coincide exactly with the bright lines emitted by the vapour of the metal itself. In the case of the single metal iron, about 60 bright lines have been determined as belonging to it. When the light from the incandescent vapour of iron, obtained by passing electric sparks between two iron wires, is allowed to pass through one-half of a fine slit, and the light of the sun through the other half, the spectra from both sources of light may be placed together ; and when this is done it is found that for every bright line of the iron spectrum there is a dark line of the solar spectrum. Reduced to actual calculation, this means that the chances are more than 1,000,000,000,000;000,000 to 1 that iron is in the atmosphere of the sun. Comparing the spectra of other metals in the same manner, Professor Kirchhof, to whose genius we owe this splendid generalisation, finds iron, calcium, magnesium, sodium, chromium, and other metals, to be constituents of the solar atmosphere, but as yet he has been unable to detect gold, silver, mercury, aluminium, tin, lead, arsenic, or antimony.

I can imitate in a way more precise than that hitherto employed, the solar constitution here supposed. I place

in the electric lamp a cylinder of carbon about half an
inch thick; on the top, and round about the edge of
the cylinder, I place a ring of sodium, leaving the central
portion of the cylinder clear. I bring down the upper
coal point upon the middle of the cylinder's upper surface,
thus producing the ordinary electric light. The proximity
of this light to the sodium is sufficient to volatilise the
latter, and thus I surround my little central sun with an
atmosphere of sodium vapour, as the real sun is surrounded
by its photosphere. In the spectrum of this light you see
the yellow band is absent.

The energy of solar emission has been measured by
Sir John Herschel at the Cape of
Good Hope, and by M. Pouillet in
Paris. The agreement between the
measurements is very remarkable.
Sir John Herschel finds the direct
heating effect of a vertical sun, at
the sea level, to be competent to
melt 0·00754 of an inch of ice per
minute; while according to M.
Pouillet, the quantity is 0·00703 of
an inch. The mean of the determinations cannot be far from the
truth; this gives 0·00728 of an inch
of ice per minute, or nearly half an
inch per hour. Before you (fig. 100)
I have placed an instrument similar
in form to that used by M. Pouillet,
and called by him a pyrheliometer.
The particular instrument which
you now see is composed of a
shallow cylinder of steel, $a\,a$, which
is filled with mercury. Into the cylinder this thermometer $d$, is introduced, the stem of which is protected by a

FIG. 100.

piece of brass tubing. We thus obtain the temperature of the mercury. The flat end of the cylinder is to be turned towards the sun, and the surface thus presented is coated with lampblack. Here is a collar and screw, *cc*, by means of which the instrument may be attached to a stake driven into the ground, or into the snow, if the observations are made at considerable heights. It is necessary that the surface which receives the sun's rays should be perpendicular to the rays, and this is secured by appending to the brass tube which shields the stem of the thermometer, a disk, *ee*, of precisely the same diameter as the steel cylinder. When the shadow of the cylinder accurately covers the disk, we are sure that the rays fall, as perpendiculars, on the upturned surface of the cylinder.

The observations are made in the following manner :— First, the instrument is permitted, not to receive the sun's rays, but to radiate its own heat for five minutes against an unclouded part of the firmament; the decrease of the temperature of the mercury consequent on this radiation is then noted. Next, the instrument is turned towards the sun, so that the solar rays fall perpendicularly upon it for five minutes,—the augmentation of temperature is now noted. Finally, the instrument is turned again towards the firmament, away from the sun, and allowed to radiate for another five minutes, the sinking of the thermometer being noted as before. You might, perhaps, suppose that exposure to the sun alone would be sufficient to determine his heating power; but we must not forget that during the whole time of exposure to the sun's action, the blackened surface of the cylinder is also radiating into space; it is not therefore a case of pure gain: the heat received from the sun is, in part, thus wasted, even while the experiment is going on; and to find the quantity lost, the first and last experiments are needed. In order to obtain the whole heating power of the sun, we must add to his

obser.ved heating power, the quantity lost during the time
of exposure, and this quantity is the mean of the first and
last observations.  Supposing the letter R to represent the
augmentation of temperature by five minutes' exposure to
the sun, and that $t$ and $t'$ represent the reductions of temperature observed before and after, then the whole force
of the sun, which we may call T, would be thus expressed:

$$T = R + \frac{t + t'}{2}$$

The surface on which the sun's rays here fall is known ;
the quantity of mercury within the cylinder is also known ;
hence we can express the effect of the sun's heat upon a
given area, by stating that it is competent, in five minutes,
to raise so much mercury, or so much water, so many
degrees in temperature.  Water indeed, instead of mercury,
was used in M. Pouillet's pyrheliometer.

The observations were made at different hours of the
day, and, hence, through different thicknesses of the earth's
atmosphere ; augmenting from the minimum thickness at
noon, up to the maximum at 6 P.M., which was the time
of the latest observation.  It was found that the solar
energy diminished according to a certain law, as the
thickness of the air crossed by the sunbeams increased ;
and from this law M. Pouillet was enabled to infer what
the atmospheric absorption of a beam would be, if
directed downwards to his instrument from the zenith.
This he found to be 25 per cent.  Doubtless, this absorption would be chiefly exerted upon the longer undulations emitted by the sun, the aqueous vapour of our air,
and not the air itself, being the main agent of absorption.
Taking into account the whole terrestrial hemisphere
turned towards the sun, the amount intercepted by the
atmospheric envelope is four-tenths of the entire radiation
in the direction of the earth.  Thus, were the atmosphere
removed, the illuminated hemisphere of the earth would

receive nearly twice the amount of heat from the sun that now reaches it.   The total amount of solar heat received by the earth in a year, if distributed uniformly over the earth's surface, would be sufficient to liquefy a layer of ice 100 feet thick, and covering the whole earth.

Knowing thus the annual receipt of the earth, we can calculate the entire quantity of heat emitted by the sun in a year.   Conceive a hollow sphere to surround the sun, its centre being the sun's centre, and its surface at the distance of the earth from the sun.   The section of the earth cut by this surface, is to the whole area of the hollow sphere, as 1 : 2,300,000,000; hence, the quantity of solar heat intercepted by the earth is only $\frac{1}{2300,000000}$ of the total radiation.

The heat emitted by the sun, if used to melt a stratum of ice applied to the sun's surface, would liquefy the ice at the rate of 2,400 feet an hour.   It would boil, per hour, 700,000 millions of cubic miles of ice-cold water.   Expressed in another form, the heat given out by the sun, per hour, is equal to that which would be generated by the combustion of a layer of solid coal, 10 feet thick, entirely surrounding the sun; hence, the heat emitted in a year is equal to that which would be produced by the combustion of a layer of coal, 17 miles in thickness.

These are the results of direct measurement; and should greater accuracy be conferred on them by future determinations, it will not deprive them of their astounding haracter.   And this expenditure has been going on for ages, without our being able, in historic times, to detect the loss.   When the tolling of a bell is heard at a distance, the sound of each stroke soon sinks, the sonorous vibrations are quickly wasted, and renewed strokes are necessary to maintain the sound.   Like the bell,

Die Sonne tönt nach alter Weise.

But how is its tone sustained ? How is the perennial loss

of the sun made good? We are apt to overlook the
wonderful in the common. Possibly to many of us—and
even to some of the most enlightened among us— the sun
appears as a fire, differing from our terrestrial fires only in
the magnitude and intensity of its combustion. But what
is the burning matter which can thus maintain itself? All
that we know of cosmical phenomena declares our brotherhood with the sun,—affirms that the same constituents
enter into the composition of his mass as those already
known to chemistry. But no earthly substance with
which we are acquainted—no substance which the fall of
meteors has landed on the earth—would be at all competent
to maintain the sun's combustion. The chemical energy
of such substances would be too weak, and their dissipation
would be too speedy. Were the sun a solid block of coal,
and were it allowed a sufficient supply of oxygen, to enable
it to burn at the rate necessary to produce the observed
emission, it would be utterly consumed in 5,000 years. On
the other hand, to imagine it a body originally endowed
with a store of heat—a hot globe now cooling—necessitates
the ascription to it of qualities, wholly different from those
possessed by terrestrial matter. If we knew the specific
heat of the sun, we could calculate its rate of cooling.
Assuming this to be the same as that of water—the terrestrial substance which possesses the highest specific heat—
at its present rate of emission, the entire mass of the
sun would cool down 15,000° Faht. in 5,000 years. In
short, if the sun be formed of matter like our own,
some means must exist of restoring to him his wasted
power.

The facts are so extraordinary, that the soberest hypothesis regarding them must appear wild. The sun
we know rotates upon his axis; he turns like a wheel
once in about 25 days: can it be the friction of the
periphery of this wheel against something in surrounding

space which produces the light and heat? Such a notion has been entertained. But what forms the brake, and by what agency is it held, while it rubs against the sun? The action is inconceivable; but, granting the existence of the brake, we can calculate the total amount of heat which the sun could generate by such friction. We know his mass, we know his time of rotation; we know the mechanical equivalent of heat; and from these data we deduce, with certainty, that the entire force of rotation, if converted into heat, would cover more than one, but less than two centuries of emission.* There is no hypothesis involved in this calculation.

There is another theory, which, however bold it may, at first sight, appear, deserves our earnest attention. I have already referred to it as the Meteoric Theory of the sun's heat. Solar space is peopled with ponderable objects: Kepler's celebrated statement that 'there are more comets in the heavens than fish in the ocean,' refers to the fact that a small portion only of the total number of comets belonging to our system, are seen from the earth. But besides comets, and planets, and moons, a numerous class of bodies belong to our system,—asteroids, which, from their smallness, might be regarded as cosmical atoms. Like the planets and the comets these smaller bodies obey the law of gravity, and revolve on elliptic orbits round the sun; and it is they, when they come within the earth's atmosphere, that, fired by friction, appear to us as meteors and falling stars.

On a bright night, 20 minutes rarely pass at any part of the earth's surface without the appearance of at least one meteor. At certain times (the 12th of August and the 14th of November) they appear in enormous numbers. During nine hours of observation in Boston, when they

---

* Mayer Dynamik des Himmels, p. 10.

were described as falling as thick as snowflakes, 240,000 meteors were calculated to have been observed. The number falling in a year might, perhaps, be estimated at hundreds or thousands of millions, and even these would constitute but a small portion of the total crowd of asteroids that circulate round the sun. From the phenomena of light and heat, and by the direct observations of Encke on his comet, we learn that the universe is filled by a resisting medium, through the friction of which all the masses of our system are drawn gradually towards the sun. And though the larger planets show, in historic times, no diminution of their periods of revolution, this may not hold good for the smaller bodies. In the time required for the mean distance of the earth from the sun to alter a single yard, a small asteroid may have approached thousands of miles nearer to our central luminary.

Following up these reflections we should infer, that while this immeasurable stream of ponderable matter rolls unceasingly towards the sun, it must augment in density as it approaches its centre of convergence. And here the conjecture naturally rises, that that weak nebulous light, of vast dimensions, which embraces the sun—the Zodiacal Light—may owe its existence to these crowded meteoric masses. However this may be, it is at least proved that this luminous phenomenon arises from matter which circulates in obedience to planetary laws ; the entire mass constituting the zodiacal light must be constantly approaching, and incessantly raining its substance down upon the sun.

We observe the fall of an apple and investigate the law which rules its motion. In the place of the earth we set the sun, and in the place of the apple we set the earth, and thus possess ourselves of the key to the mechanics of the heavens. We now know the connection between height of fall, velocity, and heat at the surface of the earth. In the place of the earth let us set the sun, with 300,000 times the earth's mass, and, instead of a fall of a few feet, let us take cosmical elevations; we thus obtain a means of generating heat which transcends all terrestrial power.

It is easy to calculate both the maximum and the minimum velocity, imparted by the sun's attraction to an asteroid circulating round him; the maximum is generated when the body approaches the sun from an infinite distance; the *entire pull* of the sun being then expended upon it; the minimum is that velocity which would barely enable the body to revolve round the sun close to his surface. The final velocity of the former, just before striking the sun, would be 390 miles a second, that of the latter 276 miles a second. The asteroid, on striking the sun with the former velocity, would develope more than 9,000 times the heat generated by the combustion of an equal asteroid of solid coal; while the shock, in the latter case, would generate heat equal to that of the combustion of upwards of 4,000 such asteroids. It matters not, therefore, whether the substances falling into the sun be combustible or not; their being combustible would not add sensibly to the tremendous heat produced by their mechanical collision.

Here then we have an agency competent to restore his lost energy to the sun, and to maintain a temperature at his surface which transcends all terrestrial combustion. The very quality of the solar rays—their incomparable penetrative power—enables us to infer that the temperature of their origin must be enormous; but in the fall of asteroids we find the means of producing such a temperature. It may be contended that this showering down of matter must be accompanied by the growth of the sun in size; it is so; but the quantity necessary to produce the observed calorific emission, even if accumulated for 4,000 years, would defeat the scrutiny of our best

instruments. If the earth struck the sun it would utterly
vanish from perception, but the heat developed by its
shock would cover the expenditure of the sun for a
century.

To the earth itself apply considerations similar to those
which we have applied to the sun. Newton's theory of
gravitation, which enables us, from the present form of
the earth, to deduce its original state of aggregation,
reveals to us, at the same time, a source of heat powerful
enough to bring about the fluid state—powerful enough
to fuse even worlds. It teaches us to regard the molten
condition of a planet as resulting from the mechanical
union of cosmical masses, and thus reduces to the same
homogeneous process, the heat stored up in the body of
the earth, and the heat emitted by the sun.

Without doubt the whole surface of the sun displays
an unbroken ocean of fiery fluid matter. On this ocean
rests an atmosphere of glowing gas—a flame atmosphere,
or photosphere. But gaseous substances, when compared with solid ones, emit, even when their temperature
is very high, only a feeble and transparent light. Hence it
is probable that the dazzling white light of the sun comes
through the atmosphere, from the more solid portions of
the surface.*

There is one other consideration connected with the
permanence of our present terrestrial conditions, which is
well worthy of our attention. Standing upon one of the
London bridges, we observe the current of the Thames
reversed, and the water poured upwards twice a-day. The
water thus moved rubs against the river's bed and sides,
and heat is the consequence of this friction. The heat thus
generated is, in part, radiated into space, and there lost, as

---

* I am quoting here from Mayer, but this is the exact view now entertained by Kirchhof. We see the solid or liquid mass of the sun *through*
his photosphere.

far as the earth is concerned.  What is it that supplies
this incessant loss ?  The earth's rotation.  Let us look a
little more closely at this matter.  Imagine the moon
fixed, and the earth turning like a wheel from west to
east in its diurnal rotation.  A mountain on the earth's
surface, on approaching the moon's meridian, is, as it
were, laid hold of by the moon ; forms a kind of handle
by which the earth is pulled more quickly round.  But
when the meridian is passed the pull of the moon on
the mountain would be in the opposite direction ; it now
tends to diminish the velocity of rotation as much as it
previously augmented it ; and thus the action of all fixed
bodies on the earth's surface is neutralised.

But suppose the mountain to lie *always* to the east of
the moon's meridian, the pull then would be always exerted
against the earth's rotation, the velocity of which would be
diminished in a degree corresponding to the strength of
the pull.  *The tidal wave occupies this position* — it lies
always to the east of the moon's meridian ; the waters
of the ocean are, in part, dragged as a brake along the
surface of the earth, and as a brake they must diminish
the velocity of the earth's rotation.  The diminution,
though inevitable, is, however, too small to make itself
felt within the period over which observations on the
subject extend.  Supposing, then, that we turn a mill by
the action of the tide, and produce heat by the friction of
the millstones ; that heat has an origin totally different
from the heat produced by another pair of millstones which
are turned by a mountain stream.  The former is produced
at the expense of the earth's rotation ; the latter at the
expense of the sun's radiation, which lifted the millstream
to its source.*

Such is an outline of the Meteoric Theory of the sun's

* Dynamik des Himmels, p. 38, &c.

heat, as extracted from Mayer's Essay on Celestial
Dynamics. I have held closely to his statements, and in
most cases simply translated his words. But the sketch
conveys no adequate idea of the firmness and consistency
with which he has applied his principles. He deals with
true causes; and the only question that can affect his theory
refers to the quantity of action which he has ascribed to
these causes. I do not pledge myself to this theory, nor
do I ask you to accept it as demonstrated; still it would be
a great mistake to regard it as chimerical. It is a noble
speculation; and depend upon it, the true theory, if this,
or some form of it, be not the true one, will not appear
less wild or less astounding.*

Mayer published his Essay in 1848; five years afterwards
Mr. Waterston sketched, independently, a similar theory, at
the Hull Meeting of the British Association. The Transactions of the Royal Society of Edinburgh for 1854 contain an
extremely beautiful memoir, by Professor William Thomson,
in which Mr. Waterston's sketch is developed. He considers that the meteors which are to furnish stores of

---

* While preparing these sheets finally for press, I had occasion to look
once more into the writings of Mayer, and the effect was a revival of the
interest with which I first read them. Dr. Mayer was a working physician
in the little German town of Heilbronn, who, in 1840, made the observation
that the venous blood of a feverish patient in the tropics was redder than
in more northern latitudes. Starting from this fact, while engaged in the
duties of a laborious profession, and apparently without a single kindred
spirit to support and animate him, Mayer raised his mind to the level
indicated by the references made to his works, throughout this book. In
1842 he published his first memoir 'On the Forces of Inorganic Nature;'
in 1845, his 'Organic Motion' was published; and in 1848, his 'Celestial
Dynamics' appeared. After this, his overtasked brain gave way, and a
cloud settled on the intellect which had accomplished so much. The shade,
however, was but temporary, and Dr. Mayer is now restored. I have never
seen him, nor has a line of correspondence ever passed between us.
Modestly and noiselessly he has done his work; and having spoken of his
merits, as accident made it my duty to speak, I confidently leave to history
the care of his fame.

energy for our future sunlight, lie principally within the
earth's orbit, and that we see them there, as the Zodiacal
Light, ' an illuminated shower, or rather tornado, of stones '
(Herschel, § 897). Thus he points to the precise source of
power previously indicated by Mayer. ' In conclusion,
then,' writes Professor Thomson, ' the source of energy
from which solar heat is derived is undoubtedly meteoric.
. . . . The principal source — perhaps the sole appreciable
efficient source — is in bodies circulating round the sun at
present inside the earth's orbit, and probably seen in the
sunlight by us called " Zodiacal Light." The store of energy
for future sunlight is at present partly dynamical — that of
the motions of these bodies round the sun ; and partly
potential — that of their gravitation towards the sun.
This latter is gradually being spent, half against the resisting medium, and half in causing a continuous increase of
the former. Each meteor thus goes on moving faster and
faster, and getting nearer and nearer the centre, until
some time, very suddenly, it gets so much entangled in the
solar atmosphere as to begin to lose velocity. In a few
seconds more it is at rest on the sun's surface, and the
energy given up is vibrated across the district where it was
gathered during so many ages, ultimately to penetrate, as
light, the remotest regions of space.'

From the tables published by Prof. Thomson I extract the
following interesting data ; firstly, with reference to the
amount of heat equivalent to the rotation of the sun and
planets round their axes ; the amount, that is, which would
be generated, supposing a brake applied at the surfaces
of the sun and planets, until the motion of rotation was
entirely stopped : secondly, with reference to the amount
of heat due to the sun's gravitation—the heat, that is,
which would be developed by each of the planets in falling
into the sun. The quantity of heat is expressed in terms
of the time during which it would cover the solar emission.

| | Heat of Gravitation, equal to Solar emission for a period of | | Heat of Rotation, equal to Solar emission for a period of | |
|---|---|---|---|---|
| Sun | | | 116 years | 6 days. |
| Mercury | 6 years | 214 days | | 15 „ |
| Venus | 83 „ | 227 „ | | 99 „ |
| Earth | 94 „ | 303 „ | | 81 „ |
| Mars | 12 „ | 252 „ | | 7 „ |
| Jupiter | 32240 „ | | 14 „ | 144 „ |
| Saturn | 9650 „ | | 2 „ | 127 „ |
| Uranus | 1610 „ | | | 71 „ |
| Neptune | 1890 „ | | | |

The heat of rotation of the sun and planets, taken all together, would cover the solar emission for 134 years; while the heat of gravitation (that produced by falling into the sun) would cover the emission for 45,589 years. There is nothing hypothetical in these results; they follow directly and necessarily from the application of the mechanical equivalent of heat to cosmical masses.

Helmholtz has shown that if the solar system has ever been a nebulous mass of extreme tenuity, the mechanical force equivalent to the mutual gravitation of the particles of such a mass would be 454 times the quantity of mechanical force which we now possess in our system; $\frac{453}{454}$ths of the gravitating tendency has been already satisfied and wasted as heat. The $\frac{1}{454}$th that remains to us would, however, if converted into heat, raise the temperature of a mass of water, equal to the sun and planets in weight, 28 millions of degrees Centigrade. The heat of the lime light, it may be remarked, is estimated at 2,000° C.; of a temperature of 28,000,000° C. we can therefore form no conception. If our entire system were pure coal, by the combustion of the whole of it only $\frac{1}{3500}$th of the above enormous amount of heat would be generated.

'But,' continues Helmholtz, 'though the store of our planetary system is so immense as not to be sensibly diminished by the incessant emission which has gone on during the period of man's history, and though the time which

must elapse before a sensible change in the condition of
our planetary system can occur is totally incapable of
measurement, the inexorable laws of mechanics show that
this store, which can only suffer loss, and not gain, must
finally be exhausted. Shall we terrify ourselves by this
thought? Men are in the habit of measuring the greatness of the universe, and the wisdom displayed in it, by the
duration and the profit which it promises to their own race;
but the past history of the earth shows the insignificance of
the interval, during which man has had his dwelling here.
What the museums of Europe show us of the remains of
Egypt and Assyria we gaze upon with silent wonder, and
despair of being able to carry back our thoughts to a
period so remote. Still the human race must have existed
and multiplied for ages, before the pyramids could have
been erected. We estimate the duration of human history at 6,000 years; but vast as this time may appear to
us, what is it in comparison with the period during which
the earth bore successive series of rank plants and mighty
animals, but no men? * Periods, during which, in our own
neighbourhood (Königsberg), the amber tree bloomed, and
dropped its costly gum on the earth and in the sea; when
in Europe and North America groves of tropical palms
flourished, in which gigantic lizards, and, after them, ele-
phants, whose mighty remains are still buried in the
earth, found a home. Different geologists, proceeding from
different premises, have sought to estimate the length of
the above period, and they set it down from one to nine
millions of years. The time during which the earth
has generated organic beings is again small, compared with
the ages, during which the world was a mass of molten
rocks. The experiments of Bischof upon basalt show,
that for our globe to cool down from 2,000° to 200°

* The absence of men may be doubted. See the conclusion of Lubbock's article on the Lake Habitations of Switzerland, in the *Natural History Review.*

Centigrade, would require 350 millions of years. And
with regard to the period during which the first nebulous
masses condensed, so as to form our planetary system,
conjecture must entirely cease. The history of man, therefore, is but a minute ripple in the infinite ocean of time.
For a much longer period than that during which he has
already occupied this world, the existence of a state of
inorganic nature, favourable to man's continuance, seems to
be secured, so that for ourselves, and for long generations
after us, we have nothing to fear. But the same forces of
air and water, and of the volcanic interior, which produced
former geologic revolutions, and buried one series of
living forms after another, still act upon the earth's crust.
They, rather than those distant cosmical changes of which
we have spoken, will end the human race; and, perhaps,
compel us to make way for new and more complete forms of
life, as the lizard and the mammoth have given way to
us and our contemporaries.'

Grand, however, and marvellous as are those questions
regarding the physical constitution of the sun, they are
but a portion of the wonders connected with our luminary.
His relationship to life is yet to be referred to. The
earth's atmosphere contains carbonic acid, and the
earth's surface bears living plants; the former is the nutri-
ment of the latter. The plant apparently seizes the combined carbon and oxygen; tears them asunder, storing up
the carbon and letting the oxygen go free. By no special
force, different in quality from other forces, do plants
exercise this power,—the real magician here is the sun.
We have seen in former lectures (see Lecture V.) how heat
is consumed in forcing asunder the atoms and molecules
of solids and liquids, converting itself into potential
energy, which reappeared as heat, when the attractions of
the separated atoms were again allowed to come into play.
Precisely the same considerations which we then applied to

heat we have now to apply to light; for it is at the
expense of the solar light that the decomposition of the
carbonic acid is effected. Without the sun the reduction
cannot take place, and an amount of sunlight is consumed
exactly equivalent to the molecular work accomplished.
Thus trees are formed, thus the meadows grow, thus the
flowers bloom. Let the solar rays fall upon a surface of
sand, the sand is heated and finally radiates away as much
as it receives; let the same rays fall upon a forest, the
quantity of heat given back is less than that received, for
the energy of a portion of the sunbeams is invested in the
building of the trees.* I have here a bundle of cotton,
which I ignite; it bursts into flame and yields a definite
amount of heat; precisely that amount of heat was
abstracted from the sun, in order to form that bit of cotton.
This is a representative case;—every tree, plant, and
flower, grows and flourishes by the grace and bounty of the
sun.

But we cannot stop at vegetable life; for this is the
source, mediate or immediate, of all animal life. In the
animal body vegetable substances are brought again into
contact with their beloved oxygen, and they burn within us,
as a fire burns in a grate. This is the source of all animal
power; and the forces in play are the same, in kind, as
those which operate in inorganic nature. In the plant
the clock is wound up, in the animal it runs down. In
the plant the atoms are separated, in the animal they
re-combine. And as surely as the force which moves a
clock's hands is derived from the arm which winds up the
clock, so surely is all terrestrial power drawn from the
sun. Leaving out of account the eruptions of volcanoes,
and the ebb and flow of the tides, every mechanical action
on the earth's surface, every manifestation of power, organic and inorganic, vital and physical, is produced by the

* Mayer ' Die organische Bewegung,' p. 39.

sun.*  His warmth keeps the sea liquid, and the atmosphere
a gas, and all the storms which agitate both are blown by
the mechanical force of the sun.  He lifts the rivers and the
glaciers up the mountains; and thus the cataract and the
avalanche shoot with an energy derived immediately from
him.  Thunder and lightning are also his transmuted
strength.  Every fire that burns and every flame that glows
dispenses light and heat which originally belonged to the
sun.  In these days, unhappily, the news of battle is familiar
to us, but every shock, and every charge, is an application,
or misapplication, of the mechanical force of the sun.
He blows the trumpet, he urges the projectile, he bursts
the bomb.  And remember, this is not poetry, but rigid
mechanical truth.  He rears, as I have said, the whole
vegetable world, and through it the animal; the lilies of
the field are his workmanship, the verdure of the meadows,
and the cattle upon a thousand hills.  He forms the muscle,
he urges the blood, he builds the brain.  His fleetness is in
the lion's foot; he springs in the panther, he soars in the
eagle, he slides in the snake.  He builds the forest and
hews it down, the power which raised the tree, and which
wields the axe, being one and the same.  The clover
sprouts and blossoms, and the scythe of the mower swings,
by the operation of the same force.  The sun digs the
ore from our mines, he rolls the iron; he rivets the plates,
he boils the water; he draws the train.  He not only
grows the cotton, but he spins the fibre and weaves the
web.  There is not a hammer raised, a wheel turned, or a
shuttle thrown, that is not raised, and turned, and thrown
by the sun.  His energy is poured freely into space, but our
world is a halting place where this energy is conditioned.
Here the Proteus works his spells; the selfsame essence

---

* The germ, and much more than the germ, of what is here stated is
to be found in a paragraph in Sir John Herschel's *Outlines of Astronomy*,
published in 1833.

takes a million shapes and hues, and finally dissolves into its primitive and almost formless form.  The sun comes to us as heat; he quits us as heat; and between his entrance and departure the multiform powers of our globe appear. They are all special forms of solar power—the moulds into which his strength is temporarily poured, in passing from its source through infinitude.

Presented rightly to the mind, the discoveries and generalizations of modern science constitute a poem more sublime than has ever yet been addressed to the intellect and imagination of man.  The natural philosopher of to-day may dwell amid conceptions, which beggar those of Milton.  So great and grand are they, that in the contemplation of them, a certain force of character is requisite to preserve us from bewilderment.  Look at the integrated energies of our world,—the stored power of our coal fields; our winds and rivers; our fleets, armies, and guns.  What are they?  They are all generated by a portion of the sun's energy, which does not amount to $\frac{1}{2,300,000,000,000}$th of the whole.  This, in fact, is the entire fraction of the sun's force intercepted by the earth, and, in reality, we convert but a small fraction of this fraction, into mechanical energy.  Multiplying all our powers by millions of millions, we do not reach the sun's expenditure.  And still, notwithstanding this enormous drain, in the lapse of human history we are unable to detect a diminution of his store.  Measured by our largest terrestrial standards, such a reservoir of power is infinite; but it is our privilege to rise above these standards, and to regard the sun himself as a speck in infinite extension,— a mere drop in the universal sea.  We analyse the space in which he is immersed, and which is the vehicle of his power.  We pass to other systems and other suns, each pouring forth energy like our own, but still without infringement of the law, which reveals immutability in

F F

the midst of change, which recognises incessant transference and conversion, but neither final gain nor loss. This
law generalises the aphorism of Solomon, that there is
nothing new under the sun, by teaching us to detect everywhere, under its infinite variety of appearances, the same
primeval force. To Nature nothing can be added; from
Nature nothing can be taken away; the sum of her energies
is constant, and the utmost man can do in the pursuit of
physical truth, or in the applications of physical knowledge, is to shift the constituents of the never-varying total,
and out of one of them to form another. The law of conservation rigidly excludes both creation and annihilation.
Waves may change to ripples, and ripples to waves,—
magnitude may be substituted for number, and number
for magnitude,—asteroids may aggregate to suns, suns
may resolve themselves into floræ and faunæ, and floræ and
faunæ melt in air,—the flux of power is eternally the
same. It rolls in music through the ages, and all terrestrial energy,—the manifestations of life, as well as the
display of phenomena, are but the modulations of its
rythm.

# APPENDIX TO LECTURE XII.

For various reasons I am anxious that this book should embrace all that I have written with regard to the relationship of Dr. Mayer to the Dynamical Theory of Heat. Here, in the first place, follows an abstract of a Lecture on Force, given at the Royal Institution on the evening of Friday June 6, 1862, and published in the Proceedings of the Institution and in the 'Philosophical Magazine.'

## ON FORCE.

The existence of the International Exhibition suggested to our Honorary Secretary the idea of devoting the Friday evenings after Easter of the present year to discourses on the various agencies on which the material strength of England is based. He wished to make iron, coal, cotton, and kindred matters, the subjects of these discourses: opening the series by a discourse on the Great Exhibition itself; and he wished me to finish the series by a discourse on 'Force' in general. For some months I thought over the subject at intervals, and had devised a plan of dealing with it; but three weeks ago I was induced to swerve from this plan, for reasons which shall be made known towards the conclusion of the discourse.

We all have ideas more or less distinct regarding force; we know in a general way what muscular force means, and each of us would less willingly accept a blow from a pugilist than have his ears boxed by a lady. But these general ideas are not now sufficient for us; we must learn how to express numerically the exact mechanical value of the two blows; this is the first point to be cleared up.

A sphere of lead weighing 1 lb. was suspended at a height of 16 feet above the theatre floor. It was liberated, and fell by gravity. That weight required exactly a second to fall to the

earth from that elevation, and the instant before it touched the
earth it had a velocity of 32 feet a second.   That is to say, if at
that instant the earth were annihilated, and its attraction annulled,
the weight would proceed through space at the uniform velocity
of 32 feet a second.

Suppose that, instead of being pulled downward by gravity, the
weight is cast upward in opposition to the force of gravity—with
what velocity must it start from the earth's surface in order to
reach a height of 16 feet ?   With a velocity of 32 feet a second.
This velocity imparted to the weight by the human arm, or by
any other mechanical means, would carry the weight up to the
precise height from which it had fallen.

Now the lifting of the weight may be regarded as so much mechanical work.   I might place a ladder against the wall, and carry
the weight up to a height of 16 feet ;  or I might draw it up to this
height by means of a string and pulley ;  or I might suddenly jerk
it up to a height of 16 feet.   The amount of work done in all
these cases, as far as the raising of the weight is concerned, would
be absolutely the same.   The absolute amount of work done
depends solely upon two things :  first of all, on the quantity of
matter that is lifted :  and secondly, on the height to which it is
lifted.   If you call the quantity or mass of matter $m$, and the
height through which it is lifted $h$, then the product of $m$ into $h$, or
$m\,h$, expresses the amount of work done.

Supposing, now, that instead of imparting a velocity of 32
feet a second to the weight we impart twice this speed, or 64 feet
a second.   To what height will the weight rise ?   You might be
disposed to answer, ' To twice the height ; ' but this would be
quite incorrect.   Both theory and experiment inform us that the
weight would rise to four times the height ; instead of twice 16,
or 32 feet, it would reach four times 16 or 64 feet.   So also, if we
treble the starting velocity, the weight would reach nine times
the height ; if we quadruple the speed at starting, we attain
sixteen times the height.   Thus, with a velocity of 128 feet a
second at starting, the weight would attain an elevation of 256
feet.   Supposing we augment the velocity of starting seven times,
we should raise the weight to 49 times the height, or to an elevation of 784 feet.

Now the work done—or, as it is sometimes called, the
*mechanical effect* — as before explained, is proportional to the

height, and as a double velocity gives four times the height, a treble velocity nine times the height, and so on, it is perfectly plain that the mechanical effect increases as the square of the velocity. If the mass of the body be represented by the letter $m$, and its velocity by $v$, then the mechanical effect would be represented by $m\,v^2$. In the case considered, I have supposed the weight to be cast upward, being opposed in its upward flight by the resistance of gravity ; but the same holds true if I send the projectile into water, mud, earth, timber, or other resisting material. If, for example, you double the velocity of a cannon ball, you quadruple its mechanical effect. Hence the importance of augmenting the velocity of a projectile, and hence the philosophy of Sir William Armstrong in using a 50 lb. charge of powder in his recent striking experiments.

The measure then of mechanical effect is the mass of the body multiplied by the square of its velocity.

In firing a ball against a target the projectile, after collision, is often found hissing hot. Mr. Fairbairn informs me that in the experiments at Shoeburyness it is a common thing to see a flash of light, even in broad day, when the ball strikes the target. And if I examine my lead weight after it has fallen from a height I also find it heated. Now here experiment and reasoning lead us to the remarkable law that the amount of heat generated, like the mechanical effect, is proportional to the product of the mass into the square of the velocity. Double your mass, other things being equal, and you double your amount of heat; double your velocity, other things remaining equal, and you quadruple your amount of heat. Here then we have common mechanical motion destroyed and heat produced.

I take this violin bow and draw it across this string. You hear the sound. That sound is due to motion imparted to the air, and to produce that motion a certain portion of the muscular force of my arm must be expended. We may here correctly say, that the mechanical force of my arm is converted into music. And in a similar way we say that the impeded motion of our descending weight, or of the arrested cannon-ball, is converted into heat. The mode of motion changes, but it still continues motion ; *the motion of the mass is converted into a motion of the atoms of the mass* ; and these small motions, communicated to the nerves, produce the sensation which we call

heat.  We, moreover, know the amount of heat which a given
amount of mechanical force can develope.  Our lead ball, for example, in falling to the earth generated a quantity of heat sufficient
to raise the temperature of its own mass three-fifths of a Fahrenheit degree.  It reached the earth with a velocity of 32 feet a
second, and forty times this velocity would be a small one for a
rifle bullet; multiplying ⅗ths by the square of 40, we find that
the amount of heat developed by collision with the target would,
if wholly concentrated in the lead, raise its temperature 960°.
This would be more than sufficient to fuse the lead.  In reality,
however, the heat developed is divided between the lead and the
body against which it strikes; nevertheless, it would be worth
while to pay attention to this point, and to ascertain whether rifle
bullets do not, under some circumstances, show signs of fusion.

From the motion of sensible masses, by gravity and other
means, the speaker passed to the motion of atoms towards each
other by chemical affinity.  A collodion balloon filled with a
mixture of chlorine and hydrogen was hung in the focus of a
parabolic mirror, and in the focus of a second mirror, 20 ft. distant,
a strong electric light was suddenly generated; the instant the
light fell upon the balloon, the atoms within it fell together with
explosion, and hydro-chloric acid was the result.  The burning of
charcoal in oxygen was an old experiment, but it had now a
significance beyond what it used to have; we now regard the act
of combination on the part of the atoms of oxygen and coal
exactly as we regard the clashing of a falling weight against the
earth.  And the heat produced in both cases is referable to a
common cause.  This glowing diamond, which burns in oxygen
as a star of white light, glows and burns in consequence of the
falling of the atoms of oxygen against it.  And could we measure
the velocity of the atoms when they clash, and could we find
their number and weight, multiplying the weight of each atom
by the square of its velocity, and adding all together, we should
get a number representing the exact amount of heat developed
by the union of the oxygen and carbon.

Thus far we have regarded the heat developed by the clashing
of sensible masses and of atoms.  Work is expended in giving
motion to these atoms or masses, and heat is developed.  But we
reverse this process daily, and by the expenditure of heat execute
work.  We can raise a weight by heat; and in this agent we

possess an enormous store of mechanical power.  This pound of coal, which I hold in my hand, produces by its combination with oxygen an amount of heat which, if mechanically applied, would suffice to raise a weight of 100 lbs. to a height of 20 miles above the earth's surface.  Conversely, 100 lbs. falling from a height of 20 miles, and striking against the earth, would generate an amount of heat equal to that developed by the combustion of a pound of coal.  Wherever work is done by heat, heat disappears. A gun which fires a ball is less heated than one which fires blank cartridge.  The quantity of heat communicated to the boiler of a working steam-engine is greater than which could be obtained from the re-condensation of the steam after it had done its work ; and the amount of work performed is the exact equivalent of the amount of heat lost.  Mr. Smyth informed us in his interesting discourse, that we dig annually 84 millions of tons of coal from our pits.  The amount of mechanical force represented by this quantity of coal seems perfectly fabulous.  The combustion of a single pound of coal, supposing it to take place in a minute, would be equivalent to the work of 300 horses ; and if we suppose 108 millions of horses working day and night, with unimpaired strength, for a year, their united energies would enable them to perform an amount of work just equivalent to that which the annual produce of our coal-fields would be able to accomplish.

Comparing the energy of the force with which oxygen and carbon unite together, with ordinary gravity, the chemical affinity seems almost infinite.  But let us give gravity fair play ; let us permit it to act throughout its entire range.  Place a body at such a distance from the earth that the attraction of the earth is barely sensible, and let it fall to the earth from this distance.  It would reach the earth with a final velocity of 36,747 feet in a second, and on collision with the earth the body would generate about twice the amount of heat generated by the combustion of an equal weight of coal.  We have stated that by falling through a space of 16 feet our lead bullet would be heated three-fifths of a degree ; but a body falling from an infinite distance has already used up 1,299,999 parts out of 1,300,000 of the earth's pulling power, when it has arrived within 16 feet of the surface ; on this space only $\frac{1}{1300000}$ths of the whole force is exerted.

Let us now turn our thoughts for a moment from the earth

towards the sun. The researches of Sir J. Herschel and M. Pouillet have informed us of the annual expenditure of the sun as regards heat, and by an easy calculation we ascertain the precise amount of the expenditure which falls to the share of our planet. Out of 2,300 million parts of light and heat the earth receives one. The whole heat emitted by the sun in a minute would be competent to boil 12,000 millions of cubic miles of ice-cold water. How is this enormous loss made good? Whence is the sun's heat derived, and by what means is it maintained? No combustion, no chemical affinity with which we are acquainted would be competent to produce the temperature of the sun's surface. Besides, were the sun a burning body merely, its light and heat would assuredly speedily come to an end. Supposing it to be a solid globe of coal, its combustion would only cover 4,600 years of expenditure. In this short time it would burn itself out. What agency, then, can produce the temperature and maintain the outlay? We have already regarded the case of a body falling from a great distance towards the earth, and found that the heat generated by its collision would be twice that produced by the combustion of an equal weight of coal. How much greater must be the heat developed by a body falling towards the sun! The maximum velocity with which a body can strike the earth is about 7 miles in a second; the maximum velocity with which it can strike the sun is 390 miles in a second. And as the heat developed by the collision is proportional to the square of the velocity destroyed, an asteroid falling into the sun with the above velocity would generate about 10,000 times the quantity of heat generated by the combustion of an asteroid of coal of the same weight.

Have we any reason to believe that such bodies exist in space, and that they may be raining down upon the sun? The meteorites flashing through our air are small planetary bodies, drawn by the earth's attraction, and entering our atmosphere with planetary velocity. By friction against the air they are raised to incandescence and caused to emit light and heat.* At certain seasons of the year they shower down upon us in great numbers. In Boston 240,000 of them were observed in nine hours. There is no reason to suppose that the planetary system is limited to ' vast masses of enormous weight;' there is

---

* To Mr. Joule, as stated in Lecture I., we owe this hypothesis.

every reason to believe that space is stocked with smaller masses, which obey the same laws as the large ones. That lenticular envelope which surrounds the sun, and which is known to astronomers as the Zodiacal light, is probably a crowd of meteors; and moving as they do in a resisting medium they must continually approach the sun. Falling into it, they would be competent to produce the heat observed, and this would constitute a source from which the annual loss of heat would be made good. The sun, according to this hypothesis, would be continually growing larger; but how much larger? Were our moon to fall into the sun it would develope an amount of heat sufficient to cover one or two years' loss; and were our earth to fall into the sun a century's loss would be made good. Still, our moon and our earth, if distributed over the surface of the sun, would utterly vanish from perception. Indeed, the quantity of matter competent to produce the necessary effect would, during the range of history, produce no appreciable augmentation in the sun's magnitude. The augmentation of the sun's attractive force would be more appreciable. However this hypothesis may fare as a representant of what is going on in nature, it certainly shows how a sun might be formed and maintained by the application of known thermo-dynamic principles.

Our earth moves in its orbit with a velocity of 68,040 miles an hour. Were this motion stopped, an amount of heat would be developed sufficient to raise the temperature of a globe of lead of the same size as the earth 384,000 degrees of the Centigrade thermometer. It has been prophesied that 'the elements shall melt with fervent heat. The earth's own motion embraces the conditions of fulfillment; stop that motion, and the greater part, if not the whole, of her mass would be reduced to vapour. If the earth fell into the sun, the amount of heat developed by the shock would be equal to that developed by the combustion of 6,435 earths of solid coal.

There is one other consideration connected with the permanence of our present terrestial conditions which is well worthy of our attention. Standing upon one of the London bridges, we observe the current of the Thames reversed, and the water poured upward twice a-day. The water thus moved rubs against the river's bed and sides, and heat is the consequence of this friction. The heat thus generated is in part radiated into space, and then

lost, as far as the earth is concerned.   What is it that supplies
this incessant loss ?   The earth's rotation.   Let us look a little
more closely at the matter   Imagine the moon fixed, and the
earth turning like a wheel from west to east in its diurnal
rotation.   Suppose a high mountain on the earth's surface ; on
approaching the moon's meridian, that mountain is, as it were,
laid hold of by the moon, and forms a kind of handle by which
the earth is pulled more quickly round.   But when the meridian
is passed, the pull of the moon on the mountain would be in the
opposite direction ; it now tends to diminish the velocity of
rotation as much as it previously augmented it ; and thus the
action of all fixed bodies on the earth's surface is neutralised.
But suppose the mountain to lie *always* to the east of the moon's
meridian : the pull then would be always exerted against the
earth's rotation, the velocity of which would be diminished in a
degree corresponding to the strength of the pull.   *The tidal wave
occupies this position* — it lies always to the east of the moon's
meridian, and thus the waters of the ocean are in part dragged as
a brake along the surface of the earth ; and as a brake they must
diminish the velocity of the earth's rotation.   The diminution,
though inevitable, is, however, too small to make itself felt
within the period over which observations on the subject extend.
Supposing then that we turn a mill by the action of the tide, and
produce heat by the friction of the millstones ; that heat has an
origin totally different from the heat produced by another mill
which is turned by a mountain stream.   The former is produced
at the expense of the earth's rotation, the latter at the expense of
the sun's radiation.

The sun, by the act of vaporization, lifts mechanically all the
moisture of our air.   It condenses and falls in the form of rain—
it freezes and falls as snow.   In this solid form it is piled upon
the Alpine heights, and furnishes materials for the glaciers of the
Alps.   But the sun again interposes, liberates the solidified liquid,
and permits it to roll by gravity to the sea.   The mechanical
force of every river in the world, as it rolls towards the ocean, is
drawn from the heat of the sun.   No streamlet glides to a lower
level without having been first lifted to the elevation from which
it springs by the mighty power of the sun.   The energy of winds
is also due entirely to the sun ; but there is still another work
which he performs, and his connection with which is not so ob-

vious. Trees and vegetables grow upon the earth, and when burned they give rise to heat, and hence to mechanical energy. Whence is this power derived? You see this oxide of iron, produced by the falling together of the atoms of iron and oxygen; here also is a transparent gas which you cannot now see — carbonic acid gas—which is formed by the falling together of carbon and oxygen. These atoms thus in close union resemble our lead weight while resting on the earth; but I can wind up the weight and prepare it for another fall, and so these atoms can be wound up, separated from each other, and thus enabled to repeat the process of combination. In the building of plants carbonic acid is the material from which the carbon of the plant is derived; and the solar beam is the agent which tears the atoms asunder, setting the oxygen free, and allowing the carbon to aggregate in woody fibre. Let the solar rays fall upon a surface of sand; the sand is heated, and finally radiates away as much heat as it receives; let the same beams fall upon a forest, the quantity of heat given back is less than the forest receives, for the energy of a portion of the sunbeams is invested in building up the trees in the manner indicated. Without the sun the reduction of the carbonic acid cannot be effected, and an amount of sunlight is consumed exactly equivalent to the molecular work done. Thus trees are formed; thus the cotton on which Mr. Bazley discoursed last Friday is formed. I ignite this cotton, and it flames; the oxygen again unites with its beloved carbon; but an amount of heat equal to that which you see produced by its combustion was sacrificed by the sun to form that bit of cotton.

But we cannot stop at vegetable life, for this is the source, mediate or immediate, of all animal life. The sun severs the carbon from its oxygen; the animal consumes the vegetable thus formed, and in its arteries a reunion of the severed elements take place, and produce animal heat. Thus, strictly speaking, the process of building a vegetable is one of winding up; the process of building an animal is one of running down. The warmth of our bodies, and every mechanical energy which we exert, trace their lineage directly to the sun. The fight of a pair of pugilists, the motion of an army, or the lifting of his own body up mountain slopes by an Alpine climber, are all cases of mechanical energy drawn from the sun. Not, therefore, in a poetical, but in a purely mechanical sense, are we children of the sun. Without

food we should soon oxidize our own bodies.   A man weighing
150 lbs. has 64 lbs. of muscle; but these, when dried, reduce
themselves to 15 lbs.   Doing an ordinary day's work, for 80 days,
this mass of muscle would be wholly oxidized.   Special organs
which do more work would be more quickly oxidized : the heart,
for example, if entirely unsustained, would be oxidized in about a
week.   Take the amount of heat due to the direct oxidation of a
given amount of food; a less amount of heat is developed by this
food in the working animal frame, and the missing quantity is the
exact equivalent of the mechanical work which the body accomplishes.

I might extend these considerations : the work, indeed, is done
to my hand—but I am warned that I have kept you already too long.
To whom, then, are we indebted for the striking generalisations of
this evening's discourse ?   All that I have laid before you is the
work of a man of whom you have scarcely ever heard.   All that I
have brought before you has been taken from the labours of a
German physician, named Mayer.   Without external stimulus,
and pursuing his profession as town physician in Heilbronn, this
man was the first to raise the conception of the interaction of
natural forces to clearness in his own mind.   And yet he is
scarcely ever heard of in scientific lectures ; and even to scientific
men his merits are but partially known.   Led by his own beautiful
researches, and quite independent of Mayer, Mr. Joule published his first Paper on the ' Mechanical Value of Heat,' in 1843 ;
but in 1842 Mayer had actually calculated the mechanical
equivalent of heat from data which a man of rare originality alone
could turn to account.   From the velocity of sound in air Mayer
determined the mechanical equivalent of heat.   In 1845 he published his Memoir on ' Organic Motion,' and applied the mechanical Theory of Heat in the most fearless and precise manner to
vital processes.   He also embraced the other natural agents in
his chain of conservation.   In 1853 Mr. Waterston proposed,
independently, the Meteoric Theory of the sun's heat, and in 1854
Professor William Thomson applied his admirable mathematical
powers to the developement of the theory ; but six years previously the subject had been handled in a masterly manner by
Mayer, and all that I have said on this subject has been derived
from him.   When we consider the circumstances of Mayer's life,
and the period at which he wrote, we cannot fail to be struck with

astonishment at what he has accomplished. Here was a man of
genius working in silence, animated solely by a love of his subject,
and arriving at the most important results, some time in advance
of those whose lives were entirely devoted to Natural Philosophy.
It was the accident of bleeding a feverish patient at Java in 1840
that led Mayer to speculate on these subjects. He noticed that
the venous blood in the tropics was of a much brighter red than
in colder latitudes, and his reasoning on this fact led him into the
laboratory of natural forces, where he has worked with such
signal ability and success. Well, you will desire to know what
has become of this man. His over-tasked mind gave way—
a result felt to be quite possible in his own case by many a
great scientific worker — and he was sent to an asylum. In a
biographical dictionary of his country it is stated that Mayer
died in the asylum : but this is incorrect. He recovered ; and,
I believe, is at this moment a cultivator of vineyards in Heilbronn.

While preparing for publication my last course of Lectures on
Heat, I wished to make myself acquainted with all that Mayer
had done in connection with this subject. I accordingly wrote
to two gentlemen who above all others seemed likely to give me
the information which I needed. Both of them are Germans,
and both particularly distinguished in connection with the Dynamical Theory of Heat. Each of them kindly furnished me with
the list of Mayer's publications, and one of them was so friendly
as to order them from a bookseller, and to send them to me.
This friend, in his reply to my first letter regarding Mayer,
stated his belief that I should not find anything very important
in Mayer's writings ; but before forwarding the memoirs to me
he read them himself. His letter, accompanying the first of these
papers, contains the following words : — ' I must here retract the
statement in my last letter, that you will not find much matter of
importance in Mayer's writings ; I am astonished at the multitude
of beautiful and correct thoughts which they contain ; ' and he
goes on to point out various important subjects, in the treatment
of which Mayer had anticipated other eminent writers. My second
friend, in whose own publications the name of Mayer repeatedly
occurs, and whose papers containing these references were translated some years ago by myself, was, on the 10th of last month,
unacquainted with the thoughtful and beautiful essay by Mayer,
entitled ' Beiträge zur Dynamik des Himmels ; ' and in 1854,

when Professor William Thomson developed in so striking a manner the meteoric theory of the sun's heat, he was certainly not aware of the existence of that essay, though from a recent article in 'Macmillan's Magazine' I infer that he is now aware of it. Mayer's physiological writings have been referred to by physiologists — by Dr. Carpenter, for example — in terms of honourable recognition. We have hitherto, indeed, obtained fragmentary glimpses of the man, partly from physicists and partly from physiologists; but his total merit has never yet been recognised, as it assuredly would have been had he chosen a happier mode of publication. I do not think a greater disservice could be done to a man of science than to overstate his claims : such overstatement is sure to recoil to the disadvantage of him in whose interest it is made. But when Mayer's opportunities, achievements, and fate, are taken into account, I do not think that I shall be deeply blamed for attempting to place him in that honourable position which I believe to be his due.

Here, however, are the titles of Mayer's papers, the perusal of which will correct any error of judgment into which I may have fallen regarding their author. 'Bemerkungen über die Kräfte der unbeleten Natur,' Liebig's Annalen, 1842, vol. xlii. p. 231; 'Die Organische Bewegung in ihrem Zusammenhange mit dem Stoff-wechsel ; ' Heilbronn, 1845 ; 'Beiträge zur Dynamik des Himmels,' Heilbronn, 1848; 'Bemerkungen über das Mechanische Equivalent der Wärme,' Heilbronn, 1851.

J. T.

With reference to this Lecture, Mr. Joule published the following letter in the August number of the 'Philosophical Magazine: '—

## NOTE ON THE HISTORY OF THE DYNAMICAL THEORY OF HEAT.

BY J. P. JOULE, LL.D., F.R.S.

*To the Editors of the Philosophical Magazine and Journal.*

GENTLEMEN,—Will you permit me to trouble your readers with a few remarks on the subject of my friend Professor Tyndall's Lecture at the Royal Institution, reported in your last Number ? In this Lecture he enforces the claims of M. Mayer, a philosopher whose merit has perhaps been overlooked by some of our English physicists, and unaccountably so by his fellow-countrymen. I myself was only imperfectly acquainted with his papers when, in good conscience and with the materials at command, I gave a sketch of the history of the Dynamical Theory of Heat, in my paper published in the Philosophical Transactions for 1850. M. Mayer's merit consists in having announced, apparently without knowledge of what had been done before, the true Theory of Heat. This is no small merit, and I am the last person who would wish to detract from it. But to give to Mayer, or indeed to any single individual, the undivided praise of propounding the Dynamical Theory of Heat, is manifestly unjust to the numerous contributors to that great step in physical science. Two centuries ago, Locke said that ' Heat is a very brisk agitation of the insensible parts of the object, which produces in us that sensation from whence we denominate the object hot; so that what in our sensation is *heat*, in the object is nothing but *motion*.' In 1798, Rumford, inquiring into the source of heat developed in the boring of cannon, observed that it was ' extremely difficult, if not quite impossible, to form any distinct idea of anything capable of being excited and communicated, in the manner the heat was excited and communicated in these experiments, except it be motion.' In 1812, Davy wrote: ' The immediate cause of the phenomena of heat, then, is motion, and the laws of its communication are precisely the same as the laws of the communication of motion;'* and he confirmed his views by that original and most interesting experiment in which he melted ice by friction.† In 1839, Séguin published a work entitled *De l'Influence des Chemins de Fer*. He shows that the theory generally adopted would lead to the absurd conclusion that a finite quantity of heat can produce an indefinite quantity of mechanical action, and remarks (p. 328), ' Il me paraît plus naturel de supposer qu'une certaine quantité de calorique disparaît dans l'acte même de la production de la force ou puissance mécanique, et réciproquement.' At p. 383 he remarks: ' La force mécanique qui apparaît pendant l'abaissement de température d'un gaz comme de tout autre corps qui se dilate, est la mesure et la représentation de cette diminution de chaleur.' In p. 389

---

* Elements of Chemical Philosophy, p. 94.
† My morning Lectures had rendered all this familiar.—J. T.

he gives a Table of the quantity of mechanical effect produced corresponding to the loss of temperature of steam on expanding. From this it appears that 1° Cent. corresponds with 363 kilogrammes raised to the height of 1 metre. At p. 403 he states: ' Je bornerai là mes réflexions sur un sujet dont chacun saura apprécier l'importance. Du calorique qui est employé par l industrie à produire de la force, et aux usages domestiques, une faible partie seulement est utilisée; une autre quantité bien plus considérable, et qui pourrait suffire à créer d'immenses valeurs et à augmenter d'autant la richesse nationale, se trouve absolument perdue.' From the above extracts, it will be seen that a great advance had been made before Mayer wrote his paper in 1842. Mayer discourses to the same effect as Séguin, but at greater length, with greater perspicuity, and with more copiousness of illustration. He adopts the same hypothesis as the latter philosopher, viz.: that the heat evolved on compressing an elastic fluid is exactly the equivalent of the compressing force, and thus arrives at the same equivalent, viz., 365 kilogrammes per 1° Cent.

It must be remarked that, at the time Séguin and Mayer wrote, there were no known facts to warrant the hypothesis they adopted. There was no reason to assert that the heat evolved by compressing a gas was even approximately the equivalent of the compressing force. This being the case may account for the inattention of the scientific world to these writings. The Dynamical Theory of Heat certainly was not established by Séguin and Mayer. To do this required experiment; and I therefore fearlessly assert my right to the position which has been generally accorded to me by my fellow physicists as having been the first to give a decisive proof of the correctness of this theory.

In saying this I do not wish to claim any monopoly of merit. Even if Rumford, Mayer, and Séguin had not produced their works, justice would still compel me to share with Thomson, Rankine, Helmholtz, Holtzman,* Clausius, and others, whose labours have not only given developements and applications of the Dynamical Theory which entitle them to merit as well as their predecessors in these enquiries, but who have contributed most essentially in supporting it by new proofs.

* The name of this philosopher ought to be added to those mentioned in the Preface as the builders of the Dynamical Theory of Heat.—J. T.

Permit me to remark, in conclusion, that I applied the Dynamical Theory to vital processes in 1843;* and that in 1847, in a popular lecture, published in the 'Manchester Courier,' I explained the phenomena of shooting stars, and also stated that the effect of the earth falling into the sun would be to increase the temperature of that luminary.† Since that time Thomson, by his profound investigations, has made the Dynamical Theory oi' Heat, as applied to cosmical phenomena, his own.

I sincerely trust that, by the foregoing remarks, I have done no injustice to Mayer, especially as I grieve to hear that sickness has removed him (I hope for only a short time) from the science to which he has contributed with so much ability. The reproduction of some of his papers in the 'Philosophical Magazine,' particularly that 'On the Forces of Inorganic Nature,' would, I am sure, interest many of your readers, and enable them to fully appreciate his just claims.

I remain, Gentlemen,<br>Yours respectfully,<br>J. P. JOULE.

---

I was in Switzerland when this letter appeared, and immediately after my return I published the following letter to Mr. Joule. (Phil. Mag. Sept. 1862.)

MY DEAR JOULE,

ON my return from Switzerland, two days ago, I became acquainted with the note which you have published in the last Number of the 'Philosophical Magazine.' Would you allow me to make the following remarks in connection with the subject of it?

During the spring of the present year I gave, at the Royal Institution, a Course of Lectures 'On Heat, regarded as a kind of Motion.' During the early portion of the course I had engaged a short-hand writer to report the lectures, with a view to their subsequent publication; and from this gentleman's notes of my second Lecture I make the following extract, which refers to the

* Phil. Mag. § 3. vol. xxiii. p. 442.
† Ibid. vol. xxxii. p. 350, and Manchester Courier, May 12, 1847.

mechanical Theory of Heat :—' It is to Mr. Joule, of Manchester, that we are almost wholly indebted for the experimental treatment of this subject. With his mind firmly fixed upon a principle, and undismayed by the coolness with which his first labours appear to have been received, he persisted for years in his attempts to prove the invariability of the relation between heat and ordinary mechanical force. He placed water in a suitable vessel, agitated it by paddles moved by measurable forces, and determined the elevation of temperature; he did the same with mercury and sperm oil. He also caused disks of cast iron to rotate against each other, and measured the heat produced by their friction. He urged water through capillary tubes, and measured the heat thus generated. The results of his experiments leave no doubt upon the mind that under all circumstances the absolute amount of heat produced by the expenditure of a definite amount of mechanical force is fixed and invariable.' Such has been my language regarding you; and to it I still adhere. I trust you find nothing in it which indicates a desire on my part to question your claim to the honour of being the experimental demonstrator of the equivalence of heat and work.

It was not my object in the Lecture to which you refer to give a history of the mechanical Theory of Heat, but simply to place a man of genius, to whom the fates had been singularly unkind, in a position in some measure worthy of him. I was quite aware of all that you have stated regarding Locke, Rumford, Davy, and others: you might have added Bacon to your list—probably no great generalization was ever established without having first simmered in the minds of many thinkers. But the writings of Mayer form an epoch in the history of this subject; and I certainly should not feel disposed to retract a single sentence that I have written in his favour. I believe he deserves more praise than I have given him. It was he who first used the term ' equivalent' in the precise sense in which you have applied it ; he calculated the mechanical equivalent of heat from data which, as I have said, ' a man of rare ingenuity alone could turn to account;' and his calculation is in striking accordance with your own experimental determinations*. You worked independently of Mayer, and in a totally different way. You brought the

---

* The corrected specific heat of air being made use of.

mechanical theory to the test of experiment, and in this way proved its truth.

Mayer calculated correctly the mechanical equivalent of heat; but you say that, at the time he wrote, there were no known facts to warrant the hypothesis which he adopted. If by this you mean to say that he made a haphazard guess, which had no basis of physical probability, I cannot agree with you. The known constitution of an elastic fluid is, in my opinion, quite sufficient to justify Mayer's proceeding. His hypothesis was this : Let the quantity of heat required to raise the temperature of gas, preserved *at a constant volume*, $t°$, be $x$, and let the heat required to raise the same gas, under *constant pressure*, $t°$, be $x + y$. The weight raised by the expanding gas in the latter case being P, and the height to which it is raised $h$, then, according to Mayer,

$$y = \text{P} \times h;$$

that is to say, the excess of heat imparted in the latter case is precisely equivalent to the mechanical work performed.

It is undoubtedly implied in this equation that the quantity of heat $y$ is expended wholly in *external* work, and that none of it has been consumed in overcoming internal molecular attractions. This, I think, on the face of it is an extremely probable hypothesis — so probable, indeed, as to amount, in my estimation, almost to a certainty. Clausius makes the same assumption with no better authority than Mayer ; and I believe (for I here trust my memory merely) that the assumption has been completely verified by the experiments of the very philosophers who once questioned it. 'The law,' says Mayer, ' " Heat = Mechanical effect," is independent of the nature of an elastic fluid, which only serves as the apparatus by means of which the one force is converted into the other.'

The law of Mariotte was an old principle when Mayer wrote ; and the fact of its holding good for gases generally renders the conclusion exceedingly probable that, in yielding to compression, the attractions of the gaseous molecules were insensible ; otherwise it is hardly conceivable that the same results could have been obtained with gases so differently constituted : the attractions of the hydrogen atoms, for example, would in all probability be different from those of oxygen. Mayer was further justified in his hypothesis, as to the absence of interior work in the case of a true gas, by the experiments of Œrsted and Despretz, which showed that the law of Mariotte was departed from by the liquefiable gases — the amount of departure depending on the proximity of the gas to its point of condensation. Where, therefore, no departure from the law had been observed (in the case of air for instance), Mayer, I submit, was perfectly warranted in assuming that the molecular attractions were insensible, and that the quantity of heat ($y$) before referred to was entirely expended in raising the weight, and had its true mechanical equivalent in the weight so raised.

With reference to the application of the mechanical Theory of Heat to cosmical phenomena, if it were not a liberty, I would ask whether you have ever read the Essay of Mayer entitled, ' Beiträge zur Dynamik des Himmels ' ?   If so, then I have good reason to suspect my competence to come to a correct conclusion as to what constitutes a scientific right.

Knowing that the original memoirs of Mayer would be the true court of appeal in connection with this subject, I some months ago urged the responsible editor of the ' Philosophical Magazine ' to publish translations of them.   This I hope he will do ; for I quite agree with you in thinking that they would interest many of the readers of the magazine.   Let me add, in conclusion, that I do not think the public estimate of your labours can be in the least affected by any recognition which may be accorded to Mayer.   There is room for both of you on this grand platform.   Certainly, had Mayer never written a syllable on the mechanical Theory of Heat, I should not deem your work a whit nobler than I now hold it to be.

Believe me, yours, &c.,

JOHN TYNDALL.

ROYAL INSTITUTION :
    *August* 1862.

The public is now in possession of all that I have written with reference to the claims of Dr. Mayer.   The whole of it being placed thus together will facilitate future reference.

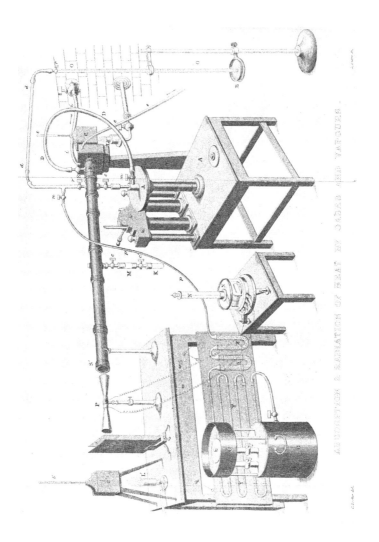

The material originally positioned here is too large for reproduction in this reissue. A PDF can be downloaded from the web address given on page iv of this book, by clicking on 'Resources Available'.

# INDEX.

# INDEX.

LONDON

PRINTED BY SPOTTISWOODE AND CO.

NEW-STREET SQUARE

Printed in the United States
By Bookmasters